MW00451823

Power Quality Primer

Power Quality Primer

Barry W. Kennedy

McGraw-Hill

New York San Francisco Washington, D.C. Auckland Bogotá
Caracas Lisbon London Madrid Mexico City Milan
Montreal New Delhi San Juan Singapore
Sydney Tokyo Toronto

Library of Congress Cataloging-in-Publication Data

Kennedy, Barry W.
Power quality primer. / Barry W. Kennedy.
p. cm.
Includes bibliographical references and index.
ISBN 0-07-134416-0
1. Electric power system stability. 2. Electric power systems—Electric losses. 3. Electric power systems—Protection 4. Electric utilities—Quality control. I. Title.

TK1010 K46 2000
621.31'2—dc21

00-033257

McGraw-Hill

A Division of The ***McGraw-Hill*** *Companies*

2 3 4 5 6 7 8 9 0 MMN/MMN 0 6 5 4 3 2 1

ISBN 0-07-134416-0

The sponsoring editor for this book was Scott Grillo, the editing supervisor was Gerry Fahey, and the production supervisor was Pamela Pelton. It was set in New Century Schoolbook by Joanne Morbit of McGraw-Hill's Professional Book Group composition unit, Hightstown, N.J.

Printed and bound by Maple-Vail Book Manufacturing Group.

This book is printed on recycled, acid-free paper containing a minimum of 50% recycled, de-inked fiber.

I dedicate this book to the Lord Jesus Christ for giving me the discipline to write each day and to my wife, Helen, for her emotional support and getting up early each morning to fix my breakfast so I could write.

Contents

FOREWORD

Deregulation of the electric power industry is making the quality of the power delivered a topic of increasing importance. It is no longer just an issue for a technical group within the utility that investigates unusual problems of interaction between the power system and customer facilities. It is a problem related to basic system design issues, system maintenance issues, investments that are required to protect equipment within customer facilities, and the implementation of new technologies. Unfortunately, no one has figured out who should be responsible for this power quality.

We are creating a new industry structure where there are many different entities with different responsibilities. The objective is to achieve deregulation of the electric power generation and realize the benefits of efficiency and innovation that result from competition. This is a good objective and consumers should benefit from this approach to generating power. However, there is still the problem of getting the power from these generators to the consumers. This involves a portion of the electric utilities that will still have to be regulated because putting in redundant systems for the transmission and distribution functions will never be the optimum approach for society. Regulation of the transmission and distribution portions of electric utilities ("the line companies") will have to consider power quality in some form. Regulators are already looking at the issues of reliability and consumers have already experienced reliability impacts associated with the new structure. Reliability is really just one part of the overall power quality issue—many other aspects of power quality can also have important impacts on customer operations. Many of these are quite complicated and involve interaction between the transmission system, distribution system, and even customer facilities.

What are these power quality concerns and how should we address them? When you are finished with this book, you should have a basic understanding of the important concerns. You will also understand that there is no simple method of dealing with them. Different customers have different needs with respect to power quality. It would not be fair to increase the costs of distributing power to all customers to meet the power quality requirements of the most sensitive customers. These leads to the concept of differentiated levels of power quality, or "custom power." There should be many opportunities for individual contracts that define the power quality to be delivered for specific customers and regulation of the transmission and distribution companies should support this concept.

First of all, we need standards that define the basic requirements, the responsibilities of the different parties, and the methods of characterizing the power quality so that everyone starts at the same point. There are many different standards efforts under way, both in North America and internationally. This book will help you understand some of these important activities so that you can keep track of the developments and provide your input where it is appropriate.

Barry Kennedy had been involved in the power quality area for a number of years. His background in energy efficiency is particularly appropriate because many of the same devices that are appropriate for improving the energy efficiency of a facility also have important impacts on the power quality levels and concerns. For instance, adjustable-speed motor drives save energy and provide important advantages in controlling processes but they also can be very sensitive to small variations in the voltage supplied and they can introduce harmonic distortion which may affect other loads on the system. Our policies in promoting energy efficiency technologies must also address the associated power quality issues. Barry recognized this many years ago and managed a project for EPRI and BPA to develop a workbook that addresses these concerns. It is still one of the best references in the industry on this topic.

Barry's perspective on the problem should be valuable for many people. This book is designed to complement more advanced books on power quality issues and should become an important reference for everyone's power quality library. It should provide a basic understanding for the wide variety of people that may now be impacted by power quality issues—utility engineers, regulators, all types of customers, equipment manufacturers,and even politicians. In this sense, it fills a very important need for the whole industry.

Mark McGranahan
Electrotek Concepts

PREFACE

The term *power quality* seems ambiguous. It means different things to different people. So, what is power quality? Is power quality a problem or a product? It depends on your perspective. If you are an electrical engineer, power quality expert, or electrician, you may tend to look at power quality as a problem that must be solved. If you are an economist, power marketer, or purchaser of electrical power, you may look at power as a product and power quality as an important part of that product. Whatever your background, if you are involved in the sale or purchase of electrical power, you will benefit from this book.

I have designed this book to be a technical book for both a nontechnical and a technical audience. Electric utility staff can use this book as a reference. I have written it to help them understand how to compete in the new deregulated, competitive utility industry—not just on the price of electricity but through better customer service and power quality. It will also help utility engineers to provide better customer service to their customers. It will provide the consumer of electricity with important guidelines on how they can get better customer service and power quality from their servicing utility.

Four factors cause an increased need to solve and prevent power quality problems: (1) the increased use of power quality–sensitive equipment, (2) the increased use of equipment that generates power quality problems, (3) the increased interconnectedness of the power system, and (4) the deregulation of the power industry. All of these factors influence utilities' ability to compete with each other to gain new—and keep existing—customers. They also affect the consumers' end users of electricity ability to succeed at their business. Utilities can cause end users to experience costly disruption of production. I discuss each one of these factors in more detail in Chap. 1.

Traditionally, utilities avoided involvement in power quality problems that occurred on a customer's system, and only got involved when their customers complained about power quality problems. Utilities would only react to customer complaints. When they received a confusing complaint, they would first try to determine the cause of the problem and who caused it. Utilities in a competitive environment have found that this reactive approach makes customers unhappy. In today's electricity market, most utilities want to keep their customers happy and satisfied. At the same time, many utility customers expect high-quality power and cannot afford the cost and bother of power quality problems. Consequently, utilities have discovered that a proactive approach to power quality problems works better in satisfying and keeping their customers happy. Utilities and their customers have found the need to look at each other's side of the revenue meter when encountering power quality problems. Even though utilities cause many power quality problems, such as voltage sags, recent studies by research organizations, like the Electric Power Research Institute (EPRI), have found that utility customers cause 80 to 90 percent of their own power quality problems.

Many utilities and their customers have discovered the importance of solving and preventing power quality problems. They have found the need to prevent the cost of lost production caused by poor power quality. They have learned that they need to understand the cause and effect of power quality problems in order to prevent them. More and more utilities are working with their customers on the "other side" of the meter to help them solve power quality problems. Yes, both providers and purchasers of electricity need to know the causes and solutions to power quality problems. I have designed this book to provide both suppliers and consumers of electricity not only with a clear understanding of the cause and effect of power quality problems but also the solutions to those problems as well.

Several books about power quality are available, but none is dedicated solely to providing the reader with the solutions to power quality problems or help them to understand how to sell or buy power high in quality. Other books define the technical problems and solutions associated wth power quality, while this book is a power quality primer that will help both the provider and consumer of electrical energy to cope with the customer service and power quality impact of the deregulated electric utility industry. My goal in writing this book is to help providers and users of electricity to understand the basics of power quality. If you are new to power quality, you will find that Chap. 2 provides you with the necessary fundamentals on power quality theory, power quality variations, and power quality solutions. If you are familiar with power quality, you will find the later chapters on new power quality monitoring and diagnostic tools informative and valuable. And for both the beginner and advanced provider and user of electricity, Chapt. 3 provides a clear explanation of all the many and often confusing international and national power quality standards.

Power quality standards have been changing over the years. They will become even more important as the utility industry restructures and becomes more competitive. Chapter 3 discusses the various power quality standards developed by IEEE, IEC, EPRI, and other organizations. In addition to understanding power quality standards, you need to know how to solve power quality problems.

Chapter 4 outlines how to solve power quality problems. The various types of power conditioning equipment available on the market are presented, along with an explanation of how they can solve your power quality problem.

Because poor wiring and grounding cause many power quality problems (80 to 90 percent), Chapter 5 is devoted to identifying and solving wiring and grounding power quality problems.

I organized this book into logical steps to help you obtain easily the information you need to either develop a power quality program or just to understand power quality in general. I explain how to use this book in Chap. 1.

Whether you are a power quality expert or end user experiencing power quality problems, you need to decide how to use the many types of meters on the market today for measuring power quality. What meter should you use to solve your—or your client's—power quality problem? Just like doctors and their patients need to understand how to use instruments for diagnosing health problems, power quality experts and their clients need to understand how to use instruments for diagnosing power quality problems. You'll learn to understand how they work and how to use them from this book. Chapter 6 shows you the inner workings of power quality meters and how to apply them to solve your power quality problem. Not only will I show you how to use power quality meters but also how to determine what type of power quality meter is best for your situation.

Many utilities and their customers are looking for permanent power quality monitoring systems that allow them to respond quickly to power quality problems. Several systems are available today. Chapter 6 explains these systems and how to use them. Once you have determined the cause of power quality problems, you need to know what solution best fits your power quality problem. To help you to determine the various solutions available, I cover the technology and equipment being used to prevent or solve power quality problems in Chap. 4.

Many power quality problems are so complex that a computer simulation is required to solve them. Not only do computer simulations provide an opportunity to look at alternative technical solutions but they also allow you to evaluate the economics of various solutions. Then you can pick the solution that solves your problem with the least amount of cost. The various tools for power quality simulations and how to use them, as well as the steps in performing a power quality survey, are discussed in Chap. 7.

Often, the analysis of power quality problems requires extensive experience in looking at the power quality signature obtained from your monitoring equipment and the ability to recognize the cause of the problem. This is not unlike a doctor looking at laboratory results and being able to identify the medical problem that is making you sick. Obtaining the necessary experience can be expensive and time consuming. Computer diagnostic tools developed by EPRI and others can be real assets in identifying and isolating a particular power quality problem. I present these tools and how to use them in Chap. 7.

Deregulation and restructuring of the electric utility industry will have an effect on power quality. What will the restructured utility industry look like and how will it effect power quality? Deregulation's effect and other future trends, described in Chap. 9 provide you with the structure of the new utility industry and how it will affect power quality.

Changes in technologies and the structure of the power industry will make the need for continuous power quality training essential. Chapter 9 is devoted to the type of training available today from utilities, EPRI, IEEE, and consultants, along with the steps for developing your power quality training if you so choose.

Once you have examined the various components that make up a power quality program, you are ready to develop your own power quality program. Your program could include any of the modules I discuss in Chap. 9. I have tried to present to you with the parts of a power quality program in modules to allow you to create a program that meets your needs.

The cost of power quality problems and solutions needs to be evaluated not only from the utility's perspective but also from the utility customer's perspective. The economics of power quality programs need to be evaluated as well. In Chap. 8 I show you how to evaluate the cost of power quality problems, solutions, and programs.

Can power quality be treated as a business? Whether you wish to sell power quality as a service bundled with your power rates or as a separate unbundled service, you need to develop a business plan. In Chap. 9 I discuss how various companies are treating power quality as a business and how to write the technical components of a business plan for a power quality, industrial, or commercial organization.

Deregulation and restructuring the electric utility industry will make the use of power quality contracts imperative. These contracts will not only involve agreements between the servicing utility and their customers but also between power consumers, transmission and distribution companies, power quality servicing companies and their customers, and between generators and their customers. I discuss how to write these contracts in Chap. 9.

When the utility industry becomes deregulated, users of electricity will need to evaluate their supplier of power not only on the basis of the power cost but also on power quality. How to evaluate providers of power quality services, and transmission and distribution services, is the subject of Chap. 9.

Research and development of new tools for diagnosing and solving power quality problems is constantly changing. New technologies that result in power quality problems will require new methods for diagnosing them. I discuss the status of research and development in Chap. 9.

What are the future trends in technology and organizational structure in the power industry? How will these trends affect the end user of electricity and the utility industry? I look into my crystal ball and present in Chap. 9 what I think will be the important trends in power quality.

In order to help you sort through the jargon and technical language in the power quality and electric utility industry, I provide an extensive glossary of terms and abbreviations. Often you might have a desire to do further research on power quality. To meet that need, I provide a bibliography that includes references to several Internet web sites that deal with power quality.

You, the users and providers of electricity, benefit from the selection and operation of a power system that provides power that is high in power quality. You have the data on the cost of power quality problems and the cost of power quality solution to make the decision that benefits you and your customers. With this book, you have the knowledge and methods for evaluating the cost effectiveness of power quality solutions that meets your needs to serve your customer and save money.

Acknowledgments

The author would like to thank the following individuals who provided help during preperation of the manuscript and production: Wayne Beaty, Roger Dugan, Gerry Fahey, John Fenker, Tom Key, Alex McEachern, Mark McGranahan, David Muller, Dan Sabin, and Frances-Crystal Wilson.

Barry W. Kennedy

Credits

Table 3.2 reprinted with permission from IEEE Std. 1159-1995 "IEEE Recommended Practice for Monitoring Electric Power Quality" Copyright © 1995, by IEEE. Tables 3.7 and 3.8 and quote on page 83 reprinted with permission from IEEE Standard 519-1992 "IEEE Recommended Practices and Requirements for Harmonic Control in Electrical Power Systems" Copyright © 1992 by IEEE. Page 87, Equation 3.2; Table 4.1; quotes on pages 151, 152, and 194; and Figs. 5.29, 5.30, and 7.7 reprinted with permission from IEEE Standard 1100-1992 "IEEE Recommended Practices for Powering and Grounding Sensitive Electronic Equipment. (*The Emerald Book*)" Copyright © 1993, by IEEE. Quote on page 76 reprinted with permission from IEEE Standard 493-1990 "IEEE Recommended Practice for the Design of Reliable Industrial and Commercial Power Systems (*The Gold Book*)" Copyright © 1991, by IEEE. Definition on page 89 reprinted with permission from IEEE Standard C62.47-1992 "IEEE Guide on Electrostatic Discharge (ESD): Characterization of the ESD Environment" Copyright © 1993, by IEEE. Quote on page 166 reprinted with permission from IEEE Standard 142-1991 "IEEE Recommended Practice for Grounding of Industrial and Commercial Power Systems (*IEEE Green Book*)" Copyright © 1992, by IEEE. Equation on pages 248–250 reprinted with permission from IEEE Standard 446-1995 "IEEE Recommended Practice for Emergency and Standby Power Systems for Industrial and Commercial Applications (*The Orange Book*)" Copyright © 1996 by IEEE. Form on Table 8.2 reprinted with permission from IEEE Standard 1346-1998 "IEEE Recommended Practice for Evaluating Electric Power System Compatibility With Electronic Process Equipment" Copyright © 1998, by IEEE.

IEEE disclaims any responsibility or liability resulting from the placement and use in the described manner.

Chapter

1

Introduction

It's Friday. Your boss gave you a deadline to have that report done by close of business. You're almost done with the report. So you don't bother to save it. Then your computer "freezes." You're upset. You take a deep breath, say a prayer, and reboot your computer. You've lost several hours of work. You may have lost a promotion and certainly a chance to impress your boss. You decide to work overtime and vow to back up your material more often. You're not alone. What may have been an annoyance to you and your boss multiplied many times has become a costly problem throughout the United States and the world. In many cases where offices and factories have become dependent on the smooth operation of computers, a single outage can be very costly. For example, a glass plant in 1993 estimated that an interruption of power of less than a tenth of a second can cost as much as $200,000, while for a computer center that experienced a 2-second interruption, it can cost $600,000 and a loss of 2 hours of data processing. According to *Science* ("Editorial: Magnetic Energy Storage," October 7, 1994), costs due to power fluctuations in the United States range from $12 to $26 billion. Consequently, the United States market for power quality services and equipment has grown to over $5 billion in 1999. Figure 1.1 shows how the cost of power quality disturbances have increased over the last 30 years.

Electrical power engineers have always been concerned about power quality. They see power quality as anything that affects the voltage, current, and frequency of the power being supplied to the end user, i.e., the ultimate user or consumer of electricity. They are intimately familiar with the power quality standards that have to be maintained. They deal with power quality at all levels of the power system, from the generator to the ultimate consumer of electrical power. They are not the only ones who need to be aware of power quality.

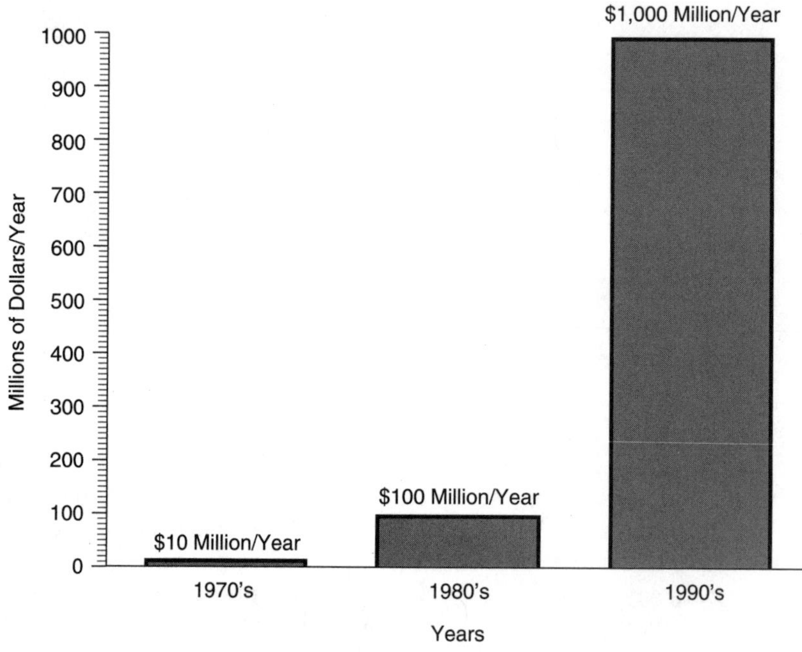

Figure 1.1 Increase in the cost of power quality in the United States. (*Courtesy of www.powerqualityinc.com*)

They share their concern with other professionals who sell and buy electrical power as well as those who sell and buy electricity-consuming appliances and equipment. They see that the market has expanded to include suppliers and consumers of equipment that mitigates power quality problems. That is why, as an electrical engineer, I see a need to communicate to others the importance of understanding power quality and power quality problems.

Power quality problems occur when the alternating-voltage power source's 60-Hz (50-Hz in Europe) sine wave is distorted. In the past, most power-consuming equipment tolerated some distortion. Today, highly sensitive computers and computer-controlled equipment require a power source of higher quality and more reliability than standard, less sensitive electricity-consuming equipment of the past, like motors and incandescent lights. Figure 1.2 illustrates how a voltage sine wave can become distorted.

The undistorted alternating-voltage sine wave repeats itself every cycle. The time required to complete one cycle is called a *period*. Because it repeats itself it is referred to as a *periodic wave*. The flow of electrons is called *current* and is measured in amperes. Current times voltage equals electrical power. Our beating heart pumps

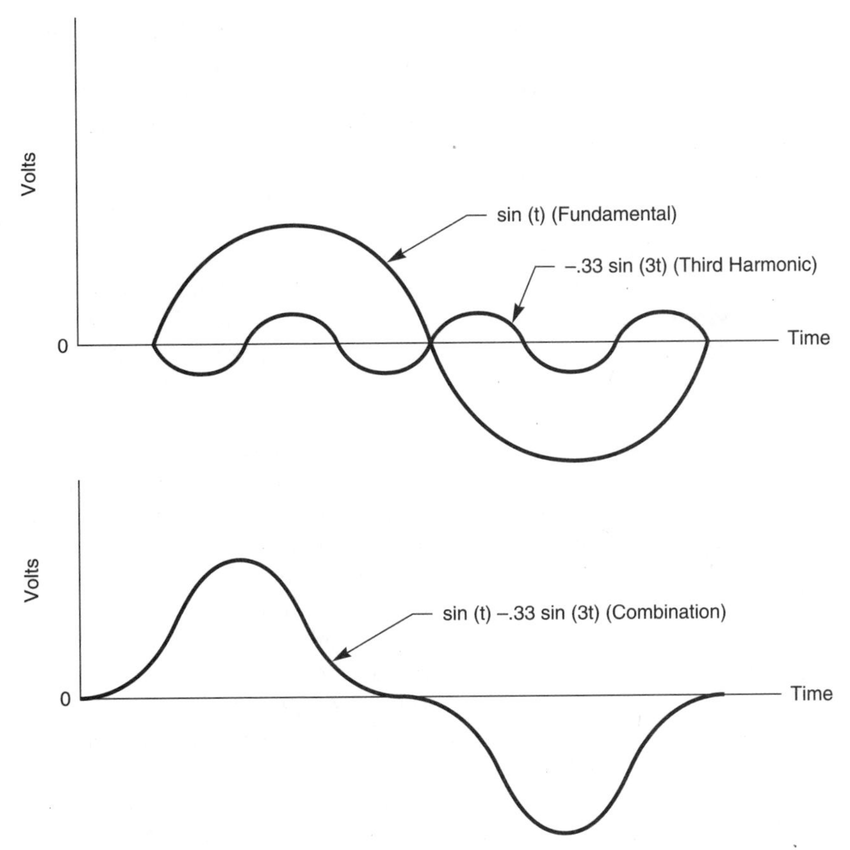

Figure 1.2 Distorted voltage sine wave.

blood that produces a periodic wave that can be seen on a heart monitor. The flow of electrons in a conductor is analogous to the flow of blood in an artery. The transmission and distribution systems that deliver electrons to the consumer are somewhat analogous to the arteries and veins that deliver blood to the vital organs of the body. Blood pressure is like voltage or the potential for the current to flow to the consumer. Voltage is a force or pressure and is measured in volts. The frequency of the heartbeat is like the frequency of electrical power. And the organs of the body are the various types of electrical loads distributed throughout the electrical power system. In the supply of electrical power, frequency is measured in Hertz (Hz; cycles per second). The United States uses 60-Hz power while Europe and Asia use 50-Hz power (by comparison the human heart normally beats at about 75 beats per minute). Figure 1.3 shows the similarity of a heart monitor to a power quality monitor for an electrical power system.

As loads have become more sensitive to variations in the quality of power, the definition of power quality has become important but somewhat confusing. This has caused utilities and their customers to take a look at the definition of power quality.

Power Quality Definition

Power quality can be defined from two different perspectives, depending on whether you supply or consume electricity. Power quality at the generator usually refers to the generator's ability to generate power at 60 Hz with little variation, while power quality at the transmission and distribution level refers to the voltage staying within plus or minus 5 percent. Gerry Heydt in *Electric Power Quality* defines power quality as "the measure, analysis, and improvement of bus voltage, usually a load bus voltage, to maintain that voltage to be a sinusoid at rated voltage and frequency." The type of equipment being used by the end user affects power quality at the end-user level. Roger Dugan, Mark McGranaghan, and Wayne Beaty in *Electrical Power Systems Quality* define a power quality problem as "any power problem manifested in voltage, current, or frequency deviations that results in failure or missed operation of utility or end user equipment." Figure 1.4 illustrates the different meanings of power quality. Economists and power marketers see power as a product and power quality as a measure of the quality of that product. The definition of power quality becomes even more unclear when the roles of utility and customer become blurred as the utility industry is restructured and deregulated. Because of the changing roles of the utility and the customer, I will try to present power quality from a power system standpoint rather than an ownership point of view. The evolution of the power system and the types of loads it serves is the major cause of an increased need for power quality.

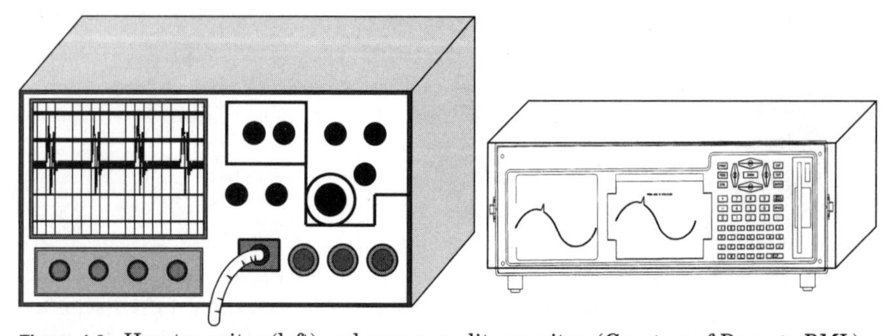

Figure 1.3 Heart monitor (left) and power quality monitor. (*Courtesy of Dranetz-BMI.*)

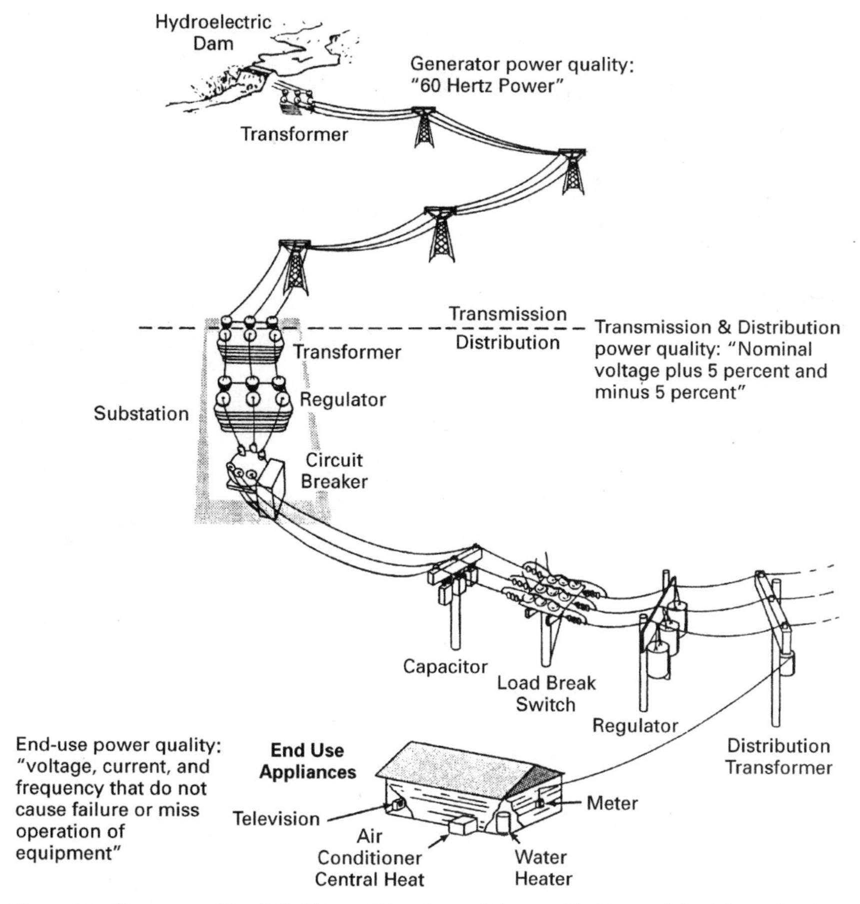

Figure 1.4 Power quality definitions. (*Courtesy of Bonneville Power Administration.*)

Need for Power Quality

Historically, power quality and reliability were synonymous. In the early days of the development of the power system, electrical engineers were mainly concerned about "keeping the lights on." They designed the power system to withstand outages by using lightning arresters, breakers and disconnect switches, and redundancy. The main concern was to prevent the frequency of the power system from deviating from 60 Hz during outages. Various devices were utilized to maintain the reliability of the power system. For example, if an outage of a major transmission line caused a large load to be dropped, there was concern about the generator "running away" and the frequency increasing above acceptable limits. Then, the whole power system would collapse. Large "dynamic brakers" consisting of many stainless

Figure 1.5 Dynamic braking resistors. (*Courtesy of Bonneville Power Administration.*)

steel wires were utilized to keep the generators from spinning out of control. These "giant toasters" are pictured in Figure 1.5.

Electrical engineers have always been concerned about the possibility of an outage of a transmission line or substation causing a cascading effect. This cascading effect would cause the various parts of the system to fall like dominos. This is what happened during the New York blackout of 1965. The failure of a relay in Canada to operate caused this particular blackout. Since then, electrical engineers have made great efforts in analyzing weaknesses in the system, using high-speed computers to perform steady-state power flow studies and transient stability studies.

Even with all these efforts, major outages have occurred in various parts of the world. For example, in 1997, the West Coast experienced a major blackout caused by a tree growing into a high-capacity 500-kV line on the Bonneville Power System in the Pacific Northwest. A contributing factor was that one of the major dams on the Columbia River was generating electricity at less than full capacity in order to allow salmon to migrate up the river to spawn. Even more recently, a large outage occurred in Canada and Northeastern United States because of extended cold weather and icing on power lines. In 1998, in New Zealand, a nationwide power outage occurred as a result of extremely hot weather and an inadequate power system. These are all examples of the need for reliable power. The need for reliable electrical power continues to grow throughout the world as the use of electricity increases.

However, brownouts (an extended reduction in voltage of more than 10 percent) and blackouts (total loss of all electrical power for more than a minute) make up only 4.7 percent of the total disturbances that may occur on a power system. Short-term changes in voltage called *transients* account for the other 95.3 percent. Power quality problems caused by transients have become an increasing concern since the 1980s.

The emphasis has shifted from concern about the reliability at the transmission and distribution level in the 1980s to concern about power quality at the end-user level. The biggest cause of this shift is the growing computer use since the 1980s. This is because computers are more sensitive to deviations in power quality.

Sensitive loads

Computers and microprocessors have invaded our homes, offices, hospitals, banks, airports, and factories. It is hard to imagine any industry today that is not impacted by computers and microprocessors. Microprocessors have even become a part of today's toys and consumer appliances. Figure 1.6 shows examples of microprocessor-controlled equipment that can be affected by poor power quality.

Why do computers cause loads to be more sensitive? The brains of all computers are integrated circuit (IC) chips. They are the source of this sensitivity, which has increased over the last 25 years as more transistors have been placed on a micro chip. The number of transistors on a chip has increased significantly from the two transistors on the first microchip invented in 1958 to 7.5 million on Intel's Pentimum II microchip in 1995, as illustrated in Figure 1.7 (mips refers to millions of instructions per second). In fact, the computer industry has observed that each new chip contains roughly twice as much capacity as its predecessor and each chip is released within 18 to 24 months of the previous chip. This principle has become known as Moore's law and was named after an Intel founder, Gordon Moore, who made this observation in a 1965 speech.

As computer chip manufacturers seek to increase the density of electrical components on a chip, the chips become even more sensitive to changes in the electrical power supply. The density of these components in a very small package causes computers to have a low tolerance for voltage deviations. They are prone to current flowing from one conductor to another if the insulation is damaged. As more components are jammed in a small area, they will tend to generate more insulation-damaging heat. Figure 1.8 shows the density of the electrical components in an IC.

In addition, computers use the on and off voltages and the timing provided by the power supply to store and manipulate data in the microprocessor. Any deviations from the voltage that is specified can cause the data to be corrupted or erased. This is what often causes your computer to "freeze up." These disturbances affect not only your personal computer, but also any industrial or commercial office process

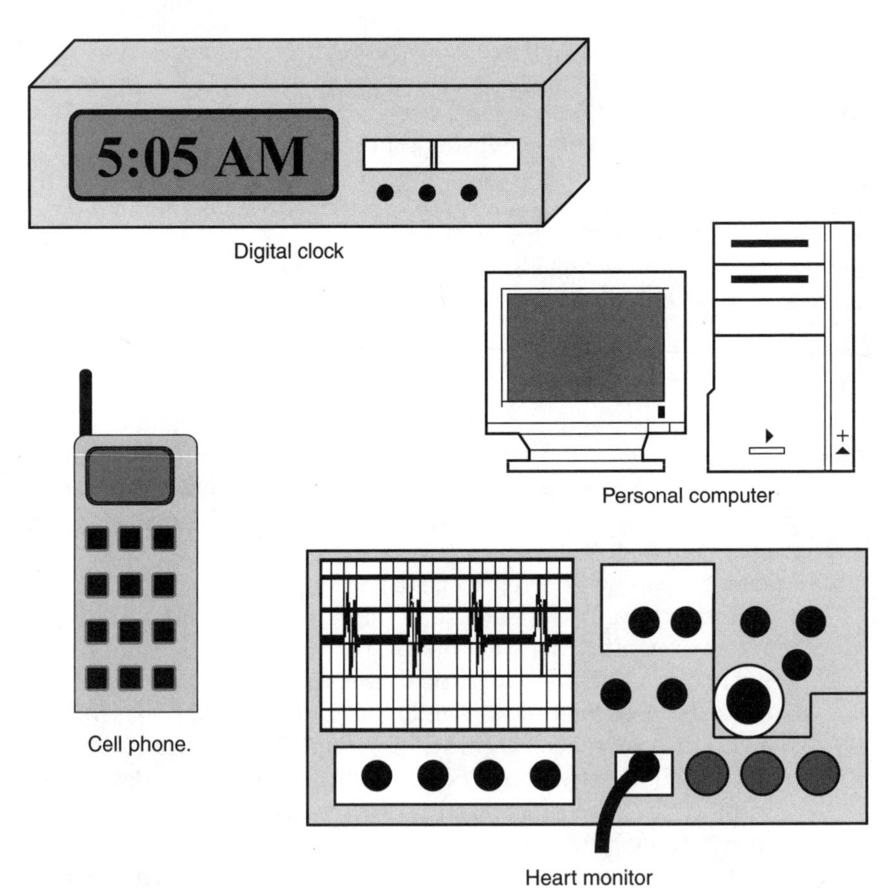

Figure 1.6 Examples of microprocessor-controlled equipment.

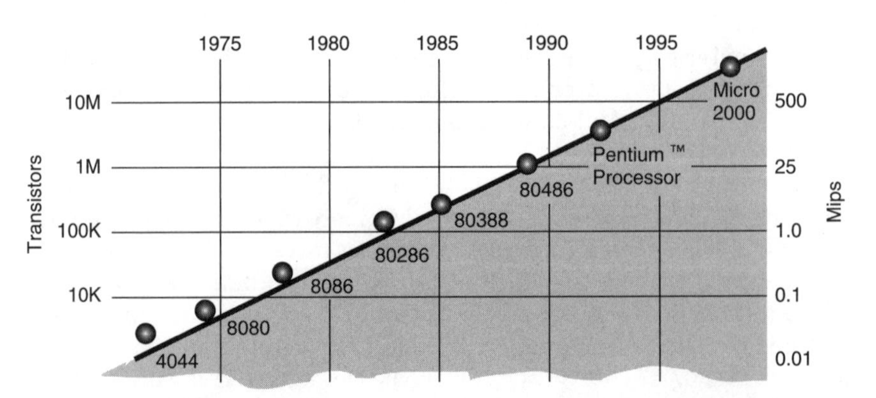

Figure 1.7 Graph of increased integrated circuit density. (*Courtesy of Intel Corp., copyright Intel Corp. 2000).*

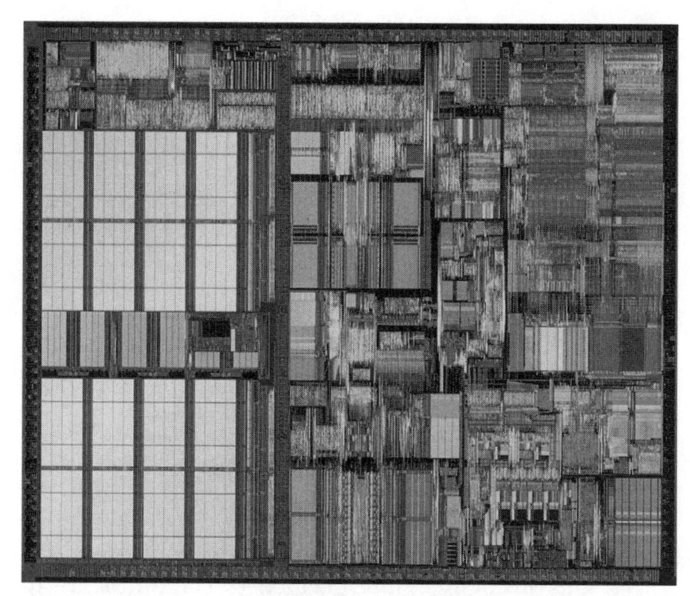

Figure 1.8 Integrated circuit components. (*Courtesy of Intel Corp., copyright Intel Corp. 2000.)*

that uses microprocessors. These include electronically controlled devices, such as adjustable-speed drives, scanners, cash registers in grocery stores, fax and copy machines in offices, telecommunication equipment, and medical equipment.

Power quality has probably not deteriorated over time, but instead the equipment requirements for higher power quality have increased in the 1990s. In the past, most equipment could tolerate a voltage disturbances of ±5 percent of nominal voltage. For example, nonelectronic equipment, like motors, incandescent lights, and resistance heaters, could tolerate decreases and increases in voltage of 6 V on a 120-V receptacle. Table 1.1 from the American National Standards Institute (ANSI) 8411 shows the voltage tolerances in the secondary system, i.e., 120 V in a residence and 480 V in a factory, of the end user.

Even though more equipment have become more voltage-sensitive, most electricians show very little concern about power quality. Often their only concern is with safety and that the wiring and grounding meet National Electrical Code (NEC) standards. The NEC standards deal with personal safety and fire protection and not with the fact that microprocessors use on and off logic voltages of 0.5 to 1 V. Someone needed to develop standards that deal with voltages disturbances on the power system that cause the logic voltage in the microprocessor to either dip below or rise above these levels. Otherwise, an erroneous data signal could be sent to the microprocessor and cause data to be corrupted and computers to freeze up. Something had to be done.

TABLE 1.1 ANSI C84.1 Secondary Voltage Standards

Type of secondary system measure		Normal conditions, V	Contingency conditions, V
120/208 V, 3 phase, 4 wire	Phases to neutral	126–114	127–110
	Phase to phase	218–197	220–191
		126–114	
120/208 V, 3 phase, 3 wire	Line to neutral	127–110	
	Line to line	254–220	252–228
277/480 V, 3 phase, 4 wire	Phase to neutral	291–263	293–254
	Phase to phase	504–456	508–440

SOURCE: Reprinted from ANSI C84.1 by permission from National Electrical Manufacturers Association. Copyright © 1996 National Electrical Manufacturers Association.

The Computer and Business Equipment Manufacturers Association (CBEMA) recognized this problem. They decided to communicate to electrical utilities the kinds of voltage variations that sensitive microprocessors could not tolerate. The association developed the so-called CBEMA curve. The United States Department of Commerce published in 1983 Federal Information Processing Standards (FIPS) Publication 94, containing the CBEMA curve. The CBEMA curve in Figure 1.9 shows the susceptibility limits for computer equipment.

The Information Technology Industry Council (ITIC) replaced the Computer and Business Equipment Association. The ITIC has created its own curve that illustrates the tolerances of voltage variations of microprocessors. Figure 1.10 shows the new ITIC curve. The ITIC plans to revise even this graph. Chapter 3, "Power Quality Standards," discusses this graph in more detail. While the computer and utility industries were trying to respond to the increased sensitivity of microprocessors to voltage variations, they were confronted by another problem: Utility customers, i.e., end users, were using equipment that in itself caused power quality problems. For example, more and more utility customers were using equipment that caused nonlinear loads.

Nonlinear loads

In the last decade, industrial end users of electricity have bought and installed the latest technology for saving energy in their factories. Utilities, state, and federal government agencies have even provided financial incentives to encourage the use of energy-saving devices, like adjustable-speed drives.

Adjustable-speed drives have become one of the most popular technologies for saving energy in factories and some commercial facilities.

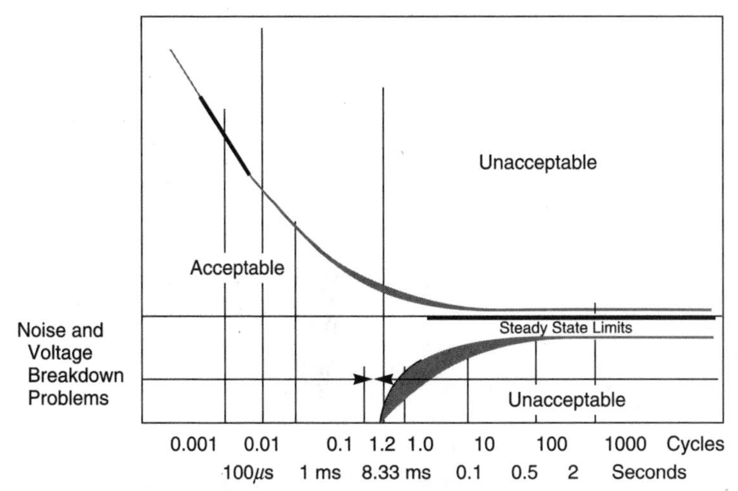

Figure 1.9 Simplified CBMA curve. (*Courtesy of the Information Technology Industry Council.*)

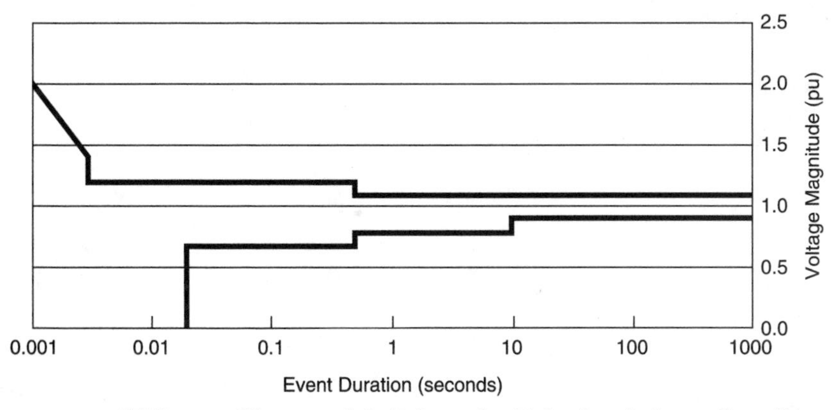

Figure 1.10 ITIC curve. (*Courtesy of the Information Technology Industry Council.*)

These devices use the latest electronic controls to control the speed of motors to match the requirements of the load. However, they have been a source of trouble. They trip off inadvertently. They cause nearby transformers to overheat and trip off. What is causing this to happen? The adjustable-speed drives produce nonlinear loads. Nonlinear loads, such as adjustable-speed drives, electronic ballasts for fluorescent lamps, and power supplies for welding machines, as shown in Figure 1.11, have become sources of poor power quality. What are nonlinear loads and how do they cause poor power quality?

Nonlinear loads are simply any piece of equipment or appliance that increases and reduces its consumption of electricity over time in a nonlinear fashion. With nonlinear loads the current and voltage do not follow each other linearly. In Article 100 of the NEC, a nonlinear load

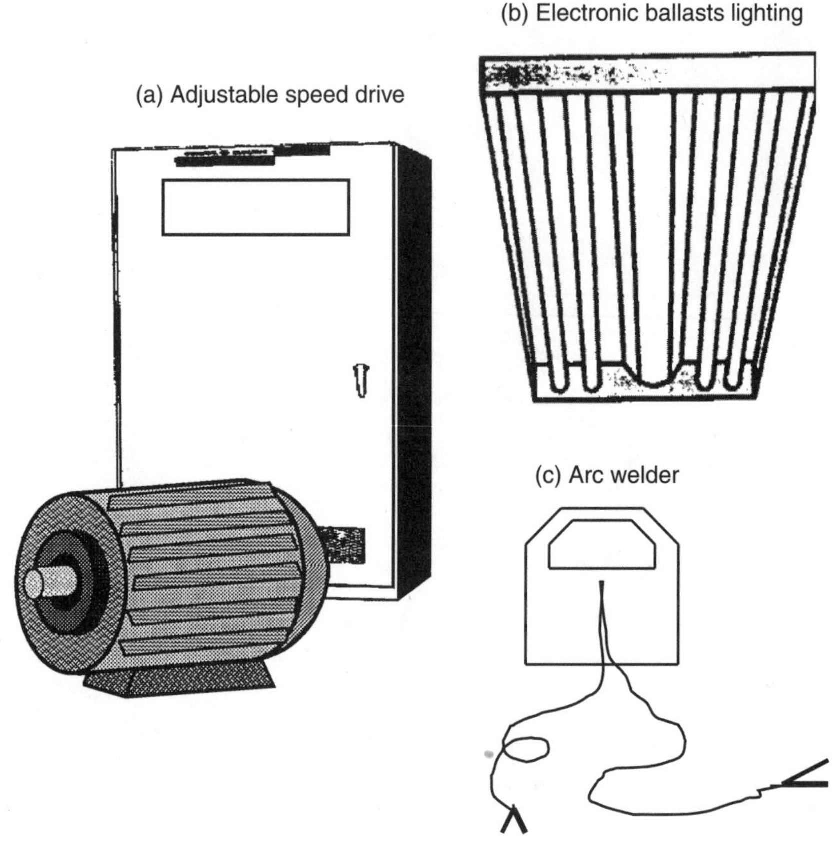

Figure 1.11 Examples of nonlinear loads.

is defined as "a load where the waveshape of the steady state current does not follow the waveshape of the applied voltage." This usually occurs when the load is not a pure resistance, capacitance, or inductance, but instead contains electronic components to control the function of the equipment to meet the requirements of the load. Often the nonlinearity of the load results in the generation of harmonics that cause overheating of electrical equipment. Figure 1.12 shows how harmonics add to the fundamental 60-Hz power and cause overheating.

Programs to improve the efficiency of production have resulted in the use of nonlinear equipment such as adjustable-speed drives, fluorescent lighting, induction heating, electron beam furnaces, static power converters, and power-factor-improving shunt capacitors. These devices often generate or amplify existing harmonic currents that distort the voltage wave. These voltage distortions can be transmitted to the utility's system and from the utility's system to nearby interconnected end

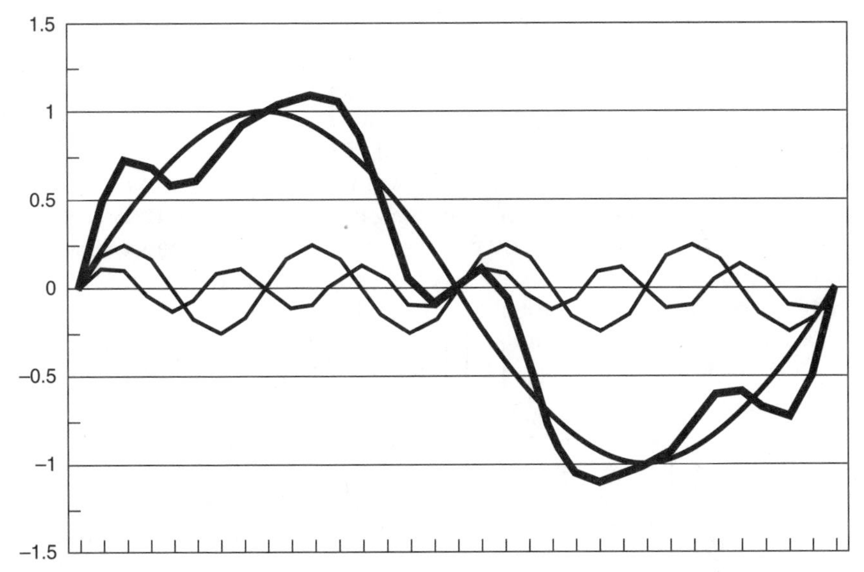

Figure 1.12 Harmonics effect.

users. In addition, increased use of arc furnaces causes voltage flicker, i.e., dips, that in turn cause lights to flicker and irritate people.

New types of loads that generate harmonic voltage distortion are becoming more common, such as electron beam furnaces for melting titanium and induction furnaces for processing aluminum. A large inrush current, as much as 6 times normal current, is required to start up large horsepower motors. This large inrush current causes the voltage to sag (dip). Chapter 2, "Power Quality Characteristics," explains in more detail how these loads cause power quality problems.

All these types of loads result in one customer causing power quality problems for another customer. Utilities cannot afford to allow such problems to continue; they affect the utilities' and their customers' competitiveness. Utilities need to identify the customers causing a power quality problem and require them to fix it. Utilities and their customers also need to have procedures that prevent power quality problems. They need to have power quality contracts that require the end user causing power quality problems to be responsible for fixing them. Chapter 8, "Future Trends," explains how to write power quality contracts. The need for power quality has become more complicated as power systems have become more interconnected.

Interconnected power systems

As utilities have increased the number of interconnections in their power systems to meet growing loads and reliability standards, they have built a increasingly complex and interconnected power system in the United States and throughout the world. The increasing interconnectedness of the power systems often results in the power quality problems of one utility or end user causing another utility or end user to have power quality problems. This is why it has become more difficult to isolate the cause of a power quality problem. For example, an end user's facility can cause a power quality problem and transmit the problem to the servicing utility power system, which then transmits the problem on another utility's power system to another end user's facility. Harmonics and flicker are good examples of power quality problems that are transferred from one utility to another through interconnected power systems. Figure 1.13 shows how the high-voltage transmission system of a utility is interconnected with its own distribution system or the distribution system of another utility serving homes, offices, and factories.

In the past, in many parts of the United States and throughout the world, one utility provided generation, transmission, and distribution services to its customers. This is called a full-service, vertically integrated utility. One utility will no longer provide all these services when the electric utility industry becomes deregulated and restructured. Different companies will supply generation, transmission, and distribution services. How will the restructuring and deregulation of

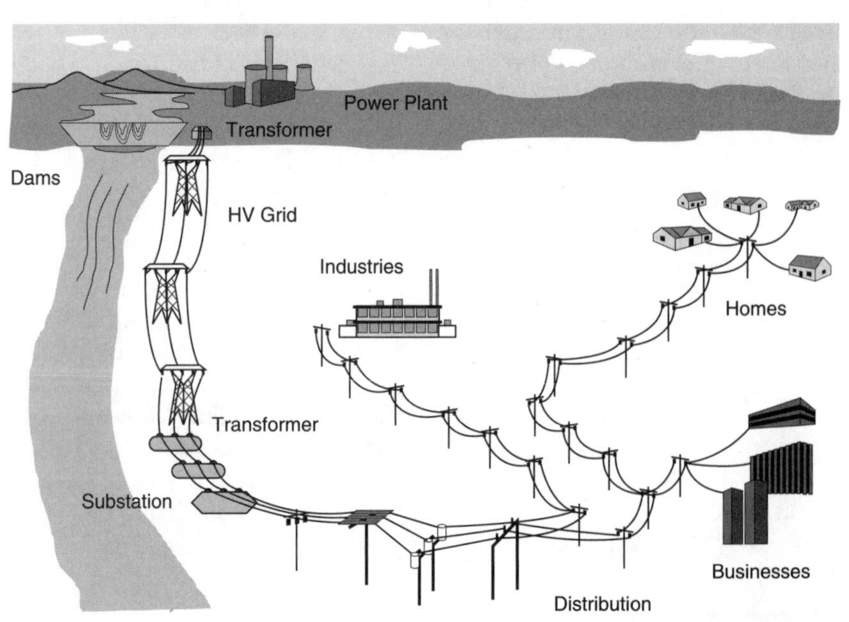

Figure 1.13 Interconnection of utility power systems (*Courtesy of Bonneville Power Administration.*)

the electric utility industry affect its ability to deliver quality power to its customers? Who will the utility customer contact when it has a power quality problem? Deregulation will have a complicating effect on the utility customer.

Deregulation

The restructuring and deregulation of the utility industry will cause many customers to choose utilities that can supply high-quality as well as low-cost power. Consequently, utilities will be able to retain existing customers and attract prospective new customers if they are able to demonstrate that they can deliver power with high quality. Utilities with power quality programs, including power quality monitoring and site surveys, will be better able to convince existing and prospective customers that they see power quality not as a problem but as an opportunity to provide customer service and help their customers be more competitive. Chapter 6, "Power Quality Measurement Tools," discusses the various types of power quality monitoring systems available today and how to use them to prevent and solve power quality problems on both sides of the meter. Chapter 7, "Power Quality Surveys," shows how to plan, conduct, and analyze power quality surveys. The utility customer sees that it can be more competitive if it has assurance that its power supply is high in quality and reliability. How will utilities and their customers deal with increasing power quality problems as the utility industry becomes deregulated and more competitive? Experience with the deregulation of the utility industry in various parts of the world can help answer this question.

Deregulation has been in effect for several years in many parts of the world, including the United Kingdom, Australia, New Zealand, and South America. Deregulation in the United States is a relatively new phenomenon. In fact, many United States utilities are purchasing deregulated foreign utilities in order to get experience in how to compete in the upcoming deregulated utility market. The deregulation process began in the United States with the passage of the 1992 Energy Policy Act. Passage of this act was soon followed by the Federal Energy Regulatory Commission (FERC) introducing on April 7, 1995, the Notice of Proposed Rulemaking (Mega-NOPR).

With Mega-NOPR, FERC requires utilities to provide open transmission access and separate their power business from their transmission and distribution (T&D) business. This has an effect on the electric utility industry in the United States. In the past, so-called vertically integrated monopolies dominated the electrical utility industry. This means that utilities owned generation, transmission, and distribution facilities and provided electrical energy to designated franchised customers. They were guaranteed a customer base and a profit by the various state regulatory commissions throughout the United States. The adoption of Mega-NOPR has caused the state legislatures and regulatory agencies to pass laws and rules to deregulate the electric utility industry.

The process of deregulation is progressing steadily. Many states are passing legislation to break up the utility monopolies. They are trying to encourage competition by allowing end users to choose electrical suppliers. Several states have begun the process of deregulating the utility industry. For example, in California, the California Public Utilities Commission (CPUC) has proposed to implement deregulation with a phased approach. The CPUC allowed the large industrial end users to choose their suppliers of electricity on January 2, 1996. The CPUC plans allow the various segments of the electrical utility market to participate in the deregulation process according to the following schedule: small industrial end users in 1997, commercial customers in 1998, and residential customers in 2002.

In a deregulated environment, utilities will be divided into separate companies. The generation companies will be called GENCOs. The transmission companies will be called TRANSCOs. The distribution companies will be called DISTCOs or DISCOs, while the companies providing unbundled energy services will be called ESCOs. Most utilities will become TRANSCOs or DISTCOs, while the great majority of public utility districts, municipalities, and cooperatives will become DISTCOs. The TRANSCOs' and DISTCOs' primary and sometimes only source of revenue will come from GENCOs. GENCOs will pay TRANSCOs and DISTCOs for the right to "wheel" (transmit electricity on someone else's power system) on their transmission and distribution systems.

The GENCOs will expect reliable and high-power-quality T&D systems. A reliable and high-power-quality T&D system offers many benefits, including making the TRANSCOs and DISTCOs more competitive, reducing the threat of end users building their own generation capability, and satisfying regulators that the T&D system is high in power quality.

The TRANSCOs and DISTCOs will most likely continue to be regulated monopolies. It is expected that the formation of an independent system operator (ISO) will be necessary to coordinate the operation of the various T&D systems. TRANSCOs and DISTCOs will probably find that the regulators will set standards on power quality. In the United States the regulators will probably adopt standards developed by the Institute of Electrical and Electronics Engineers (IEEE) and Electric Power Research Institute (EPRI). That means TRANSCOs and GENCOs will need power quality monitoring systems to assure they are adhering to those standards. Otherwise the TRANSCOs and DISTCOs will not be able to show their customers and regulators that they are not affecting the quality of power. This should provide an incentive for TRANSCOs and DISTCOs to improve the quality of power. Figure 1.14 illustrates how the utility industry will change when it becomes deregulated.

Chapter 8, "Future Trends," discusses in more detail deregulation of the utility industry and how it will affect power quality and the roles of the utilities and their customers in providing and receiving quality power. Other stakeholders besides the utilities and their customers participate in the power quality industry.

Who's Involved in the Power Quality Industry?

The primary participants in preventing and solving power quality problems include the utility, the end user, and the equipment manufacturer. In addition to these three primary participants, the power quality industry includes several other participants, including the power conditioning equipment manufacturers, standards organizations like IEC, IEEE, and ANSI, research organizations like EPRI and PEAC, consultants, monitoring and measuring equipment manufacturers, and architect/engineer facility designers. All these organizations need to work together to ensure that the end users get the power

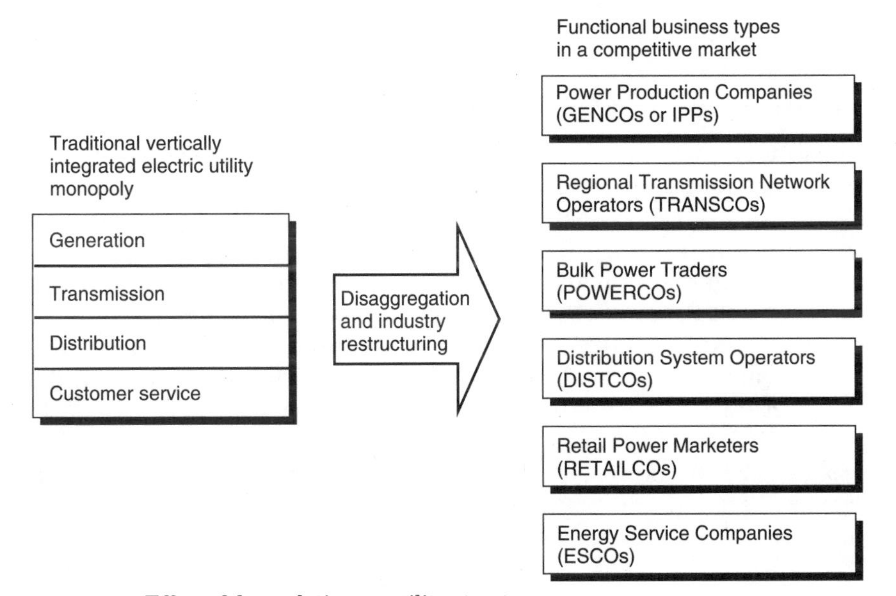

Figure 1.14 Effect of deregulation on utility structure.

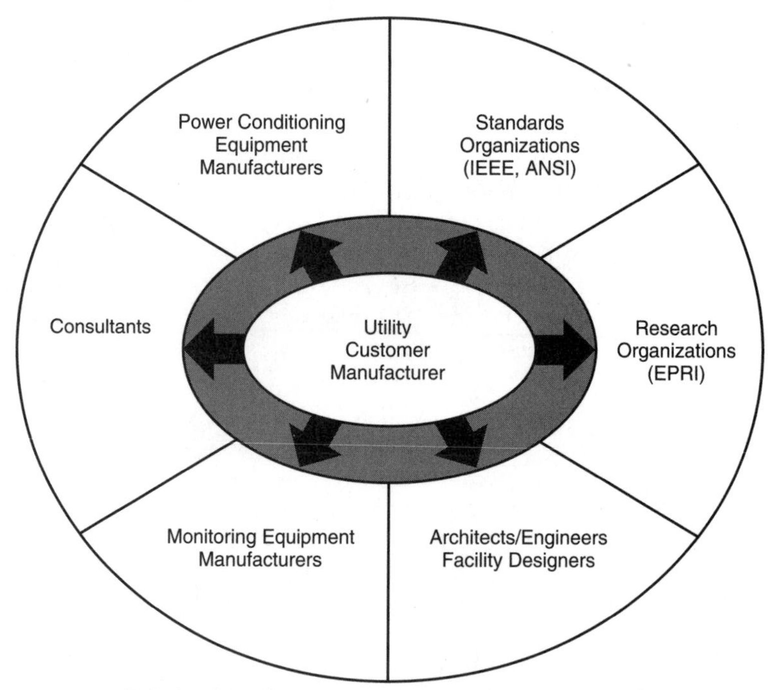

Figure 1.15 Relationship of organizations involved in power quality.

quality they need to operate their equipment. Figure 1.15 illustrates how these various organizations need to work together. Each chapter in the book discusses them in more detail.

Research and development organizations

Government agencies, universities, and manufacturing industries contribute in varying degrees to power quality research and development. EPRI has been a major contributor to power quality research and development. In the 1970s, the electric utility industry founded EPRI as a nonprofit research arm of the electric utility industry. Member utilities fund EPRI from their revenues. It has been in the forefront of research to study and solve power quality problems. It has developed Power Quality Service Centers throughout the United States and the Power Electronic Application Center (PEAC) in Knoxville, Tenn., to provide information and training on the use of EPRI power quality products. EPRI has developed power quality studies, guidebooks, training, mitigation hardware, and diagnostic software. This book describes the various EPRI power quality products and services; each chapter of the book deals with the various contributions EPRI has made in power quality. Chapter 9, "Future Trends," describes EPRI's contributions to power quality research and development. EPRI has, as well, helped develop power quality index standards. In addition, the Institute of Electrical and Electronic Engineers (IEEE) and the International Electrotechnical Commission (IEC) have contributed significantly to the development of power quality standards.

Standards organizations

The power quality standards organizations seem like a confused mix of alphabet soup with the various acronyms such as IEEE, IEC, ANSI, NEC, etc. The Glossary at the end of the book reduces this confusion by explaining these and other acronyms used in this book and in the power quality industry. Chapter 3, "Power Quality Standards," clarifies and explains how all these organizations work together to set standards and guidelines for the power quality industry. It is important to understand how these organizations develop and deliver standards so that you know what standard to refer to in solving your particular power quality problems.

In the United States, several organizations have developed power quality standards. The IEEE is the most prominent of these organizations. The IEEE is the largest professional organization in the world. It has over the years developed standards for the electrical and electronic industries. It has been especially active in developing power quality standards. Table 1.2 describes the various IEEE power quality standards as well as those IEEE power quality standards adopted by the American National Standards Institute (ANSI), identified as ANSI/IEEE standards. In addition to the IEEE power quality standards, Table 1.2 presents power quality standards developed by the National Electrical Manufacturers Association (NEMA), National Fire Protection Association (NFPA), the National Institute for Standards and Technology (NIST), and the Underwriters Laboratories (UL). Chapter 3, "Power Quality Standards," describes in more detail the IEEE and other United States organizations and their power quality standards, as well as international power quality standards organizations, like the International Electrotechnical Commission (IEC).

Consultants

The number of consultants in the power quality industry have grown considerably in the 1990s, as power quality problems have become more prevalent. They have traditionally been involved in providing power quality training, although universities and utilities have become more involved in training recently. Many consultants and utilities are providing training on how to solve power quality problems. Consultants have provided the tools that are necessary to diagnose

TABLE 1.2 United States Power Quality Standards Synopsis

Title/Subject	Standards
Industrial Electric Power Systems (*Red Book*)	ANSI/IEEE 141
Industrial & Commercial (I&C) Power System Ground (*Green Book*)	ANSI/IEEE 142
Commercial Electric Power Systems (*Gray Book*)	ANSI/IEEE 241
I&C Power System Protection (*Buff Book*)	ANSI/IEEE 242
I&C Power System Analysis (*Brown Book*)	ANSI/IEEE 399
I&C Power System Emergency Power (*Orange Book*)	ANSI/IEEE 446
I&C Power System Reliability (*Gold Book*)	ANSI/IEEE 493
Control of Noise in Electronic Controls	ANSI/IEEE 518
Harmonics in Power Systems	ANSI/IEEE 519
Electric Systems in Healthcare Facilities (*White Book*)	ANSI/IEEE 602
Energy Management in I&C Facilities (*Bronze Book*)	ANSI/IEEE 795
Interconnection Practices for Photovoltaic Systems	ANSI/IEEE 929
Interfacing Dispersed Storage and Generation	ANSI/IEEE 1001
Test Procedures for Interconnecting Static Power Converters	ANSI/IEEE 1035
Grounding of Power Station Instrumentation and Control	ANSI/IEEE 1050
Guides and Standards on Surge Protection	ANSI C62
Voltage Ratings for Power Systems and Equipment	ANSI C84.1
Guides and Standards for Relay and Overcurrent Protection	ANSI C37
Transformer Derating for Supplying Nonlinear Loads	ANSI C57.110
Electromagnetic Compatibility	ANSI C63.18
Wire Line Communication Protection in Power Stations	IEEE P487
Power and Ground Sensitive Electronic Equip. (*Emerald Book*)	IEEE 1100
Monitoring and Definition of Electric Power Quality	IEEE 1159
Guide on Equipment Sensitive to Momentary Voltage Disturbances	IEEE 1250
Guide on Compatibility for ASDs and Process Controllers	IEEE P1346
Uninterruptible Power Supply Specification	NEMA UPS
National Electric Code	NFFA 70
Protection of Electronic Computer Data Processing Equipment	NFFA 75
Lightning Protection Code for Buildings	NFFA 78
Electric Power for ADP Installations	NIST 94
Overview of Power Quality and Sensitive Electrical Equipment	NIST SP678
Standard for Safety of Transient Voltage Surge Suppressors	UL 1449

power quality problems, such as power quality monitoring systems discussed in Chapter 6, "Power Quality Measurement Tools," and power quality computer simulations and diagnostic tools discussed in Chapter 7, "Power Quality Surveys." They are highly involved in research and development as well as in helping utilities and their end-user customers solve power quality problems.

End-user equipment manufacturers

End-user equipment manufacturers include manufacturers of motors, adjustable-speed drives, lighting, computers, capacitors, transformers, and any other type of electricity-consuming equipment. They play a definite role in determining the level of power quality required by the end user. When they design and build equipment, they determine the sensitivity or robustness of their equipment to power quality variations.

Purchasers of electrical equipment need to be aware of the power quality robustness of the equipment they plan to purchase. "Robust" means equipment that has less sensitivity to power quality variations. European equipment tends to be more robust than equipment made in the United States. This is especially true of computers. Some manufacturers of equipment that generate harmonics, like fluorescent lights and adjustable-speed drives, install harmonic filters in their equipment to keep harmonics from affecting other equipment on the end user's system. The old adage of "Buyer beware" applies just as well to the purchase of electrical and electronic equipment. Buyers of electronic equipment today need to know the tolerance of the equipment they are buying. Smart buyers specify the power quality requirements of the equipment before buying it.

Monitoring-equipment manufacturers

Historically, power quality monitoring-equipment manufacturers provided monitors that were installed by the power quality engineer in the field at the point of common coupling. The point of common coupling is the point where the utility connects to the end-user customer. The power quality engineer would have to go to the site and download the data that had been measured and recorded over a period of time. The power quality engineer would take the data back to the office and analyze the data to determine the cause of the power quality problem. This approach worked when there were a few temporary meters. Recent meters allow the power quality engineer to access the data remotely through a modem and telephone line. Present and future power quality monitoring will require several meters at many sites, installed permanently to monitor the power quality for statistical or diagnostic analysis. The power quality engineer can use statistical analysis to determine the relationship between the power system configuration and the level of power quality or to show deviations from power quality standards. The power quality engineer can use diagnostic analysis to determine the source of a particular power quality problem. Chapter 6, "Power Quality Measurement Tools," presents the various types of power quality meters available today and how they measure power quality. It also shows how power quality monitoring-equipment manufacturers have developed sophisticated systems that allow the utilities and their customers to monitor the power quality status of critical loads by using the Internet and pagers.

Power conditioning equipment manufacturers

Power conditioning equipment manufacturers include manufacturers of surge suppressers, isolation transformers, filters, uninterruptible power supplies (UPSs), static VAR (volt-amperes reactive) controllers, and superconducting magnetic energy storage (SMES) devices. They have designed these devices to protect critical equipment from being damaged or not operating correctly because of a variation in power quality. A whole industry has developed to provide these devices. EPRI's 1994, Power Quality Market Assessment report concluded that the "United States power quality mitigation equipment market will grow from approximately $2 billion in 1992 to $5.6 billion in 2002." Frost and Sullivan in a June 1999 *Power Quality Assurance* magazine article indicated that the power conditioning market, including UPS, had become a $4.3 billion industry in 1999. Chapter 4, "Power Quality Solutions," explains how this equipment works and discusses who are the various suppliers of power quality mitigation equipment.

Utilities

Since electric utilities supply, transmit, and distribute electric power to residential, commercial, and industrial end users, they are intimately involved in power quality. Figure 1.16 shows the amount of energy electrical utilities sold to the industrial, commercial, and residential sectors in 1996. Their role has become more complicated today as many utilities find it necessary to get involved in power quality on both sides of the meter. As mentioned earlier, the continuing specter of deregulation of the electric utility industry will change the role of the utilities significantly. This changing role will be a theme throughout this book. One thing is certain: The role of the electric utility will continue to serve end users with electrical power of a quality that meets their needs.

End users

End users include any user of electricity. They can be categorized into residential, commercial, and industrial. Those end users most concerned about power quality have increased significantly and will continue to increase every day. It mostly depends on whether they use microprocessors. EPRI projects that over half of the electric load in the twenty-first century will use microprocessor-based equipment. Some of the end users especially concerned about power quality are hospitals with all their high-tech electronics (magnetic radiation imaging, CAT scans, and heart monitors); home-based businesses that use personal computers, fax machines, and copiers; all types of industries from potato chip to electronic chip manufacturing; retail stores with computer-controlled cash

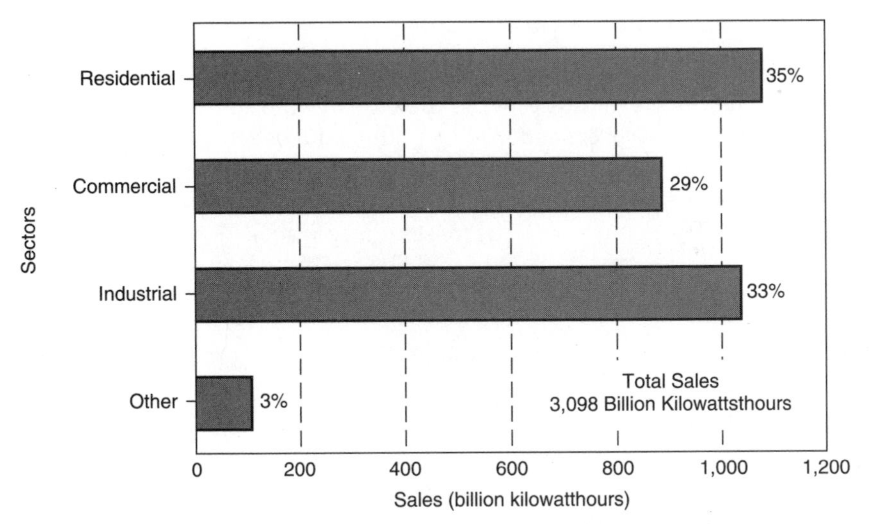

Figure 1.16 Total United States electric utility sales by sector in 1996. (*Courtesy of Energy Information Administration.*)

registers; banks with electronic automatic teller machines (ATMs), commercial office buildings and shopping centers, and even farms that use electronic equipment to milk their cows or control their irrigation systems. All the information provided throughout this book is targeted toward meeting the needs of the ultimate customer, the end user of electricity, and helping the electric utilities and manufacturers of power conditioning and power quality measuring instruments meet the needs of the customer. Both the end user and the utility are finding it necessary to seek legal advice from lawyers to protect their interests.

Lawyers

Lawyers increasingly play an important role in power quality because of legal issues. Malfunction of equipment can not only cost money but human suffering as well. A patient being treated for cancer in his arm experienced overradiation from a computer-controlled cobalt machine. He had to have his arm amputated. He sued the utility and hospital for several million dollars. Further investigation revealed that the copy machine in the room next to the cancer treatment room caused the cobalt machine to malfunction. Lawyers are also needed for preparing contracts between the various participants in power transactions. Chapter 9, "Future Trends," provides guidance in how to prepare the technical language in a power quality contract. These power quality contracts help buyers, sellers, and distributors of electricity to hold a person who causes a power quality problem responsible for paying the expense of mitigating the power quality problem.

How Much Does Power Quality Cost?

The common denominator of any power quality problem is the cost. How do you measure that cost? Is the cost measured by the loss of production? Or is the cost measured by potential legal consequences of not fixing the power quality problem? Or is the cost measured by the cost to mitigate or prevent the problem from ever occurring? Chapter 8, "Power Quality Economics," discusses these issues.

How to Use This Book

The book is organized to make it easy for a reader to find an answer to any question about power quality. The first five chapters provide background information on power quality, while the next three chapters provide ways to diagnose and solve power quality problems. Chapter 8 explains how to evaluate the economics of alternative power quality solutions and choose the most cost-effective solution. Chapter 9 explains how to treat power quality as a business from the utility perceptive and provides guidelines on how to write power quality contracts. Chapter 9 also explains how to segment the power quality market and provide the consumers of electricity with information to help them make wise choices in getting the power quality they need. It provides information on the latest power quality research and development projects. Figure 1.17 illustrates in flowchart form how the various chapters of the book build on one another.

With new and existing electronic technologies and the effect of utility deregulation, power quality becomes increasingly important. Using the techniques and knowledge described in the chapters of this book will allow all suppliers, distributors, and consumers of electricity to deal with power quality problems in a way that makes economic sense for them. The next chapter will lay the framework for understanding power quality. If you are familiar with power quality already, you can skip Chapter 2 and go on to Chapter 3, "Power Quality Standards."

References

1. Ibrahim, A. Rashid and K. Seshadri. 1995. "Power Quality for Beginners." *IAEEL Newsletter 3-4/95.* URL address: *http://www.stem.se/iaeel/IAEEL/NEWSL/1995/trefyra1995/LiTech_b-3-4-95.html.* Available from Public Utilities Board, Singapore.
2. McCluer, Stephen W. 1997. "Defining Power Quality in the Age of Solid-State Electronics." *Plant Engineering,* vol. 51, no. 7, July, pp. 81–85.
3. Sabin, Daniel D. and Ashok Sundarram. 1996. "Quality Enhances." *IEEE Spectrum,* February, pp. 34–41.
4. Douglas, John. 1985. "Quality of Power in the Electronics Age." *EPRI Journal,* pp. 7–13.
5. Vassell, Gregory S. 1990. "The Northeast Blackout of 1965." *Public Utilities Fortnightly,* vol. 136, no. 8, October 11, pp. 12–17.

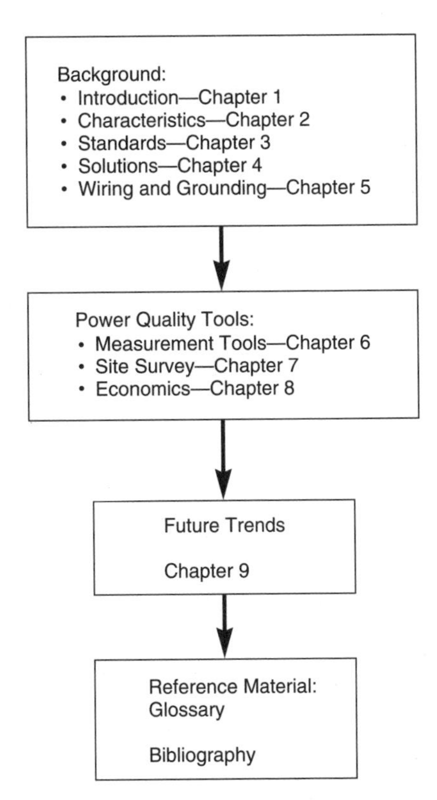

Figure 1.17 Flowchart of book chapters.

6. Casazza, John. 1998. "Blackouts: Is the Risk Increasing?" *Electrical World,* vol. 212, no. 4, April, pp. 62–64.
7. Beaty, Wayne. 1994. "Clean Power Requires Cooperative Effort." *Electric Light & Power,* vol. 72, no. 8, August, pp. 20–24.
8. Douglas, John. 1993. "Solving Problems of Power Quality." *EPRI Journal,* vol. 18, no. 8, December, pp. 8–15.
9. Gilker, Clyde. 1999. "Investigating Power Quality Problems: A Systematic Approach Is the Best Way to Track Down Power Quality Problems." *Design Extra.* URL address: *http:www.csemag.com/SquareD/topic/investigating.asp.* Available from Square D.
10. Ahuja, Anil. 1997. "Power Quality from the Bottom Up." *Consulting-Specifying Engineer,* May, pp. 68–74.
11. Fleishman, Barry J., et al. 1997. "Power Quality and Products Liability Law: Emerging Issues and Concepts." *Power Quality Assurance. URL address: http://www.powerquality.com/art0034/art1.html*
12. Hof, Robert D. 1991. "The Dirty Power Clogging Industry's Pipeline." *Business Week,* no. 3207, April 8, p. 82.

Chapter

2

Power Quality Characteristics

The invention of alternating current electricity caused a controversy that waged in the nineteenth century. The outcome of this controversy would influence the use of electricity to this day. The controversy centered on whether electricity should be delivered as direct current (dc) or alternating current (ac). On the surface, the difference between dc and ac didn't seem controversial. DC delivers electricity at a constant voltage and current over time, while ac delivers electricity at a varying voltage and current over time, as shown in Figure 2.1.

The great American inventor Thomas Edison promoted dc. By late 1887, he had built 121 central stations distributing dc power at 110 V that powered more than 300,000 of his incandescent lamps. Edison argued that dc was safer than ac. He even tested the safety of ac versus dc by electrocuting a horse. He seemed to ignore dc's inherent disadvantages.

DC can operate only at generator voltage. This is inefficient. This inefficiency of dc can best be understood by first understanding how voltage and current affect the efficiency of an electrical power system. Current is the flow of electrons in a conductor, measured in amperes and identified by the letter I. Voltage is the force or pressure that causes electrons to flow in a conductor, measured in volts, and is represented by the letter V. Electric power is measured in watts and represented by the letter P. Power is equal to the amount of voltage multiplied by the amount of current, i.e., $P = V \times I$. When electric power is used over time, it becomes electric energy and is calculated by multiplying power in watts by time in hours. Electric energy is measured in watt-hours and is represented by the letters Wh. Thus, by raising the voltage and lowering the current,

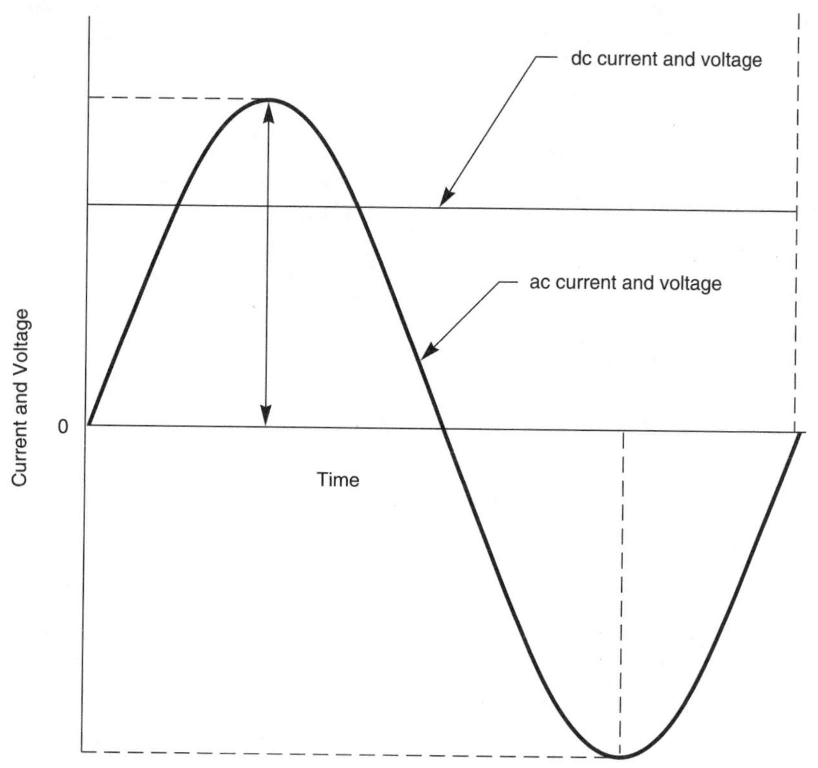

Figure 2.1 Alternating versus direct current and voltage.

the same amount of electrical energy can be transmitted with less current. The magnitude of the current determines the size of the conductor. Therefore, a high-voltage line can transmit a large amount of power with a smaller current carried on a corresponding smaller, less expensive conductor. Just as a bigger pipe is needed to transmit large volumes of water, a bigger conductor is needed to transmit large amounts of current. By increasing the pressure of the water in the pipe, you can increase the volume of water that can be transmitted without increasing the size of the pipe. The same thing is true of electricity. By increasing the voltage you can increase the amount of transmitted power without increasing the size of the conductor.

Also, losses in an electrical conductor are equal to the square of the current in the conductor multiplied by the resistance in the conductor. *Resistance* in a conductor resists the flow of electrons and requires a greater voltage or force to keep the electrons flowing. Resistance is measured in ohms (Ω) and represented by the letter R. Losses reduce the efficiency of the conductor and waste energy and money. This concept is illustrated in the following formula:

$$W = I^2R \tag{2.1}$$

where W = power loss in watts (W)
I = current in amperes (A)
R = resistance in ohms (Ω)

George Westinghouse saw the advantages of ac over dc. He promoted the use of ac. He argued that ac had an economic advantage over dc. AC allowed the generator voltage to be transformed to a higher voltage. The higher voltage would reduce losses and increase the amount of power the electrical system could transmit. Westinghouse knew that the transformer was essential to the economic advantage of ac over dc. He knew that the transformer's ability to increase voltage for economic transmission and lower voltage for safe use was essential to the economic success of ac. He knew that the higher voltage would reduce losses and increase the amount of power the electrical system could transmit. Westinghouse knew that for ac to be a viable alternative to dc, there needed to be a standard way of delivering ac and that the electrical equipment needed to be designed to accept ac. He knew that for alternating current to be practical it had to have electrical equipment that used alternating current. He found a practical use for ac with the invention of the induction motor by Nikola Tesla.

Nikola Tesla, a brilliant but eccentric physicist, discovered in 1888 that the induction motor, especially the three-phase motor, served by an ac-powered system was the most economic design. He used the same basic design of a generator to convert mechanical energy into electrical energy for the motor to convert electrical energy into mechanical energy. This required a rotating magnetic field. Tesla convinced his fellow scientists that a rotating magnetic field that produced a 60 cycle per second (hertz or Hz) alternating current was the most practical. This rotating field can easily be represented by a sinusoidal wave by using the principles of trigonometry, as shown in Figure 2.2.

Tesla designed a system that consisted of three phases. He designed the voltage to be 120° out of phase with the voltage of each phase. He also discovered that the current flowing in one phase was 120° out of phase with the current in each of the other phases. The relationship between the voltage or current of the three phases is illustrated in Figure 2.3.

From Tesla's simple design of the polyphase induction motor evolved the modern 60-Hz power system. The following formula illustrates the relationship between voltage and time for a power system operating at a frequency of 60 Hz:

$$e = E_p \sin (380.44t) \tag{2.2}$$

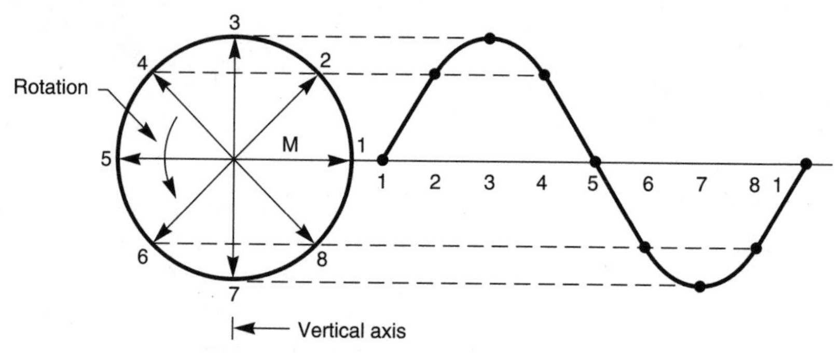

Figure 2.2 Sinusoidal wave derivation.

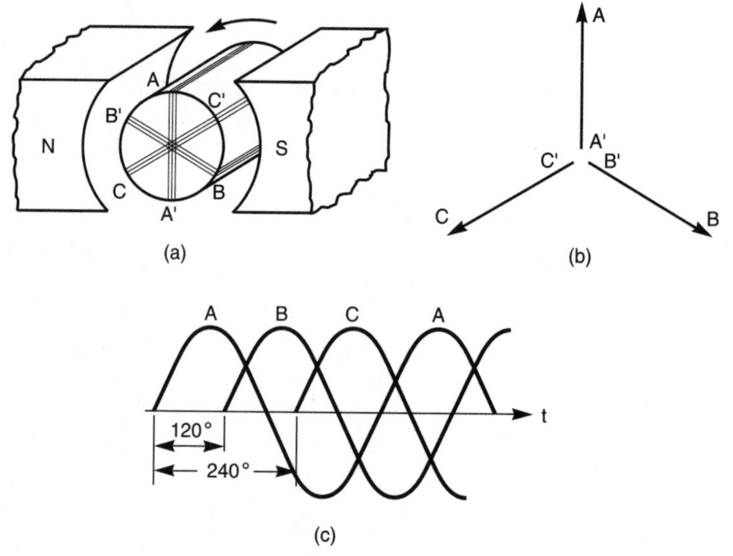

Figure 2.3 Three-phase voltage and current relationship. *(Reproduced from Schaum's Electric Circuits, copyright 1965, by Joseph A. Edminster with permission of The McGraw-Hill Companies.)*

where e = instantaneous voltage
E_p = peak voltage
t = time

The same type of formula would apply to the instantaneous current:

$$i = I_p \sin (380.44t) \tag{2.3}$$

where i = instantaneous current
I_p = peak current
t = time

An electric power system is like a hydraulic power system. The voltage is analogous to the pressure. The current is analogous to the flow of water in the pipes, while the electric power transmission and distribution system is analogous to the pipes in a hydraulic system. Figure 2.4 compares a hydraulic system to an electrical power system.

In a hydraulic system the consumers of water are concerned about the quality of water they drink. In an electric power system the consumers of electricity are concerned about the quality of power they use.

Power Quality Theory

The quality of power has often been characterized as “clean” or “dirty.” Clean power refers to power that has sinusoidal voltage and current without any distortion and operates at the designed magnitude and frequency. Dirty power describes power that has a distorted sinusoidal voltage and current or operates outside the design limits of voltage, current, and/or frequency. Natural and man-made events in the power

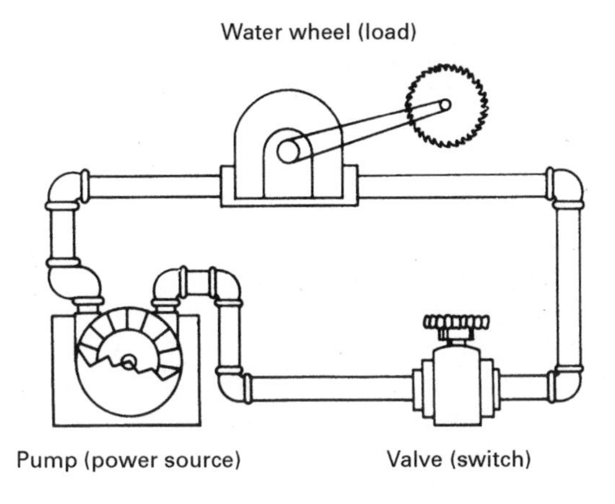

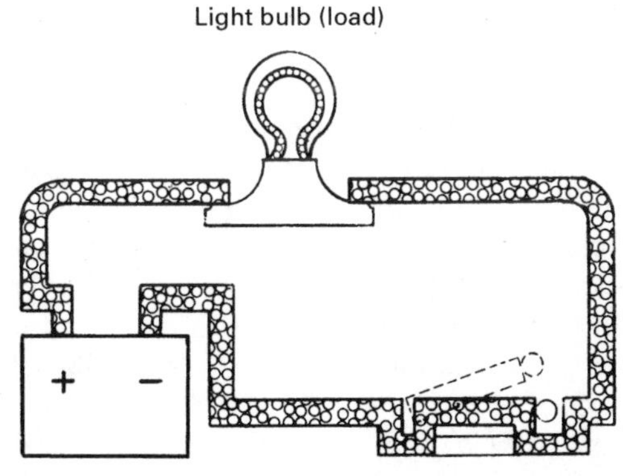

Figure 2.4 Hydraulic system versus electrical power system.

system provide sources or initiating events that cause clean power to become dirty. Categories of dirty power quality sources include power system events, nonlinear loads, and poor wiring and grounding. Examples of dirty power quality sources include lightning, adjustable-speed drives, and loose connections. Power quality experts prefer not to use the term *dirty power* but like to instead use the term *power quality problems*. Therefore, this book will use the term *power quality problems* when referring to poor power quality.

The source of a power quality problem often causes a disturbance or power quality variation. The disturbance can then affect the operation of end-user equipment. This may seem confusing. To make sense of the confusing causes and effects of power quality problems, the power quality engineer breaks down a power quality problem into three parts: sources (initiating events), causes, and effects of power quality, as shown in Figure 2.5.

In solving power quality problems, the power quality engineer uses classical problem-solving techniques. The engineer is usually contacted because some piece of equipment has failed or is not operating properly. The engineer initially asks questions and collects information about the problem before conducting an on-site power quality survey or audit of the facility. The on-site survey includes a visual inspection and electrical measurements of the affected equipment. The engineer sets up instrumentation to measure the disturbance that caused the equipment to malfunction, and collects and records data for later analysis. The engineer often categorizes the disturbance by the "signature" it leaves on power-quality-measuring instruments. A power signature refers to the wave shape of the power quality disturbance. From the power signature, the power quality engineer can determine the type of power quality problem. After diagnosing the type of power quality problem, the engineer can determine possible sources. The engineer isolates the source of the power quality problem and identifies alternative solutions to the problem. The engineer performs an economic evaluation of alternative solutions to determine the most cost-effective solution and recommends solutions to the customer.

Most books or articles about power quality problems categorize the problems by the type of disturbances. This book will categorize power quality problems by the following disturbances: voltage swells, voltage sags, various types of interruptions, overvoltage, undervoltage, harmonics, and transients. Alexander McEachern's *Handbook of Power Signatures,* published in 1989 by Basic Measuring Instruments, contains detailed examples of various power signatures. *IEEE Recommended Practice for Monitoring Electric Power Quality* (IEEE Standard 1159-1995, Copyright © 1995) provides official definitions of power quality disturbances. I have included these and other definitions of power quality, electric utility, and electronic terms in the Glossary of this book. Table 2.1

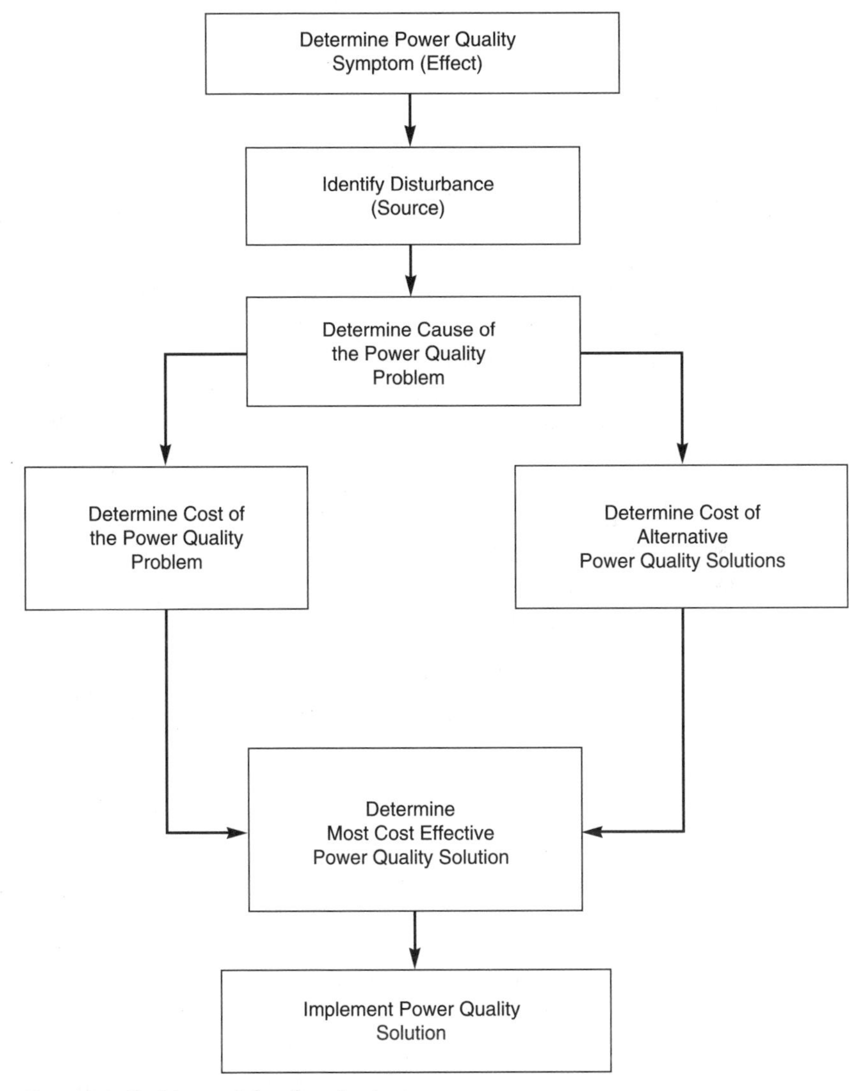

Figure 2.5 Problem-solving flowchart.

provides an overview of the causes, sources, effects, and solutions to various types of power quality problems and disturbances.

Types of Power Quality Problems

The ability to define and understand the various types of power quality problems provides the necessary background needed to prevent and solve those problems. The power quality signature, or characteristic, of

TABLE 2.1 Summary of Power Quality Problems

Example waveshape or RMS variation	Causes	Sources	Effects	Examples of power conditioning solutions
	Impulsive transients (Transient disturbance)	- Lightning - Electrostatic discharge - Load switching - Capacitor switching	- Destroys computer chips and TV regulators	- Surge arresters - Filters - Isolation transformers
	Oscillatory transients (Transient disturbance)	- Line/cable switching - Capacitor switching - Load switching	- Destroys computer chips and TV regulators	- Surge arresters - Filters - Isolation transformers
	Sags/swells (RMS disturbance)	- Remote system faults	- Motors stalling and overheating - Computer failures - ASDs shutting down	- Ferroresonant transformers - Energy storage technologies - Uninterruptible power supply (UPS)
	Interruptions (RMS disturbance)	- System protection - Breakers - Fuses - Maintenance	- Loss production - Shutting down of equipment	- Energy storage technologies - UPS - Backup generators
	Undervoltages/ overvoltages (steady-state variation)	- Motor starting - Load variations - Load dropping	- Shorten lives of motors and lightning filaments	- Voltage regulators - Ferroresonant transformers
	Harmonic distortion (steady-state variation)	- Nonlinear loads - System resonance	- Overheating transformers and motors - Fuses blow - Relays trip - Meters misoperate	- Active or passive filters - Transformers with cancellation of zero sequence components
	Voltage flicker (steady-state variation)	- Intermittent loads - Motor starting - Arc furnaces	- Lights flicker - Irritation	- Static VAR systems

the disturbance identifies the type of power quality problem. The nature of the variation in the basic components of the sine wave, i.e., voltage, current, and frequency, identifies the type of power quality problem. Voltage sags are the most common type of power quality problem.

Voltage sags (dips)

Voltage sags are referred to as *voltage dips* in Europe. IEEE defines *voltage sags* as a reduction in voltage for a short time. The duration of a voltage sag is less than 1 minute but more than 8 milliseconds (0.5 cycles). The magnitude of the reduction is between 10 percent and 90 percent of the normal root mean square (rms) voltage at 60 Hz. The rms, or effective, value of a sine wave is the square root of the average of the squares of all the instantaneous values of a cycle and is equal to 0.707 ($1/\sqrt{2}$) times the peak value of the sine wave, as shown in Figure 2.6. A more detailed discussion of rms follows in the section on harmonics.

How do voltage sags differ from other voltage reduction disturbances? Other voltage reduction disturbances often occur intermittently, like voltage flicker, while voltage sags occur once, for a short time. Figure 2.7 shows the voltage returning to normal after a 0.12-second voltage sag.

What causes voltage sags? Utilities and end users can cause voltage sags on transmission and distribution systems. For example, a transformer failure can be the initiating event that causes a fault on the utility power system that results in a voltage sag. These faults draw energy from the power system. A voltage sag occurs while the fault is on the utility's power system. As soon as a breaker or recloser clears the fault, the voltage returns to normal. Transmission faults cause voltage sags that last about 6 cycles, or 0.10 second. Distribution faults

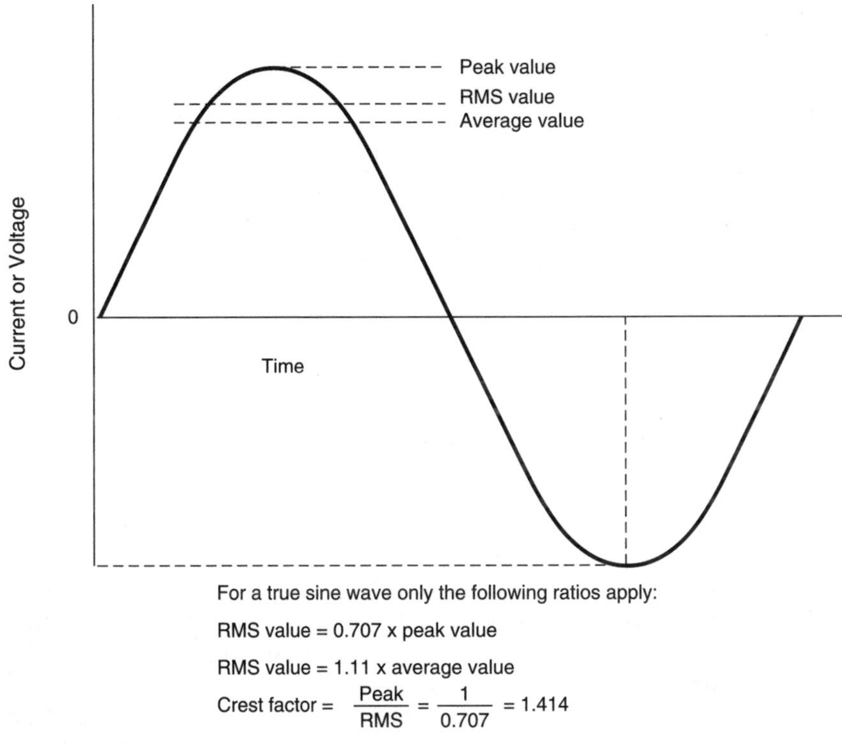

Figure 2.6 Sine wave values.

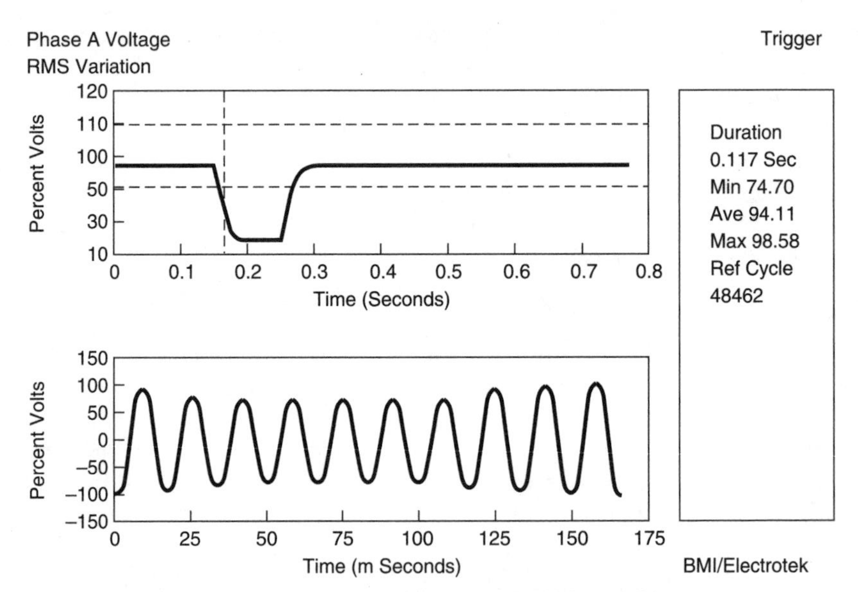

Figure 2.7 Voltage sag plot. (*Courtesy of Dranetz-BMI / Electrotek.*)

last longer than transmission faults, while large motor loads can cause voltage sag on utility's and end user's power systems.

Compared to other power quality problems affecting industrial and commercial end users, voltage sags occur most frequently. They reduce the energy being delivered to the end user and cause computers to fail, adjustable-speed drives to shut down, and motors to stall and overheat.

Solutions to voltage sag problems include equipment that protects loads that are sensitive to voltage sags. Examples of these types of equipment include ferroresonnant, i.e., constant voltage transformers; dynamic voltage restorers (DVRs); superconducting energy storage devices; flywheels; written pole motor-generator sets; and uninterruptible power supplies (UPS). Chapter 4 discusses in more detail these types of devices. Voltage swells are another type of power quality problem.

Voltage swells

Voltage swells, or momentary overvoltages, are rms voltage variations that exceed 110 percent of the nominal voltage and last for less than 1 minute. Voltage swells occur less frequently than voltage sags. Single-line to ground faults cause voltage swells. Examples of single-line to ground faults include lightning or a tree striking a live conductor. The increased energy from a voltage swell often overheats equipment and reduces its life. Figure 2.8 illustrates a typical voltage swell caused by a single-line to ground fault occurring in an adjacent phase. Figure 2.9 illustrates an example of a single-line to ground fault caused by a tree growing into a power line.

Long-duration overvoltages

Long-duration overvoltages are close cousins to voltage swells, except they last longer. Like voltage swells, they are rms voltage variations that exceed 110 percent of the nominal voltage. Unlike swells, they last longer than a minute.

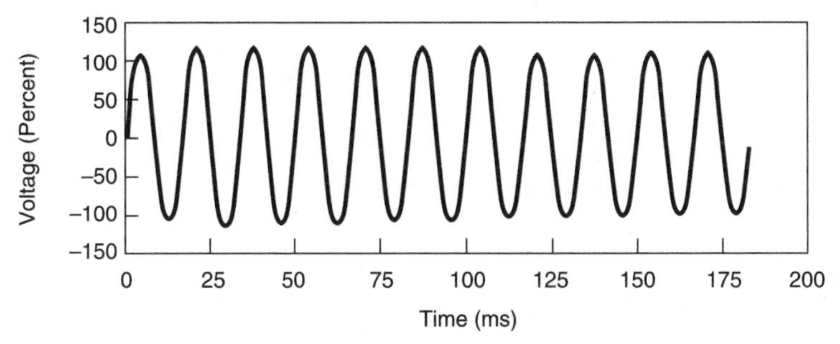

Figure 2.8 Voltage swell plot. *(Courtesy of IEEE, Std. 1159-1995. Copyright © 1995. All rights reserved.)*

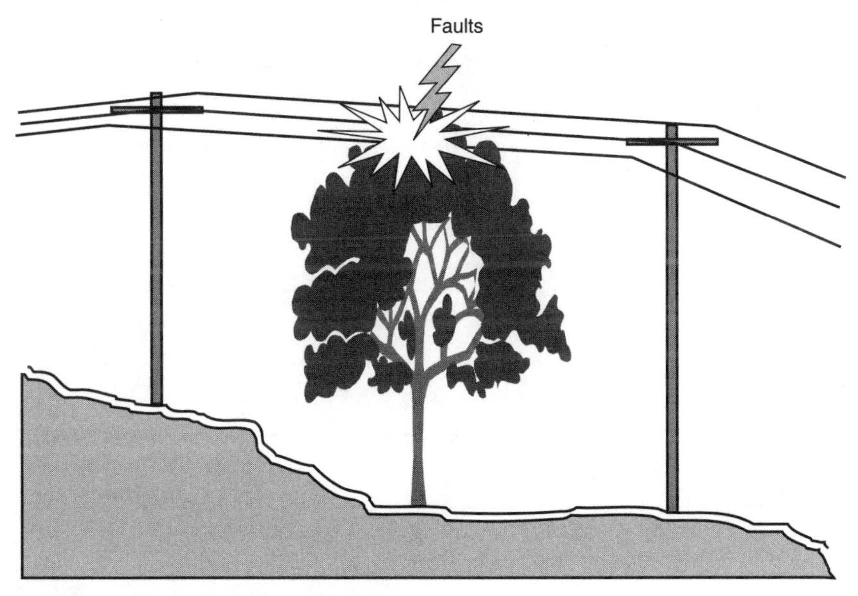

Figure 2.9 Single-line to ground fault caused by a tree.

Several types of initiating events cause overvoltages. The major cause of overvoltages is capacitor switching. This is because a capacitor is a charging device. When a capacitor is switched on, it adds voltage to the utility's system. Another cause of overvoltage is the dropping of load. Light load conditions in the evening also cause overvoltages on high-voltage systems. Another common cause of overvoltage is the missetting of voltage taps on transformers. Extended overvoltages shorten the life of lighting filaments and motors. Solutions to overvoltages include using inductors during light load conditions and correctly setting transformer taps. Figure 2.10 shows a plot of overvoltage versus time.

Undervoltages

Undervoltages occur when the voltage drops below 90 percent of the nominal voltage for more than 1 minute. They are sometimes referred to as "brownouts," although this is an imprecise nontechnical term that should be avoided. They are recognized by end users when their lights dim and their motors slow down.

Too much load on the utility's system, during very cold or hot weather, for example, or the loss of a major transmission line serving a region can cause undervoltages. Overloading inside an end user's own distribution system can cause undervoltages. Sometimes utilities deliberately cause undervoltages to reduce the load during heavy load

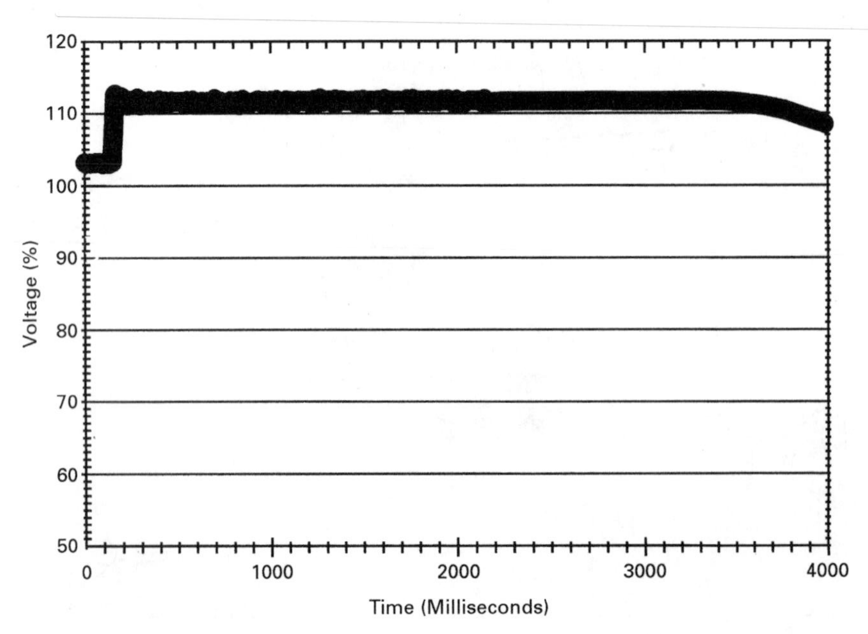

Figure 2.10 Overvoltage plot. (*Courtesy of Dranetz-BMI/Electrotek.*)

conditions. Reducing the voltage reduces the overall load, since load is voltage times current (kW = $V \times I$). Undervoltages can cause sensitive computer equipment to read data incorrectly and motors to stall and operate inefficiently. Utilities can prevent undervoltages by building more generation and transmission lines. Figure 2.11 shows a typical plot of undervoltage versus time.

Interruptions

Interruptions are a complete loss of voltage (a drop to less than 10 percent of nominal voltage) in one or more phases. *IEEE Recommended Practice for Monitoring Electric Power Quality* (IEEE Standard 1159-1995, Copyright © 1995) defines three types of interruptions. They are categorized by the time period that the interruptions occur: momentary, temporary, and long-duration interruptions.

Momentary interruptions are the complete loss of voltage on one or more phase conductors for a time period between 0.5 cycles, or 8 milliseconds, and 3 seconds. A temporary, or short-duration, interruption is a drop of voltage below 10 percent of the nominal voltage for a time period between 3 seconds and 1 minute. Long-duration, or sustained, interruptions last longer than 1 minute. Figure 2.12 shows a momentary interruption.

Loss of production in a business costs money. Any kind of interruption can result in loss of production in an office, retail market, or industrial factory. Not only does the loss of electrical service cause lost production, but the time required to restore electrical service also

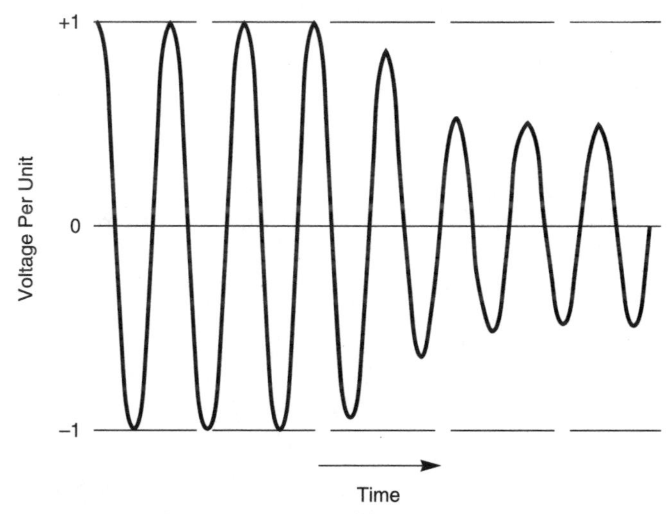

Figure 2.11 Undervoltage plot. (*Courtesy of IEEE, Std. 1159-1995, Copyright © 1995. All rights reserved.*)

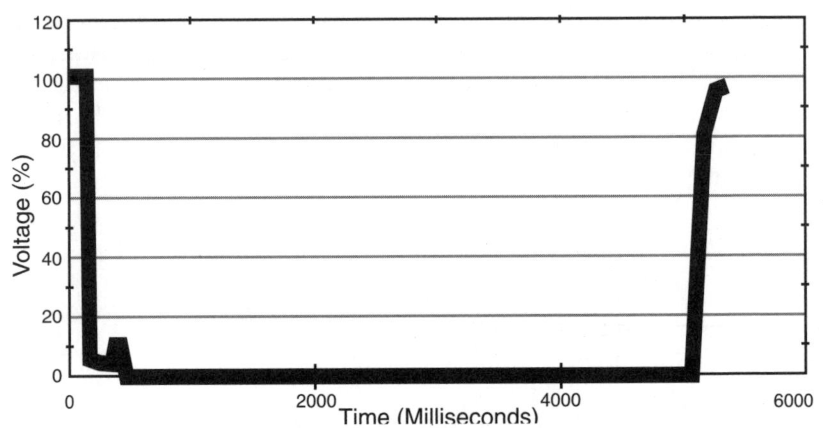

Figure 2.12 Momentary, temporary, and long-duration interruption plots. (*Courtesy of Dranetz-BMI / Electrotek.*)

causes lost production. Some types of processes cannot "ride through" even short interruptions. "Ride through" is the capability of equipment to continue to operate during a power disturbance. For example, in a plastic injection molding plant, for a short interruption of 0.5 second it takes 6 hours to restore production.

The common methods of reducing the impact of costly interruptions include on-site and off-site alternative sources of electrical supply. An end user may install on-site sources, such as battery-operated uninterruptible power supplies (UPS) or motor-generator sets, while a utility may provide an off-site source that includes two feeders with a high-speed switch that switches to the alternate feeder when one feeder fails. Chapter 4, "Power Quality Solutions," discusses these devices in more detail.

Transients

Transients can destroy computer chips and TV. Transients or surges are sometimes referred to as "spikes" in less technically correct language. A sudden increase or decrease in current or voltage characterizes them. They often dissipate quickly. There are basically two types of transients: impulsive and oscillatory.

The time it takes impulsive transients to rise to peak value and decay to normal value determines their identity. For example, page 13 of IEEE Standard 1159-1995, Copyright © 1995 describes an impulsive transient caused by a lightning stroke. In this example the transient current raises to its peak value of 2000 V in 1.2 microseconds (μs; one-millionth of a second) and decays to half its peak value in 50 μs. Resistive components of the electrical transmission and distribution system dampen (reduce) transient currents. The most frequent cause of impulsive transients is lightning strokes. Figure 2.13 illustrates an impulsive current transient caused by lightning.

What kind of device prevents damage to electrical equipment caused by impulsive transients from lightning strokes? Utilities use lightning arresters mounted on their transmission and distribution systems and in their substations, while many utility customers use transient voltage surge suppression (TVSS) or battery-operated uninterruptible power supplies in their homes, offices, or factories. If not stopped, impulsive transients can interact with capacitive components of the power system. Capacitors often cause the impulsive transients to resonant and become oscillatory transients.

Oscillatory transients do not decay quickly like impulsive transients. They tend to continue to oscillate for 0.5 to 3 cycles and reach 2 times the nominal voltage or current. Another cause of oscillatory transients, besides lightning strokes going into resonance, is switching of equipment and power lines on the utility's power system. Figure 2.14 illustrates a typical low-frequency oscillatory transient caused by the energization of a capacitor bank.

Voltage unbalance

Voltage unbalance or imbalance is the deviation of each phase from the average voltage of all three phases. It can be calculated by the formula:

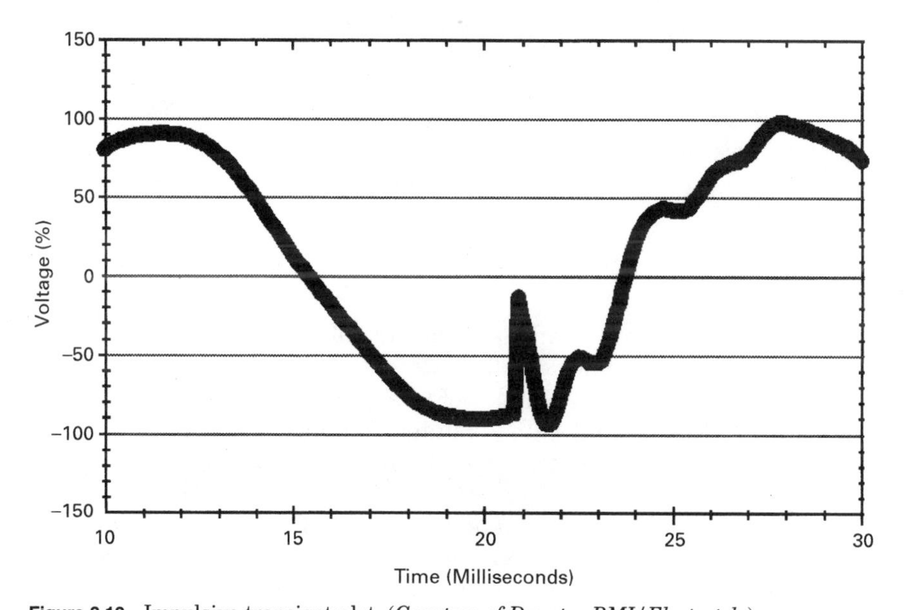

Figure 2.13 Impulsive transient plot. (*Courtesy of Drantez-BMI/Electrotek.*)

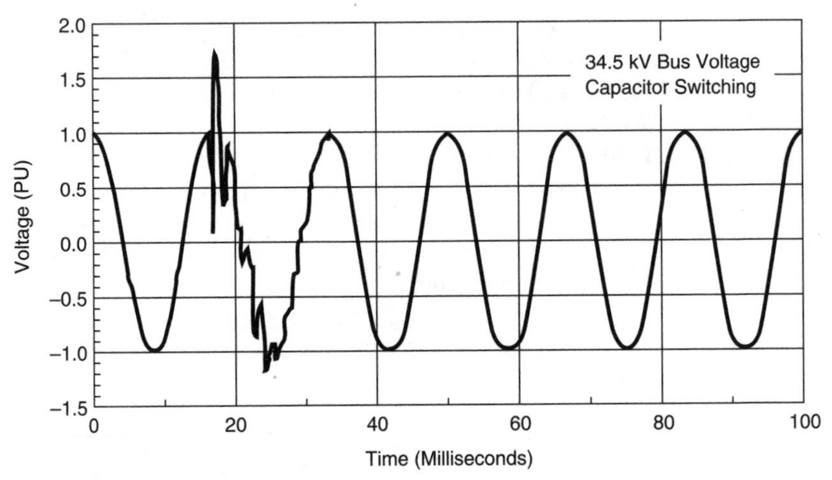

Figure 2.14 Oscillatory transient plot. (*Courtesy of Drantez-BMI/Electrotek.*)

$$\text{Voltage unbalance} = 100 \times \frac{\text{max. deviation from average voltage}}{\text{average voltage}} \tag{2.4}$$

where average voltage = (sum of voltage of each phase)/3.

Most equipment, especially motors, can tolerate a voltage unbalance of 2 percent. A voltage unbalance greater than 2 percent will cause motors and transformers to overheat. This is because a current unbalance in an induction device, like a motor or transformer, varies as the cube of the voltage unbalance applied to the terminals. Potential causes of voltage unbalance include capacitor banks not operating properly, single phasing of equipment, and connecting more single-phase loads on one phase than another. Installing monitors to measure the voltage unbalance provides the necessary data to analyze and eliminate the cause of the unbalance.

Voltage fluctuations

Voltage fluctuations are rapid changes in voltage within the allowable limits of voltage magnitude of 0.95 to 1.05 of nominal voltage. Devices like electric arc furnaces and welders that have continuous, rapid changes in load current cause voltage fluctuations. Voltage fluctuations can cause incandescent and fluorescent lights to blink rapidly. This blinking of lights is often referred to as "flicker." This change in light intensity occurs at frequencies of 6 to 8 Hz and is visible to the human eye. It can cause people to have headaches and become stressed and irritable. It can also cause sensitive equipment to malfunction. What is the solution to voltage fluctuations (flicker)?

The solution to voltage fluctuations is a change in the frequency of the fluctuation. In the case of an arc furnace, this usually involves the use of costly but effective static VAR controllers (SVCs) that control the voltage fluctuation frequency by controlling the amount of reactive power being supplied to the arc furnace. Figure 2.15 shows voltage fluctuations that produce flicker.

Harmonics

What are harmonics? Harmonics are the major source of sine waveform distortion. The increased use of nonlinear equipment have caused harmonics to become more common. Figure 2.16 shows the architecture of a standard sine wave. An analysis of the sine wave architecture provides an understanding of the basic anatomy of harmonics.

Harmonics are integral multiples of the fundamental frequency of the sine wave shown in Figure 2.16; that is, harmonics are multiples of the 60-Hz fundamental voltage and current. They add to the fundamental 60-Hz waveform and distort it. They can be 2, 3, 4, 5, 6, 7, etc., times the fundamental. For example, the third harmonic is 60 Hz times 3, or 180 Hz, and the sixth harmonic is 60 Hz times 6, or 360 Hz. The waveform in Figure 2.17 shows how harmonics distort the sine wave.

What causes harmonic currents? They are usually caused by nonlinear loads, like adjustable speed drives, solid-state heating controls, electronic ballasts for fluorescent lighting, switched-mode power supplies in

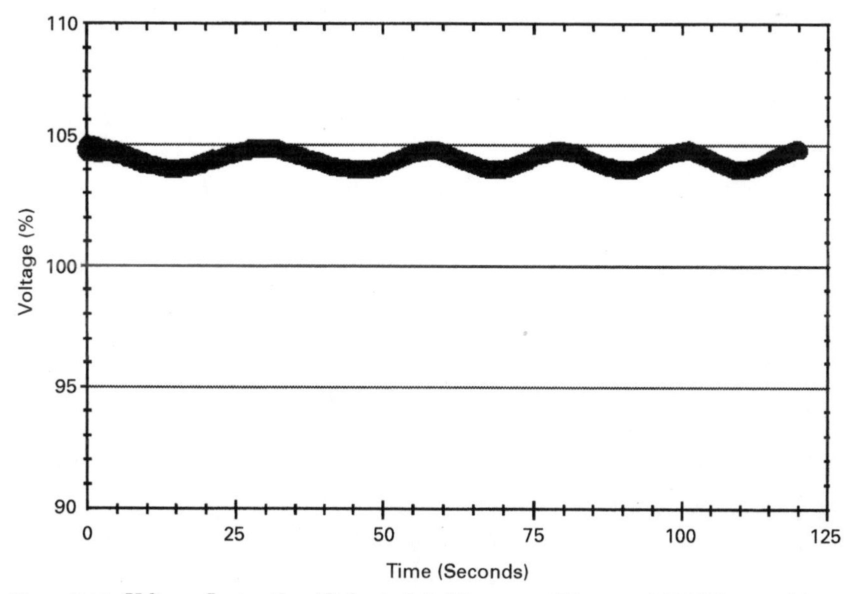

Figure 2.15 Voltage fluctuation (flicker) plot. (*Courtesy of Dranetz-BMI/Electrotek.*)

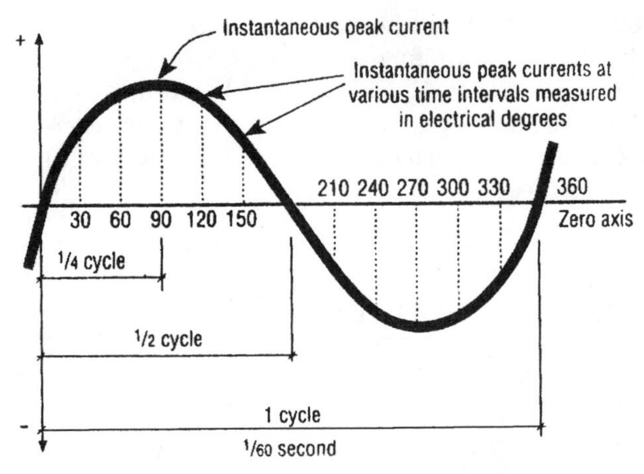

Figure 2.16 Sine wave architecture.

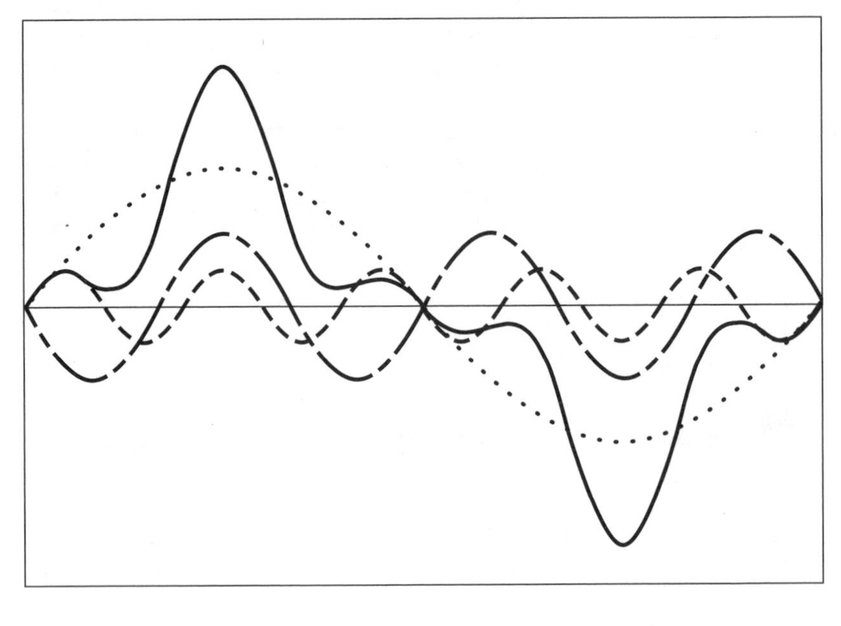

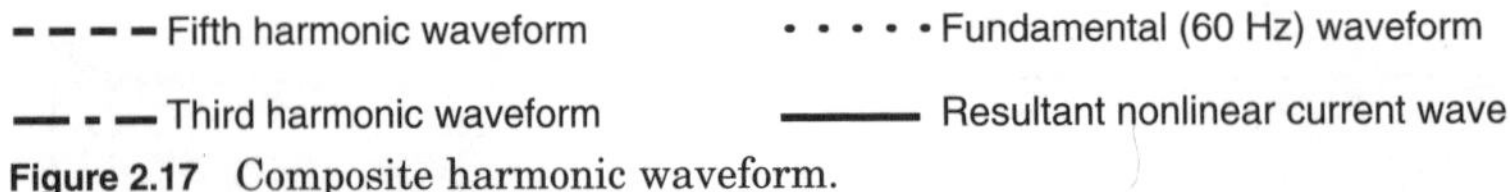

Figure 2.17 Composite harmonic waveform.

computers, static UPS systems, electronic and medical test equipment, rectifiers, filters, and electronic office machines. Nonlinear loads cause harmonic currents to change from a sinusoidal current to a nonsinusoidal current by drawing short bursts of current each cycle or interrupting the current during a cycle. This causes the sinusoidal current waveform to become distorted. The total distorted wave shape is cumulative. The resulting nonsinusoidal wave shape will be a combination of the fundamental 60-Hz sine wave and the various harmonics. Figure 2.18 illustrates the various nonlinear loads and the corresponding harmonic waveforms they generate.

Harmonic voltages result from the harmonic currents interacting with the impedance of the power system according to Ohm's law:

$$V = \frac{I}{Z} \tag{2.5}$$

where V = voltage
I = current
Z = impedance

Harmonic currents and voltages have a detrimental effect on utility and end-user equipment. They cause overheating of transformers, power cables, and motors; inadvertent tripping of relays; and incorrect measurement of voltage and current by meters. Harmonic voltages cause increased iron losses in transformers. Harmonics cause motors to experience rotor heating and pulsating or reduced torque. Table 2.2 shows the effect of harmonics on various types of equipment.

Not only can harmonics cause power quality problems on the end user or the utility serving the end user, but they can cause problems on other end users. For example, a third harmonic generated by a transformer was injected into a utility's system and transmitted to a city miles away and caused the digital clocks to show the wrong time. Because of the increased adverse effects of harmonics, the IEEE adopted a standard for harmonics in 1992. This standard is referred to as *IEEE Recommended Practices and Requirements for Harmonic Control in Electrical Power Systems* (IEEE 519-1992, Copyright © 1993). Section 6 of IEEE 519 discusses the effects of harmonics. This section describes how harmonic currents increase heating in motors, transformers, and power cables. The extent of harmonics' harmful effects is related to the ratio of harmonic current or voltage to the fundamental current or voltage. For example, IEEE 519 sets an upper current distortion limit of 5 percent to prevent overheating of transformers. The maximum overvoltage for transformers is 5 percent at rated load and 10 percent at no load. Harmonic voltages can cause increased iron losses in transformers. Harmonics reduce the torque and overheat the rotor in motors. Electronic equipment cannot tolerate more than a 5 percent harmonic voltage distortion factor, with the single harmonic being no more than 3 percent of the fundamental voltage. Higher levels of harmonics result in erratic malfunction of the electronic equipment. Harmonics can cause relays and meters to malfunction. Concern about the effects of harmonics comes not by an occasional deviation from IEEE 519 standards but by periodic frequent deviations.

Type of Load	Typical Waveform	Typical Current Distortion
Single Phase Power Supply	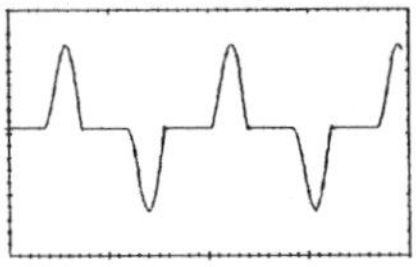	80% (high 3rd)
Semiconverter	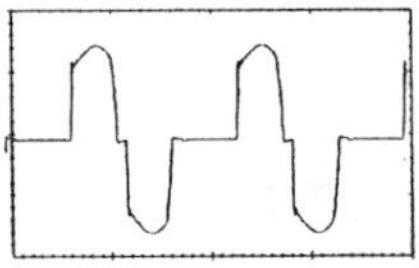	high 2nd,3rd, 4th at partial loads
6 Pulse Converter, capacitive smoothing, no series inductance	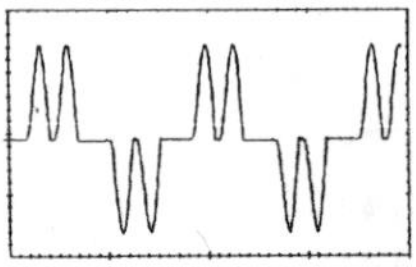	80%
6 Pulse Converter, capacitive smoothing with series inductance > 3%, or dc drive	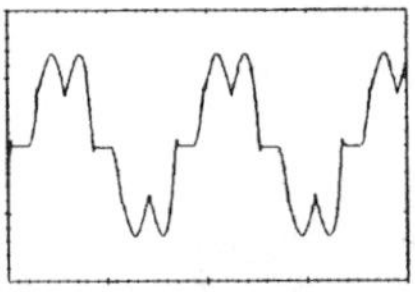	40%
6 Pulse Converter with large inductor for current smoothing	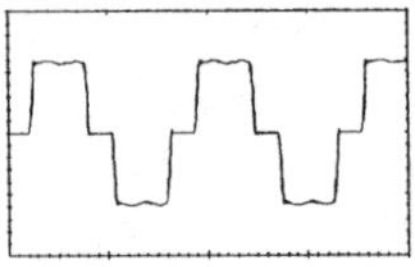	28%
12 Pulse Converter	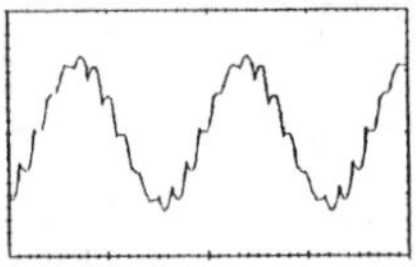	15%
ac Voltage Regulator	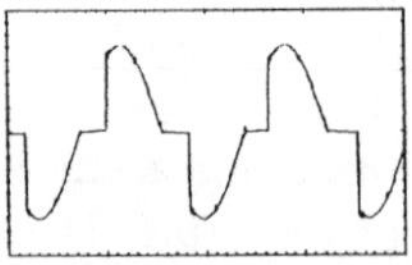	varies with firing angle

Figure 2.18 Nonlinear loads and their current waveforms. (*Courtesy of EPRI.*)

IEEE 519 sets limits on total harmonic distortion (THD) for the utility side of the meter and total demand distortion (TDD) for the end-user side of the meter. This means the utility is responsible for the voltage distortion at the point of common coupling (PCC) between the utility and the end user. Total harmonic distortion is a way to evaluate the voltage distortion effects of injecting harmonic currents into the utility's system. The formula for calculating THD (for a voltage waveform) is as follows:

TABLE 2.2 **Effects of Harmonics on Equipment**

Equipment	Harmonic effects	Results
Capacitors	- Capacitor impedance decreases with increasing frequency, so capacitors act as sinks where harmonics converge; capacitors do not, however, generate - Supply system inductance can resonate with capacitors at some harmonic frequency, causing large currents and voltages to develop - Dry capacitors cannot dissipate heat very well, and are therefore more susceptible to damage from harmonics - Breakdown of dielectric material - Capacitors used in computers are particularly susceptible, since they are often unprotected by fuses or relays - As a general rule of thumb, untuned capacitors and power switching devices are incompatible	- Heating of capacitors due to increased dielectric losses - Short circuits - Fuse failure - Capacitor explosion
Transformers	- Voltage harmonics cause higher transformer voltage and insulation stress; normally not a significant problem	- Transformer heating - Reduced life - Increased copper and iron losses - Insulation stress - Stress
Motors	- Increased losses - Harmonic voltages produce magnetic fields rotating at a speed corresponding to the harmonic frequency	- Motor heating - Mechanical vibrations and noise - Pulsating torques - Increased copper and iron losses in stator and rotor windings, from 5–10% - Reduced efficiency - Reduced life - Voltage stress on insulation of motor windings
Electromechanical induction disk relays	- Additional torque components are produced and may alter the time delay characteristics of the relays	- Incorrect tripping of relays - Incorrect readings
Circuit breakers	- Blowout coils may not operate properly in the presence of harmonic currents	- Failure to interrupt currents - Breaker failure
Watt-hour meters, overcurrent relays	- Harmonics generate additional torque on the induction disk, which can cause improper operation since these devices are calibrated for accurate operation on the fundamental frequency only	- Incorrect readings
Electronic and computer-controlled equipment	- Electronic controls are often dependent on the zero crossing or on the voltage peak for proper control; however, harmonics can significantly alter these parameters, thus adversely affecting operation	- Maloperation of control and protection equipment - Premature equipment failure - Erratic operation of static drives and robots

SOURCE: Ontario Hydro Energy Inc. (www.ontariohydroenergy.com).

$$V_{\mathrm{THD}} = \frac{\sqrt{\sum_{h=2}^{50}}}{V_1} = \sqrt{\left(\frac{V_2}{V_1}\right)^2 + \left(\frac{V_3}{V_1}\right)^2 + \cdots \left(\frac{V_n}{V_1}\right)^2} \tag{2.6}$$

where V_1 = fundamental voltage value and $V_n = V_2, V_3, V_4$, etc. = harmonic voltage value.

The THD can be used to characterize distortion in both current and voltage waves. However, THD usually refers to distortions in the voltage wave. For example, calculate the THD for a complex waveform with the following harmonic distortion as a percentage of the fundamental component for each harmonic: third harmonic distortion = 6/120 × 100% = 50%, fifth harmonic = 9/120 × 100% = 7.5%, and seventh harmonic = 3/120 × 100% = 2.5%. The THD would be calculated as follows:

$$\mathrm{THD} = \sqrt{(0.5)^2 + (0.75)^2 + (0.25)^2} = 0.093 \qquad \text{or} \qquad 0.3\%$$

This exceeds the IEEE 519 limit of 5 percent and would require some type of mitigating device, like filters, to reduce the harmonics to acceptable levels.

TDD, on the other hand, deals with evaluating the current distortions caused by harmonic currents in the end-user facilities. The definition is similar to that of THD, except that the demand current is used in the denominator of TDD instead of simply the fundamental current of a particular sample. TDD of the current I is calculated by the formula

$$\mathrm{TDD} = \frac{\sqrt{\sum_{h=1}^{h=\infty} (I_h)^2}}{I_L} \tag{2.7}$$

where I_L = rms value of maximum demand load current
h = harmonic order (1, 2, 3, 4, etc.)
I_h = rms load current at the harmonic order h

Chapter 3, "Power Quality Standards," discusses in more detail THD and TDD limits as contained in IEEE 519, 1992, Copyright © 1992.

There are several ways to reduce or eliminate harmonics. The most common way is to add filters to the electrical power system. Harmonic filters or chokes reduce electrical harmonics just as shock absorbers reduce mechanical harmonics. Filters contain capacitors and inductors in series. Filters siphon off the harmonic currents to ground. They prevent the harmonic currents from getting onto the utility's or end user's distribution system and doing damage to the utility's and other end users' equipment. There are two types of filters: static and active. Static filters do not change their value. Active filters change their value to fit the harmonic being filtered. Other ways of reducing or eliminating harmonics include using isolation transformers and detuning capacitors and designing the source of the harmonics to change the type of harmonics. Chapter 4, "Power Quality Solutions," discusses these mitigating methods in more detail.

Electrical noise

When you think of electrical noise, you may think of the audible crackling noise that emanates from high-voltage power lines. Or you may imagine the low throbbing hum of an energized transformer. This type of noise can affect your life quality as much as your power quality. In fact, one man complained that the corona noise from a nearby 500-kV power line drowned out the babbling of a brook in his backyard. When power quality experts talk about electrical noise, they do not mean these audible noises. They mean the electrical noise that is caused by a low-voltage, high-frequency (but lower than 200-Hz) signal superimposed on the 60-Hz fundamental waveform. This type of electrical noise may be transmitted through the air or wires. High-voltage lines, arcing from operating disconnect switches, start-up of large motors, radio and TV stations, switched mode power supplies, loads with solid-state rectifiers, fluorescent lights, and power electronic devices can all cause this type of noise. Electrical noise adds "hash" onto the fundamental sine wave, as shown in Figure 2.19.

Electrical noise can degrade telecommunication equipment, radio, and TV reception, and damage electronic equipment as well. How do you reduce or eliminate electrical noise?

There are two ways of solving the electrical noise problem. One solution is to eliminate the source of the electrical noise. Another way is to either stop or reduce the electrical noise from being transmitted. For example, the use of multiple conductors or installation of corona rings can reduce electrical noise from high-voltage lines. Grounding equipment and the service panel to a common point can eliminate electrical noise from ground loops. This prevents the ground wires from acting as a loop antenna and transmitting a "humming" type noise that interferes with communication signals.

The electromagnetic interference (EMI) type of noise is reduced by shielding the sensitive equipment from the source of the electrical noise. Another way of protecting sensitive equipment from EMI is to simply move the source of the EMI far enough away so that the EMI becomes too weak to affect the sensitive equipment. For example, the electromagnetic field from a tabletop fluorescent lamp near a

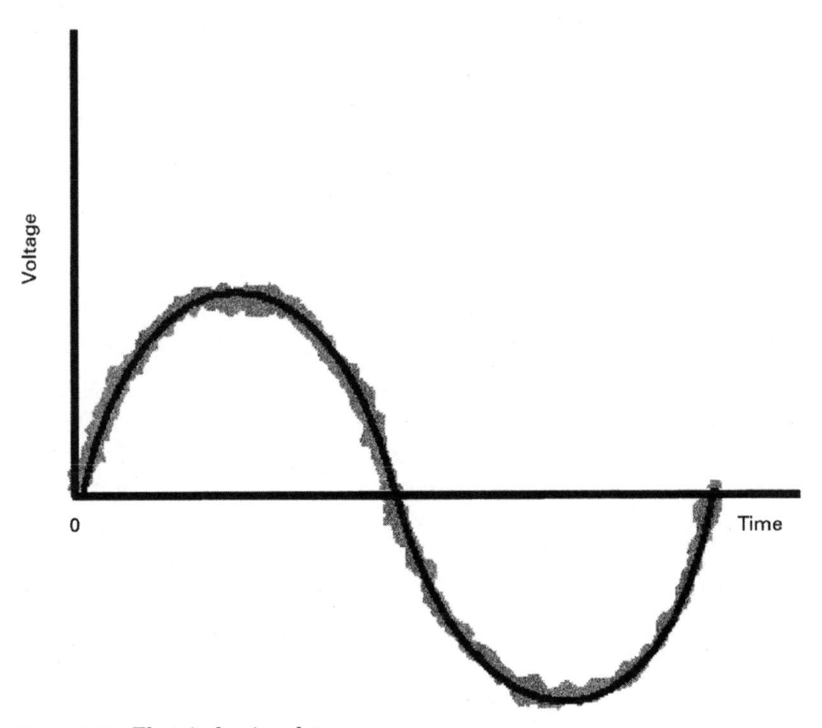

Figure 2.19 Electrical noise plot.

computer screen will cause the lines on the screen to wiggle. Move the fluorescent light far enough away, and the wiggles will stop.

Sources of Power Quality Problems

Power quality experts find it a challenge to analyze any power quality problem and determine the source of the problem. They usually measure the effect of the problem and draw on their experience to identify the type of disturbance from the measurement. Even experienced power quality experts often find it is difficult to determine the source of the power quality problem. They know they need to understand the basic reasons why different devices and phenomena cause power quality problems. One common characteristic of sources of power quality problems is the interruption of the current or voltage sine wave. This interruption results in one of the disturbances discussed at the beginning of this chapter.

The major sources of power quality problems can be divided into two categories, depending on the location of the source in relationship to the power meter. One category is on the utility side of the meter and includes switching operations, power system faults, and lightning. The other category is on the end-user side of the meter and includes nonlinear loads, poor grounding, electromagnetic interference, and static electricity. So let's first examine the characteristics of utility-caused power quality problems.

Utility side of the meter

Sources of power quality problems on the utility side of the meter involve some type of activity on the utility's electrical power system. They can be either man-made or natural events. They all involve some type of interruption of the current or voltage. The most common man-made causes are switching operations.

Utilities switch equipment on and off by the use of breakers, disconnect switches, or reclosers. Usually some type of fault on the power system causes a breaker to trip. Utilities trip breakers to perform routine maintenance. They also trip breakers to insert capacitors to improve the power factor. Lightning striking a power line or substation equipment, a tree touching a power line, a car hitting a power pole, or even an animal touching an energized line may cause the fault. The tripping of the breaker and the initiating fault can cause the voltage to sag or swell, depending on when in the periodic wave the tripping occurs. Utilities set breakers and reclosers to reclose on the fault to determine if the fault has cleared. If the fault has not cleared, the breaker or recloser trips again and stays open. Figure 2.20 shows a utility breaker.

Another type of utility activity that can cause oscillatory transients is the switching of power factor improvement capacitors. As shown in Figure 2.21, utilities use power factor improvement capacitors to improve the power factor by adding capacitive reactance to the power system. This causes the current and voltage to be in phase and thus reduces losses in the power system. When utilities insert capacitors in the power system, they momentarily cause an increase in the voltage and cause transients. Capacitors, if tuned to harmonics on the power system, can also amplify the harmonics. This is especially true if the utility and end user both switch their capacitors on at the same time.

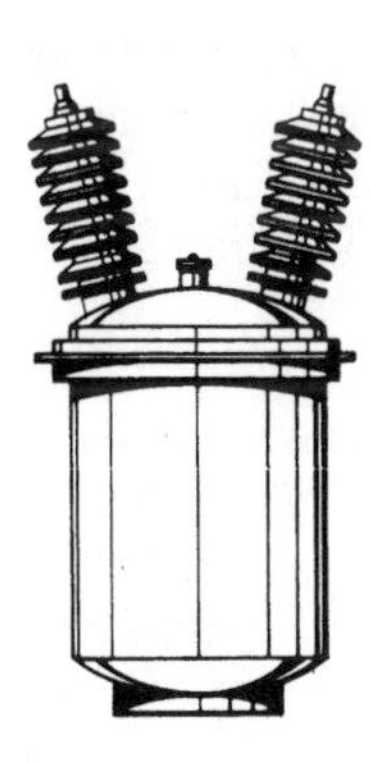

Figure 2.20 Utility breaker.

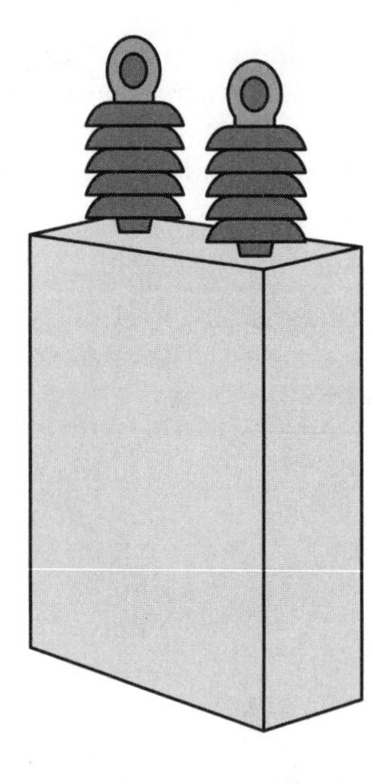

Figure 2.21 Utility power factor improvement capacitor.

Utility system faults occur on power lines or in power equipment. They are usually categorized by single-phase faults to ground, phase-to-phase faults, or three-phase faults to ground. On the utility side of the meter, the type of fault often determines the type of disturbance. On the end-user side of the meter, the type of load or wiring and grounding conditions determine the type of power quality disturbance.

End-user side of the meter

Sources of power quality problems on the end-user side of the meter usually involve a disruption of the sinusoidal voltage and current delivered to the end user by the utility. These disruptions can damage or cause misoperation of sensitive electronic equipment in not only the end-user's facilities but also in another end-user's facilities that is electrically connected. The following is a list of power quality problems caused by end users: nonlinear inrush current from the start-up of large motors, static electricity, power factor improvement capacitors amplifying harmonics, and poor wiring and grounding techniques.

Nonlinear loads. There are today many types of nonlinear loads. They include all types of electronic equipment that use switched-mode power supplies, adjustable-speed drives, rectifiers converting ac to dc, inverters converting dc to ac, arc welders and arc furnaces, electronic and magnetic ballast in fluorescent lighting, and medical equipment like MRI (magnetic radiation imaging) and x-ray machines. Other devices that convert ac to dc and generate harmonics include battery chargers, UPSs, electron beam furnaces, and induction furnaces, to name just a few. All these devices change a smooth sinusoidal wave into irregular distorted wave shapes. The distorted wave shapes produce harmonics.

Most electronic devices use switched-mode power supplies that produce harmonics. Manufacturers of electronic equipment have found that they can eliminate a filter and eliminate the power supply transformer (shown in Figure 2.22) by the use of a switched-mode power supply (shown in Figure 2.23). What is a switched-mode power supply? How does it produce harmonics? The switched-mode process converts ac to dc using a rectifier bridge, converts dc back to ac at a high frequency using a switcher, steps the ac voltage down to 5 V using a small

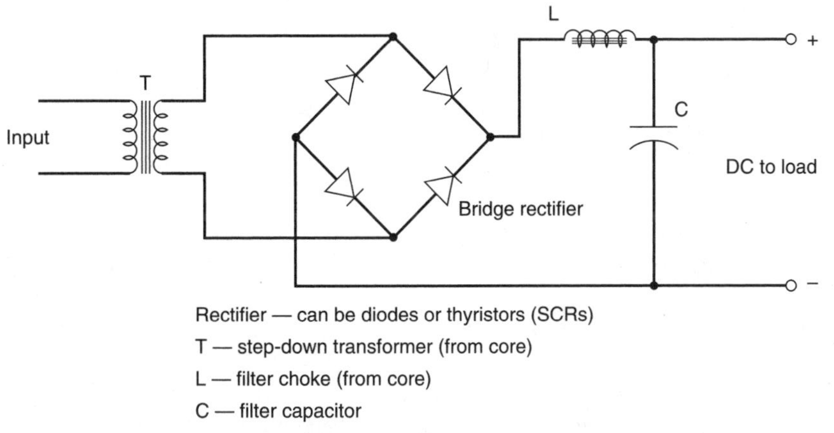

Figure 2.22 Power supply without switched mode. (Reprinted with permission from March 1988 issue of *EC&M Magazine,* Copyright 1988, Intertec Publishing Corp. All rights reserved.)

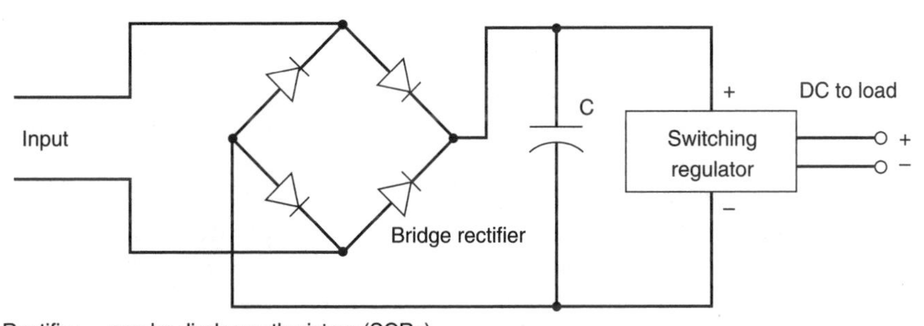

Rectifier — can be diodes or thyristors (SCRs)

C — filter capacitor

Switching regulator — high-speed switcher (20 to 100kHz, some in MHz range) with control circuitry.

Figure 2.23 Power supply with switched mode. (Reprinted with permission from March 1988 issue of *EC&M Magazine,* Copyright 1988, Intertec Publishing Corp. All rights reserved.)

transformer, and finally converts the ac to dc using another rectifier. Electronic equipment requires 5 V dc to operate. Go inside a switched-mode power supply and you'll find a switching circuit that takes stored energy from a capacitor in short pulses and delivers voltage at a frequency of 20 to 100 kHz to a transformer in the form of a square wave. The high-frequency switching requires a small and light transformer. However, the pulsed square wave distorts the sine wave and produces harmonics.

EPRI has stated that "by the year 2000, over half of all electricity produced in the United States is expected to flow through power electronic equipment." Electronic equipment in the office includes computers, copiers, printers, and fax machines.

Adjustable-speed drives save energy by adjusting the speed of the motor to fit the load. Residential heat pumps, commercial heating and ventilating systems, and factories that use motors in their processes benefit from the use of adjustable-speed drives. However, adjustable-speed drives cause harmonics by varying the fundamental frequency in order to vary the speed of the drive.

Arc furnaces use extreme heat (3000°F) to melt metal. The furnace uses an electrical arc striking from a high-voltage electrode to the grounded metal to create this extreme heat. The arc is extinguished every half-cycle. The short circuit to ground causes the voltage to dip each time the arc strikes. This causes the lights to flicker at a frequency typically less than 60 Hz that is irritating to humans. Arc furnaces also generate harmonic currents. Figure 2.24 illustrates the configuration of a one-electrode dc electric arc furnace.

Most nonlinear loads not only generate harmonics but cause low power factor. They cause low power factor by shifting the phase angle between the voltage and current. What is power factor and why is low power factor bad?

Power factor. Power factor is a way to measure the amount of reactive power required to supply an electrical system and an end-user's facility. Reactive power represents wasted electrical energy, because it does no useful work. Inductive loads require reactive power and constitute a major portion of the power consumed in industrial plants. Motors, transformers, fluorescent lights, arc welders, and induction heating furnaces all use reactive power.

Power factor is also a way of measuring the phase difference between voltage and current. Just as a rotating alternating current and voltage can be represented by a sine wave, the phase difference between voltage and current can be represented by the cosine of the phase shift angle. Figure 2.25 illustrates the relationship between power factor and the phase shift between current and voltage.

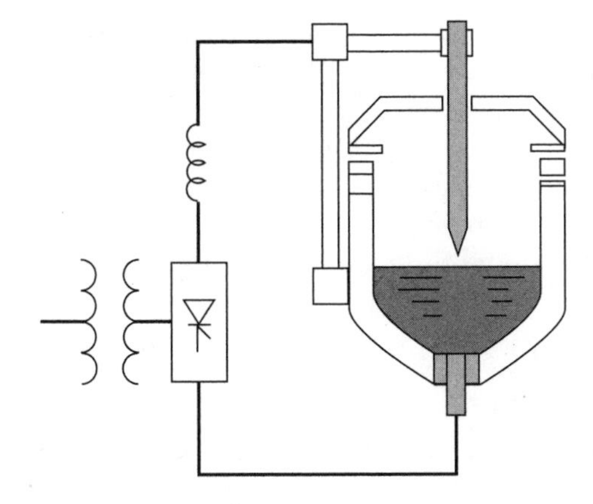

Figure 2.24 One electrode dc arc furnace. (*Courtesy of EPRI.*)

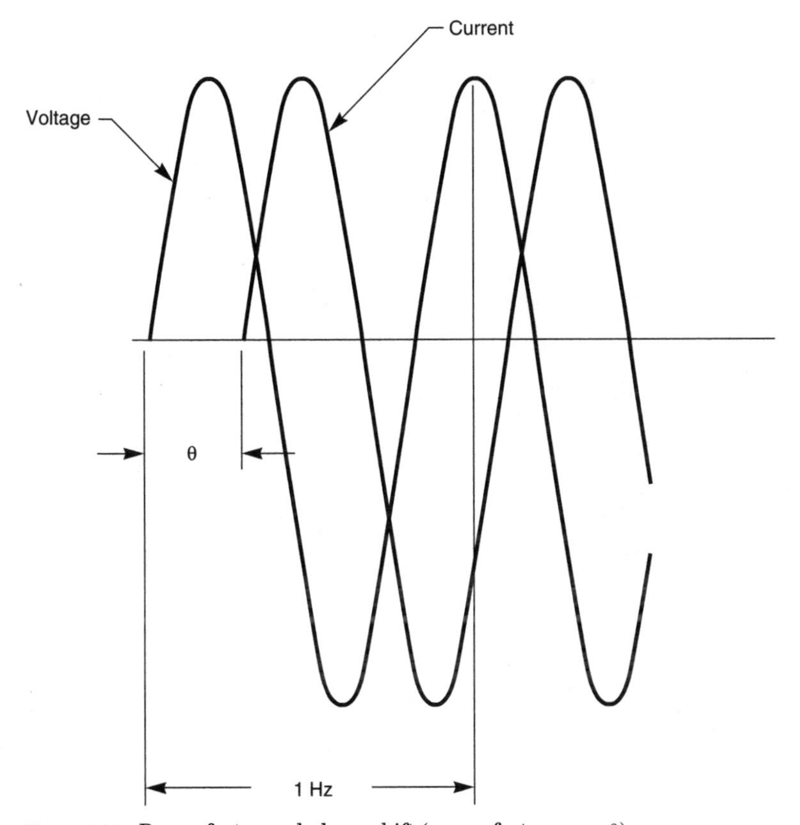

Figure 2.25 Power factor and phase shift (power factor = cos θ).

Nonlinear loads often shift the phase angle between the load current and voltage, require reactive power to serve them, and cause low power factor. Linear motor loads require reactive power to turn the rotating magnetic field in the motor and cause low power factor. Nonlinear and linear loads that cause low power factor include induction motors of all types, power electronic power converters, arc welding machines, electric arc and induction furnaces, and fluorescent and other types of arc lighting.

Power factor is defined as the ratio of active power to apparent power:

$$\text{Power factor} = \frac{\textit{active power in kW}}{\textit{apparent power in kVA}} \tag{2.8}$$

It is often represented by the power triangle shown in Figure 2.26.

Active power is the power to do useful work, such as turning a motor or running a pump, and is measured in kilowatts (kW). Electrical equipment needs active power to convert electrical energy into mechanical energy. Reactive power is the power required to provide a magnetic field to ferromagnetic equipment, like motors and transformers, and does no useful work. Reactive power is measured in kilovolt-amperes–reactive (kVAR)s. Apparent power or demand power is the total power needed to serve a load. It is measured in kilovolt-amperes (kVA) and is the vector sum of reactive and active power:

$$(kVA)^2 = (kW)^2 + (kVAR)^2 \tag{2.9}$$

Reactive power takes up capacity on the utility's and end-user's electrical distribution systems. Reactive power also increases transmission and distribution losses. Reactive power is frequently described as analogous to the foam in a beer mug. It comes with the beer and takes up capacity in the mug but does not quench the beer drinker's thirst. As can be seen from the power triangle in Figure 2.26, power factor measures the reactive efficiency of a power system. At maximum efficiency the reactive power is zero, and the power factor is unity.

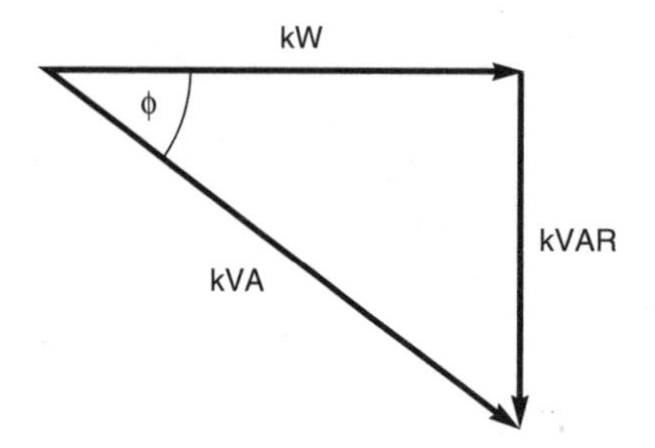

Figure 2.26 Power triangle (power factor = cos = kW/kVa).

At a power factor of unity, kVA equals kW and there is no reactive power component in the system. An example of low power factor would be a 1600-kW load requiring 2000 kVA of total power. The power factor in this case is 80 percent.

As a general rule, an electrical system using motors exhibits a low power factor. Low power factors result in overall low power system efficiency, including increased conductor and transformer losses and low voltage. Low power factor also reduces line and transformer capacity. Utilities must supply both the active and reactive power and compensate for these losses. For this reason, most utilities charge their customers a penalty for low power factor. Many utilities increase the demand charge for every percent the power factor drops below a set value, say 95 percent. However, more and more utilities are charging for kVAR-hours just like they charge for kW-hours. These charges provide utility customers an incentive to increase their power factor by the use of power factor improvement capacitors. Otherwise, the utility has to install power factor improvement capacitors on its own power system. But how do capacitors improve power factor?

It is generally more energy-efficient and cost-effective to improve the power factor of the electrical system at an industrial plant than to require generators to provide the necessary reactive requirements of the plant's loads. Improving power factor can be accomplished through the addition of shunt capacitors.

Power factor improvement capacitors. Power factor improvement capacitors improve the power factor by providing the reactive power needed by the load. They also reduce the phase shift difference between voltage and current. Like a battery, they store electrical energy. Unlike a battery, they store energy on thin metal foil plates separated by a sheet of polymer material. They release the energy every half-cycle of voltage. They cause the current to lead the voltage by 90°. This subtracts from the phase angle shift of induction loads that cause the current to lag the voltage by 90°. This is how capacitors reduce the phase shift between current and voltage and provide the magnetization that motors and transformers need to operate. Therefore, capacitors are an inexpensive way to provide reactive power at the load and increase power factor. This is illustrated in Figure 2.27.

They supply the reactive, magnetized power required by electric loads, especially industrial loads that use inductive motors. Motors with their inductive, magnetizing, reactive power cause current to "lag" behind voltage. Capacitors create "leading" current. Capacitors act in opposition to inductive loads, thereby minimizing the reactive power required. When carefully controlled, the capacitor lead can match the motor lag,

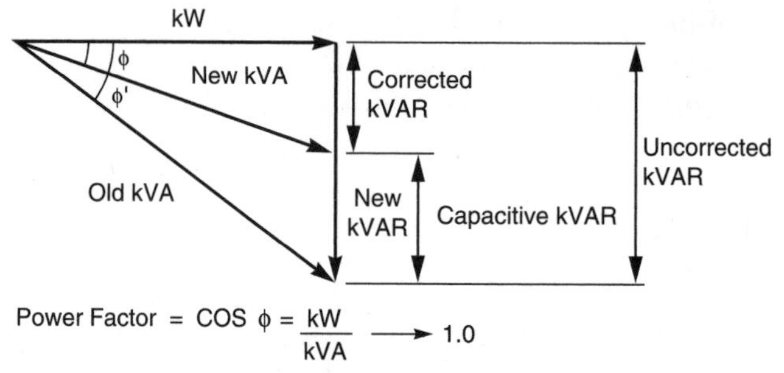

Figure 2.27 Capacitors increase power factor.

eliminate the need for reactive power, and increase the power factor toward unity.

Both fixed and dynamic shunt capacitors applied to inductive loads increase the power factor. Fixed capacitors are switched on manually and apply a constant capacitance; dynamic capacitors can be switched on automatically and adjust their capacitance according to the inductive load. Both types have advantages and disadvantages, but both types provide similar benefits. In raising the power factor, shunt capacitors release energy to the system, raise system voltage, reduce system losses and, ultimately, reduce power costs. However, capacitors have a downside. They can amplify harmonics through harmonic resonance.

Harmonic resonance. Electrical harmonic resonance occurs when the inductive reactance of a power system equals the capacitive reactance of a power system. This is a good thing at the fundamental frequency of 60 Hz and results in the current and voltage being in phase and unity power factor. However, it is not so good when it occurs at a harmonic frequency. If resonance occurs at a harmonic frequency, the harmonic current reaches a maximum value and causes overheating of transformers, capacitors, and motors; tripping of relays; and incorrect meter readings. How does resonance occur at a harmonic frequency?

The amount of inductive and capacitive reactance are dependent on the frequency of the current and voltage. Thus, resonance can occur at various harmonic frequencies. The formulas for inductive and capacitive reactance illustrate this relationship:

$$X_L = 2\pi f L \tag{2.10}$$

where X_L = inductive reactance in ohms
π = 3.14
f = frequency in cycles per second
L = induction of the power system in henries

$$X_C = \frac{1}{2\pi fC} \tag{2.11}$$

where X_C = capacitive reactance in ohms
π = 3.14
f = frequency in cycles per second
C = capacitance of the power system in farads

Capacitors can cause two types of resonance: parallel and series resonance. Since most power factor improvement capacitors are in parallel with the inductance of the power system, as shown in the schematic of a parallel resonant circuit (Figure 2.28), parallel resonance occurs most often.

When capacitive and inductive reactance connect in parallel in the power system, the magnitude of the total reactance or impedance becomes

$$X_T = \sqrt{R^2 + (X_L - X_C)^2} \tag{2.12}$$

where X_T = total reactance
R = resistance
X_L = inductive reactance = $2\pi fL$
X_C = capacitive reactance = $1/(2\pi fC)$

Harmonic resonance occurs when $X_L = X_C$ and X_T becomes a pure resistance (R) and from Ohm's law ($I = V/X_T$) the harmonic current I reaches a maximum. Therefore, the following formula determines the harmonic resonance frequency (f_{resonant}):

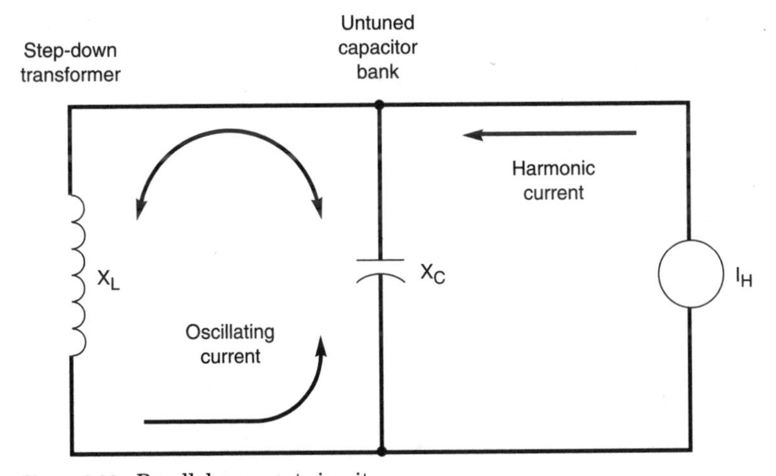

Figure 2.28 Parallel resonant circuit.

$$f_{\text{resonant}} = \frac{1}{2\pi} \frac{1}{\sqrt{LC}} \tag{2.13}$$

How do you prevent resonance? You prevent resonance by sizing and locating capacitors to avoid the harmonic resonance frequency or by using filters. A filter is simply an inductor (reactor) in series with a capacitor, as shown in Figure 2.29. Filters detune the capacitor away from the resonant frequency. Filters usually cost twice as much as capacitors. Filters also remove the effect of distortion power factor and increase the true power factor.

True power factor. True power factor is the power factor caused by harmonics and the fundamental, while the standard or displacement power factor described previously is caused by the fundamental power at 60 Hz. It is not measured by standard VAR or power factor meters. It is measured only by so-called true rms meters (see Chapter 6, "Power Quality Measurement Tools," for an explanation of true rms meters). The diagram in Figure 2.30 and the following formula define it:

$$\text{True power factor} = \frac{\text{Real power in kW}}{\text{Total power in kVA or } V_{\text{rms}} \times I_{\text{rms}}} \tag{2.14}$$

As can be seen from the diagram, the true kVA is larger than the displacement kVA because of the effect of the harmonic distortion. Even though there is no penalty associated with true power factor, it still has a detrimental effect on the power system. Low true power factor means increased losses and reduced system capacity. True power factor is

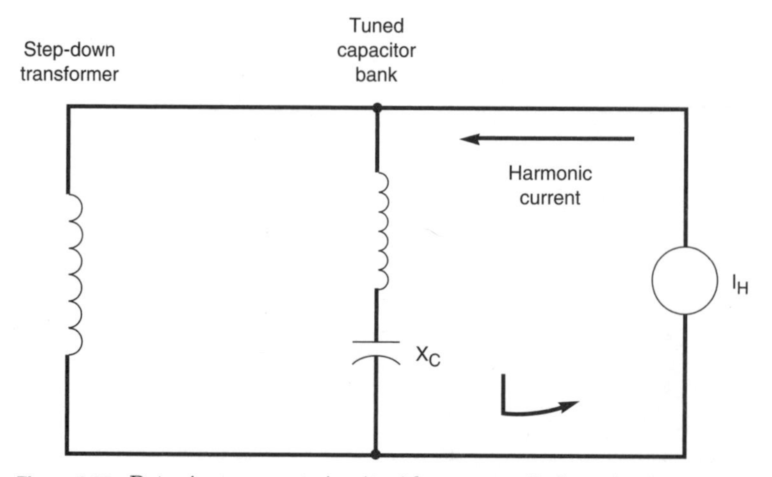

Figure 2.29 Detuning resonant circuit with a reactor (inductor).

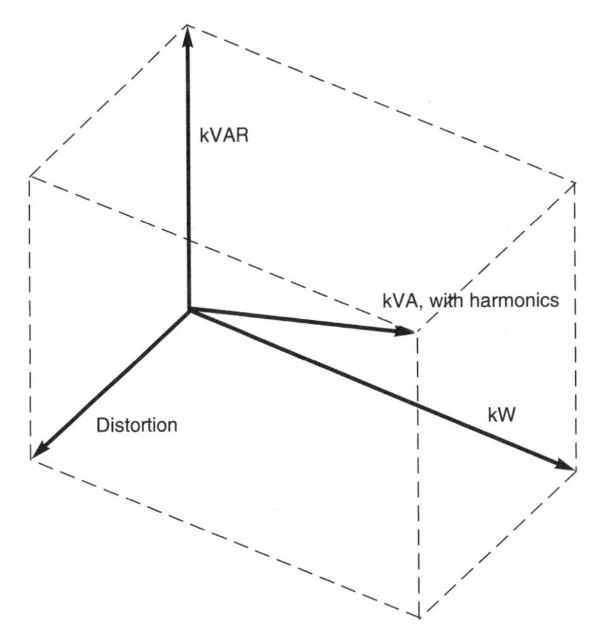

Figure 2.30 Distortion power factor power triangle.

increased not by the addition of capacitors but by the elimination of harmonics through the use of filters. The addition of capacitors can cause the true power factor to be worse by magnifying the harmonic distortion. Another cause of power quality problems is poor wiring and grounding.

Poor wiring and grounding. An EPRI survey found poor wiring and grounding in the end-user's facilities cause 80 percent of all power quality problems. Why does poor wiring and grounding cause most of the power quality problems? The National Electrical Code (NEC) determines the design of the wiring and grounding. However, the NEC, as described in Section 90-1(b), is intended to protect people from fire and electrocution, not to protect sensitive electronic equipment from damage. As a consequence there is a great need to establish guidelines for wiring and grounding that not only protects the public but prevents power quality problems.

When poor wiring and grounding cause equipment to fail, utility customers often attribute the failure to the utility. They may even buy expensive power conditioning equipment that only treats the symptom of the power quality problem and does not solve the underlining cause of the problem. They should, instead, identify the effects of poor wiring and grounding, determine the cause of the power quality problem, and find a simple way to correct the problem.

Symptoms of poor wiring and grounding include computers that lose data or stop operating; telephone systems that lose calls or are noisy; industrial processes that suddenly stop; breaker boxes that get very hot; neutral leads that catch fire; and even power conditioning equipment, like transient voltage surge suppressors (TVSSs), that catch fire. What kinds of poor wiring and grounding practices cause these problems? How can you prevent these problems from happening?

Some simple guidelines will help you identify and prevent problems caused by inadequate wiring and grounding. These guidelines can be divided into three categories: (1) wiring, (2) grounding, and (3) lightning protection.

Intermixing loads can cause power quality problems in any facility. When nonsensitive and sensitive loads are connected to the same circuit, they often interact with one another. For example, when a large motor on an elevator or an air conditioner starts, it causes a large inrush current that can cause a voltage sag. The voltage sag inside a facility has the same effect that a voltage sag has outside of the facility. It causes lights to dim and computer equipment to malfunction. The solution is to not connect nonsensitive loads that will interact with sensitive loads. Wiring sensitive loads to separate circuits connected to the main electrical service panel separates sensitive loads from nonsensitive loads.

Poor grounding can cause voltage potential differences, excessive ground loops, and interference with sensitive electronic equipment. Proper grounding not only protects people from shock but provides a reference point and a path for large currents caused by faults, like switching surges and lightning strokes. Remember reference points are critical to computers, because 5 V dc represents "1" and 0 V dc represents "0." Article 100 of the NEC defines ground as "a conducting connection, whether intentional or accidental, between an electrical circuit or equipment and the earth, or to some conducting body that serves in place of the earth." One effective method recommended by the *IEEE Green Book* for grounding equipment is a ground ring surrounding the affected area and "tied to the building steel at suitable intervals." Bonding the ground wire to the neutral wire, i.e. the white wire that is normally at or near the voltage of the ground wire, only at the service panel prevents ground loops.

Poor grounding can result in lightning destroying equipment in a home, office, or factory. Lightning surges will take the path of least resistance. Wiring and grounding should be designed to divert lightning current away from sensitive equipment to ground through lightning protection devices, such as lightning arresters and surge protectors as shown in Figure 2.31, from FIPS Publication 94.

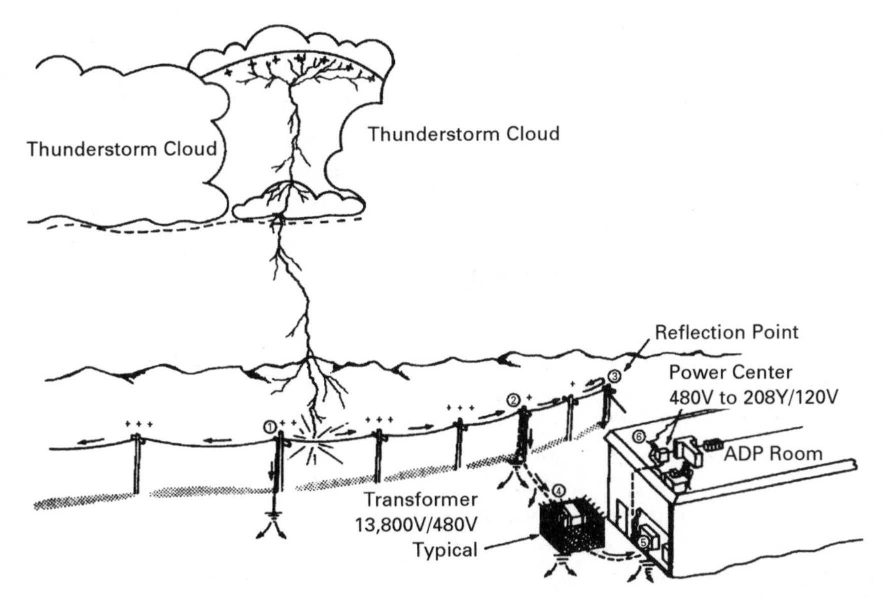

Figure 2.31 Lightning path of least resistance. (*Courtesy of National Institute of Standards and Technology.*)

Electromagnetic interference (EMI). Another source of power quality problems is electromagnetic interference (EMI). Some devices, like a large motor during start-up, emit a magnetic field that intersects with an adjacent sensitive device, like a computer or telephone. Michael Faraday's transformer law explains this phenomenon. Faraday's transformer law says that when an alternating magnetic field cuts across an adjacent conductor, it will induce an alternating current and voltage in that conductor. The induced current and voltage can damage sensitive electronic equipment or cause it to malfunction. Sensitive equipment in hospitals often experiences EMI problems. For example, in one open-heart-surgery training center, electromagnetic fields from an adjacent electrical equipment room were causing heart monitors to read incorrectly. Moving cables emitting the electromagnetic fields a safe distance from the cables feeding the heart monitors solved this problem.

Static electricity. Another cause of power quality problems is static electricity. Static electricity occurs when the rubbing of one object against another causes a voltage buildup. For example, you can build up an electric charge on your body when you rub your shoes on a carpet. A discharge of static electricity can occur when you then touch a grounded object, like another person or a metal object. Although static electricity power quality problems are infrequent, they are often overlooked. Static electricity can create voltages of 3000 V or more and damage sensitive electronic equipment. You can minimize static electricity problems by increasing the humidity, changing the carpet, clothing, and furniture to nonstatic types, and by grounding the person working on a piece of equipment to the equipment with a wrist strap.

Effects of Power Quality Problems

The effects of power quality problems are many and varied. Often a utility customer calls the utility in an attempt to determine the cause of a power quality problem. This chapter has discussed the various types of power quality problems. However, most power quality problems manifest themselves as some effect on an end-user's electrical equipment. These symptoms include motors overheating, adjustable-speed drives tripping off, computers shutting down, flickering lights, and stopped production. The effects of power quality problems can be best be understood by looking at the various types of loads that are affected by power quality problems, including computers, consumer products, lighting, meters, ferromagnetic equipment, telephones, manufacturing processes, and capacitors.

Computers and computer-controlled equipment are most subject to power quality problems. They freeze up and lose data. Most power quality problems on computers are caused by voltage variations.

Consumer products include digital clocks, microwave ovens, television sets, video cassette recorders, and stereo equipment. Most consumer products are affected by voltage sags and outages causing the electronic timer to shut down. This problem manifests itself by the blinking clock.

Lighting includes incandescent, high-intensity discharge, and fluorescent lights. Incandescent lights often dim during a voltage sag. All lighting will flicker when arc furnaces and arc welders cause the voltage to fluctuate.

Meters will give erroneous readings in the presence of harmonics.

Ferromagnetic equipment include transformers and motors. They overheat and lose life when harmonic currents increase the loading on them.

Telephones will experience noise induced by adjacent electrical equipment.

Adjustable-speed drives not only cause harmonics but are affected by them. The frequent shutdown of an adjustable-speed drive is usually an indication of excessive harmonics.

Many manufacturing processes experience frequent shutdowns due to voltage sags.

Capacitors can amplify as well as draw harmonic currents to themselves. This often causes the capacitors to fail or be tripped off-line.

Power Quality Problem-Solving Procedures

Performing a power quality survey can often prevent power quality problems. It is essential to determining the causes and solutions to an existing power quality problem. A power quality survey is a step-by-step process for identify existing and potential power quality problems. A qualified power quality expert often performs a power quality survey. However, end users can perform preliminary surveys. Chapter 7, "Power Quality Surveys," discusses in detail how to plan, conduct, and analyze a power quality survey.

Power Quality Solutions

Power quality solutions fall into two categories: prevention and diagnosing. Preventing power quality problems is preferable to trying to find a solution to a preexisting problem. It involves designing equipment so that it does not add to a potential problem. It involves wiring and grounding sensitive equipment so that electromagnetic interference or ground loops do not affect it. It involves installing power conditioning equipment, such as filters, isolation transformers, UPSs, and TVSSs, to protect sensitive equipment from damage caused by power quality problems. Chapter 4, "Power Quality Solutions," discusses in detail how equipment design and different types of power conditioning equipment solve power quality problems.

Summary

You now have a basic understanding of the steps in solving a power quality problem and how to recognize various types of power quality problems. How do you determine whether you have a problem and the extent of the problem? Power quality standards developed by recognized organizations will help you identify the problem. The next chapter will explain who develops power quality standards and the status of those standards.

References

1. Wolff, Jean-Pierre. 1998. "Power Quality: How Bad Is Bad?" *EC&M Electrical Construction & Maintenance,* vol. 97, no. 9, August, pp. 28–35.
2. Stanislawski, James J. 1994. "Power Quality under the Microscope." *Telephony,* July 18, pp. 32–55.
3. Waller, Mark. 1992. *Surges, Sags, and Spikes.* Indianapolis, Ind. PROMPT Publications.

4. "Voltage Sags & Swells." 1998. URL address: *http://www.scana.com/sce&g/business_solutions/powerquality/qcifvss.htm.*
5. Adams, R. A., et al. 1998. "Solving Customer Power Quality Problems due to Voltage Magnification," *IEEE Transactions on Power Delivery,* vol. 13, no. 4, October, pp. 1515–1520.
6. Ray, Larry. 1998. "Don't Let Sags and Interruptions Disturb You." *EC&M Electrical Construction & Maintenance,* vol. 97, no. 9, August, pp. 42–46.
7. Owen, Edward L. 1996. "Power Disturbance and Quality: Light Flicker Voltage Requirements." *Industrial Application Magazine*, vol. 2, no. 1, January/February, pp. 20–27.
8. Bingham, Rich. 1998. "All You Want to Know about Harmonics." *Power Quality Assurance,* vol. 9, no. 1, January/February, pp. 23–27.
9. Waggoner, Ray. 1994. "Electrical Noise and EMI—Part 1." *EC&M Electrical Construction & Maintenance,* vol. 93, no. 2, February, pp. 14–15.
10. Barber, Thomas L. 1996. "Handling Nonlinear Loads." *Consulting-Specifying Engineer,* pp. 59–61.
11. Bonneville Power Administration. 1991. *Reducing Power Factor Cost.* DOE/CE-0380. U.S. Department of Energy.
12. Seufert, Frederick J. 1990. "Capacitors Improve Power Factor and Reduce Losses." *EC&M Electrical Construction & Maintenance,* vol. 89, no. 8, August, pp. 63–66.
13. Waggoner, Ray. 1993. "How Harmonics Affects Power Factor—Part 1." *EC&M Electrical Construction & Maintenance,* vol. 92, no. 10, October, pp. 22–24.
14. Ballo, Jourjal. 1985. "Take Charge of Static Electricity." *Production Engineering,* vol. 32, October, pp. 66–68.
15. Holt, Mike. 1999. "Introduction to Grounding." *Power Quality Assurance,* vol. 10, no. 3, May/June, pp. 48–50.

Chapter

3

Power Quality Standards

Power quality standards are needed in the power quality industry. How can utilities deliver and their customers receive the quality of power they need without power quality standards? How can the electronics industry produce sensitive electronic equipment without power quality standards? How can the power conditioning industry produce devices that will protect sensitive electronic equipment without power quality standards? They can't.

The power quality industry recognizes that power quality standards are critical to the viability of the industry. Therefore, stakeholders in the power quality industry have developed several power quality standards in recent years. They recognize that the increased interest in power quality has resulted in the need to develop corresponding standards. They realize that the increased use of sensitive electronic equipment, increased application of nonlinear devices to improve energy efficiency, the advent of deregulation, and the increasingly complex and interconnected power system all contribute to the need for power quality standards. Standards set voltage and current limits that sensitive electronic equipment can tolerate from electrical disturbances. Utilities need standards that set limits on the amount of voltage distortion their power systems can tolerate from harmonics produced by their customers with nonlinear loads. End users need standards that set limits not only for electrical disturbances produced by utilities but also for harmonics generated by other end users. Deregulation increases the need for standards so that the offending organization causing poor quality problems is held accountable for fixing the problems. As power systems become more interconnected, contracts based on standards will be needed to protect the offended party. Standards also allow utilities to provide different levels of power quality service.

Several national and international organizations have developed power quality standards. There are a confusing number of different organizations that set power quality standards. The first step in sorting through the confusing number of standards and standards organizations is to examine the primary power quality standards organizations.

Power Quality Standards Organizations

The organizations responsible for developing power quality standards in the United States include the following: Institute of Electrical and Electronics Engineers (IEEE), American National Standards Institute (ANSI), National Institute of Standards and Technology (NIST), National Fire Protection Association (NFPA), National Electrical Manufacturers Association (NEMA), Electric Power Research Institute (EPRI), and Underwriters Laboratories (UL). Outside the United States, the primary organizations responsible for developing international power quality standards include the following: International Electrotechnical Commission (IEC), Euronorms, and ESKOM for South African standards. What is the role of each of these organizations in adopting power quality standards? What organization has the final authority in applying power quality standards? The roles of the three primary organizations involved in developing power quality standards—IEEE, ANSI, and IEC—provide answers to these questions.

Institute of Electrical and Electronic Engineers (IEEE)

The IEEE was founded in 1963 from two organizations: the American Institute of Electrical Engineers (AIEE) and the Institute of Radio Engineers (IRE). It is a not-for-profit association that has grown to be the largest professional organization in the world, with more than 330,000 members in 150 countries. Members of the IEEE have varied technical backgrounds, ranging from computer engineering, biomedical technology, and telecommunications, to electric power, aerospace, and consumer electronics, among others. It has participated in the development of electrical industry standards of all kinds, including power quality. Its membership's interest in power quality focuses on solving particular power quality problems. Its members have developed power quality standards for several years. It presently has at least four societies and dozens of committees, subcommittees, and working groups developing and revising power quality standards. In fact, in 1991, IEEE formed the Standards Coordinating Committee (SSC-22) to coordinate and oversee the myriad of IEEE power quality standards under development or revision.

IEEE power quality standards deal primarily with the power quality limits of disturbances at the point of common coupling (the point where the utility connects to its customer or end user). IEEE power quality standards have a great impact in the electrical utility industry but lack official status, while the American National Standards Institute (ANSI) has the official responsibility to adopt standards for the United States.

American National Standards Institute (ANSI)

Five engineering societies and three government agencies founded ANSI in 1918. It is a private, nonprofit organization with member organizations from the private and public sectors. It does not develop standards, but facilitates standards development by qualified groups, like the IEEE. Consequently, many officially authorized IEEE standards have the dual designation of ANSI/IEEE. It is the sole United States representative to the two major international standards organizations, the International Organization for Standardization (ISO) and the International Electrotechnical Commission (IEC).

International Electotechnical Commission (IEC)

The genesis of the IEC occurred in 1890 at the Electrical Exposition and Conference held in St. Louis during a meeting of several famous electrical pioneers. It has since evolved into an organization with membership from 43 countries. The IEC Council heads the IEC and oversees 200 technical committees, subcommittees, and working groups. IEC power quality standards working groups are concerned mainly about standards that will enhance international trade. They refer to power quality standards as so-called electromagnetic compatibility (EMC) standards. IEC's reference to power quality standards as electromagnetic compatibility standards illustrates that IEC's primary concern is the compatibility of end-user equipment with the utility's electrical supply system. Figure 3.1 shows examples of electromagnetic compatibility.

The IEC has adopted many EMC standards that seem to duplicate IEEE power quality standards. This duplication of standards has caused confusion in the power quality industry. Consequently, several power quality experts have tried to "harmonize" the IEC standards with the IEEE standards. In the meantime, the users of power quality

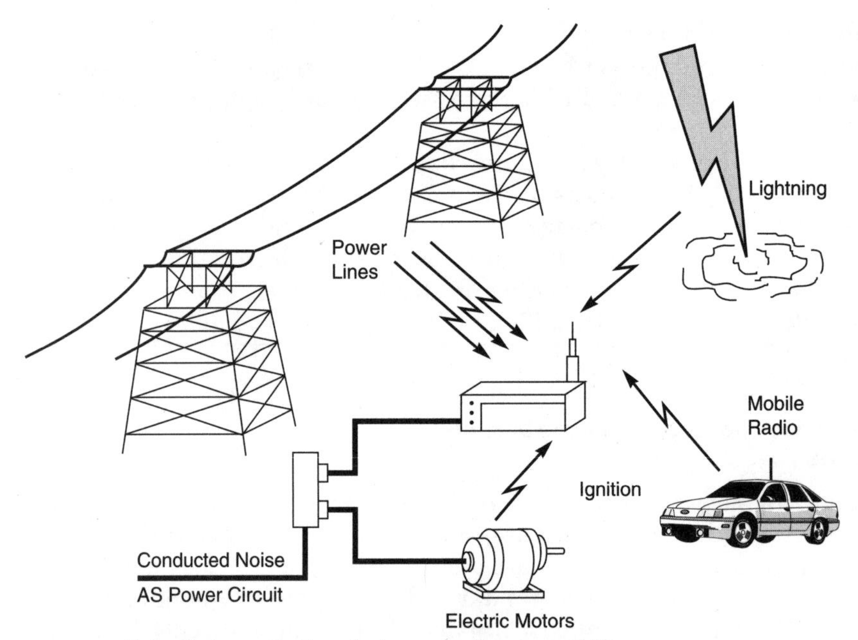

Figure 3.1 Some factors affecting electromagnetic compatibility.

standards must become familiar with both IEC and IEEE standards to determine what standards best meet their needs. In addition to the primary power quality standards organizations of ANSI, IEEE, and IEC, there are other important power quality standards organizations.

Other domestic standards organizations

In the United States, other organizations, like EPRI, UL, NEMA, NFPA, NIST, and some public utility commissions, have also developed power quality standards. For example, EPRI, the research arm of the electric utility industry, has developed reliability indices for utility distribution systems and sponsored the System Compatibility Research Project to enhance the specifications of appliances and equipment to be more compatible with their electrical environment. Underwriters Laboratories is concerned about the safety of various electrical appliances and has developed a standard for the safety of transient voltage surge suppressors, UL 1449. NEMA has set power quality standards for motors, generators, and uninterruptible power supplies (UPSs). NFPA has always been concerned about electrical standards for fire safety. Consequently, it has developed power quality standards to protect computer equipment (NFPA-75) and building lighting (NFPA-780-95) from electrical fires. The National Institute of Standards has developed an information poster on power quality

(NIST-SP768).

Even public utility commissions have adopted power quality standards. For example, the New York Public Utility Commission adopted in 1991 standards of reliability and power quality. The reliability standards address sustained interruptions lasting 5 minutes or more, while the power quality standards address all other disturbances. The New York commission did not set any numerical values but required each utility to set its own power quality standards and file an annual power quality report to the commission.

Other international standards organizations

While the IEC is the primary developer of international power quality standards, other organizations have developed their own standards. For example, ESKOM, the South African utility, has developed power quality standards based on the best of those in the United States and the rest of the world, plus new requirements that other organizations have not developed yet. These standards have allowed ESKOM to provide enhanced power quality service at a premium cost. In addition to IEC and ESKOM, The European Standards Community Standards Organization (CENELEC) has developed power quality standards called *Euronorms.* The International Union of Producers and Distributors of Electrical Energy (UNIPEDE) published, in 1995, "Measurement Guide for Voltage Characteristics." The French standards organization, Union Internationale d'Electrothermie (UIE), is preparing a power quality guide on voltage dips, short-duration interruptions, harmonics, and imbalances.

International standards tend to require more specific measurements of power quality than United States standards. International standards' purpose is to ensure electromagnetic compatibility between utilities and their customers to help commerce and business, while United States standards' purpose is usually to solve a power quality problem. Thus, international standards require more specificity than United States standards.

If you need further information on various power quality standards contact the appropriate organization. Table 3.1 provides names, addresses, and telephone numbers of organizations publishing power quality standards.

Purpose of Power Quality Standards

The purpose of power quality standards is to protect utility and end-user equipment from failing or misoperating when the voltage, current, or frequency deviates from normal. Power quality standards provide

TABLE 3.1 Organizations Publishing Power Quality Standards

Organization	Type of standards	Address
ANSI	Steady-state voltage ratings (ANSI C84.1)	American National Standards Institute 11 West 42nd St., 13th Floor New York, NY 10036 (212) 642-4900 e-mail: *ansionline@ansi.org*
CENELEC	Regional standards	European Union Standards Organization
CISPR	International standards	International Special Committee on Radio Interference
EPRI	Signature newsletter on power quality standards	Electric Power Research Institute Attn: Marek Samotyj 3112 Hillview Ave. Palo Alto, CA 94304 (650) 855-2980
IEC	International standards	International Electrotechnical Commission 31 rue de Varembc P.O. Box 131 CH-1211 Geneva 20 Switzerland 441229190265 e-mail: *info@iec.ch*
IEEE	International and United States standards Color Book Series	Institute of Electrical and Electronics Engineers 445 Hoes Lane Piscataway, NJ 08855-1331 (732) 981-0060 e-mail: *s.tatiner@ieee.org*
ITI (formerly CBEMA)	Equipment guides	Information Technology Industry Council 1250 I St. NW, Suite 200 Washington, DC 20005 (202) 737-8888 Web address: *www.itic.org*
NEMA	Equipment standards	National Electrical Manufacturers Association 1300 N 17th St., Suite 1847 Rossslyn, VA 22209 (703) 841-3258 Web address: *www.nema.org*
NFPA	Lighting protection National Electric Code	National Fire Protection Association 1 Batterymarch Park Quincy, MA 02269-0101 (800) 344-3555 Web address: *www.nfpa.org*
NIST	General Information on all standards	National Center for Standards and Certification National Institute of Standards and Technology Bldg. 820, Room 164 Gaithersburg, MD 208997 (301) 975-4040 e-mail: *ncsic@nist.gov*
UL	Safety standards for equipment	Underwriters Laboratories, Inc. 333 Pfingsten Rd. Northbrook, IL 60062-2096 (849) 272-8800 e-mail: *northbrook@ul.com*

this protection by setting measurable limits as to how far the voltage, current, or frequency can deviate from normal. By setting these limits, power quality standards help utilities and their customers gain agreement as to what are acceptable and unacceptable levels of service.

On both sides of the revenue meter, utilities and their customers need power quality standards. On the utility side of the meter, standards help utilities keep electrical disturbances from affecting their equipment and their customer's equipment. On the end-user side of the meter, end users need standards that keep user-created electrical disturbances from affecting the operation of equipment owned by the utility, other end users, and themselves. Manufacturers of sensitive electronic equipment need standards to keep their customers satisfied. Finally, the purchasers and manufacturers of power conditioning equipment need standards developed by independent organizations, like UL, to assure themselves that their power conditioning equipment does protect end-user equipment from electrical disturbances.

Types of Power Quality Standards

What types of power quality standards do utilities and their customers need to protect their equipment from damage? They need not only standards that set limits on electrical disturbances that utility and end-user equipment can withstand, but also standards that allow their various types of equipment to operate effectively. United States standards deal mostly with voltage quality, while international standards deal with compatibility limits between the electric utility power supply and the end-user equipment. There are standards on the performance of power conditioning equipment needed to prevent power quality disturbances from causing end-user equipment to misoperate. There are even standards for the equipment that measures and monitors electrical disturbances. And finally there are general standards that define power quality terms.

In order to help the power quality industry compare the results of power quality measurements from different instruments, the IEEE developed IEEE Standard 1159-1995 copyright © 1995, *Recommended Practice for Monitoring Electric Power Quality.* This standard defines various power quality terms and categorizes IEEE standards by the various power quality topics of grounding, powering, surge protection, harmonics, disturbances, life/fire safety, mitigation equipment, telecommunications equipment, noise control, utility interface, monitoring, load immunity, and system reliability. Table 3.2, from IEEE Standard 1159-1995, provides a summary of the various types of IEEE power quality standards.

How do you determine the appropriate IEC power quality standard

TABLE 3.2 United States Power Quality Standards by Topic

Topic	Relevant standards
Grounding	IEEE 446, 141, 142, 1100; ANSI/NFPA 70
Powering	ANSI C84.1; IEEE 141, 446, 1100, 1250
Surge protection	IEEE C62, 141, 142; NFPA 778; UL 1449
Harmonics	IEEE C57.110, 519, P519a, 929, 1001
Disturbances	ANSI C62.41; IEEE 1100, 1159, 1250
Life/fire safety	FIPS Pub. 94; ANSI/NFPA 70; NFPA 75; UL 1478, 1950
Mitigation equipment	IEEE 446, 1035, 1100; 1250; NEMA-UPS
Telecommunication equipment	FIPS Pub. 94; IEEE 487, 1100
Noise control	FIPS Pub. 94; IEEE 518, 1050
Utility interface	IEEE 446, 929, 1001, 1035
Monitoring	IEEE 1100, 1159
Load immunity	IEEE 141, 446, 1100, 1159, P1346
System reliability	IEEE 493

SOURCE: IEEE Standards 1159-1995

to use in a given situation? What are the various types of IEC power quality standards? Table 3.3 provides a simple user guide to IEC standards by organizing them according to general, environment, limits, testing and measurement, and installation and mitigation categories.

How do you reconcile the differences and similarities of United States and international standards? Tom Key, vice president of technology of EPRI's Power Electronic Application Center (PEAC), saw the need to compare and categorize United States and international standards by electrical disturbances. He, therefore, developed Table 3.4, which compares the United States and international standards by disturbance and the corresponding purpose of the standard.

The many United States and international power quality standards can cause much confusion. In order to sort through the confusion so that you can use these standards to solve and prevent power quality problems, this chapter categorizes them in the same way that Chapter 2 categorizes power quality characteristics. Chapter 2 categorizes the type of disturbance along with the subsequent effect of the disturbance on equipment and the corresponding mitigation solution to the power quality disturbance. Table 3.5 uses this approach to categorize the various United States power quality standards.

This method of organizing power quality standards provides you with a more logical and effective way to utilize power quality standards. First, Chapter 3 discusses both United States and international standards categorized by electrical disturbances. Next, it discusses the corresponding subcategories of standards organized by the effect

TABLE 3.3 IEC Power Quality Standards by Topic

Topic	Description	IEC number
General	-Fundamental principles -Definitions -Terminology	IEC Pub. 1000-1
Environment	-Description -Classification -Compatibility limits	IEC Pub. 1000-2
Limits	-Emission and immunity limits -Generic standards	EIC Limits 1000-3
Testing and measurement	Techniques for conducting tests	IEC Pub. 1000-4
Installation and mitigation	-Installation guidelines -Mitigation methods -Mitigation devices	IEC Guide 1000-5

TABLE 3.4 Comparison of IEEE and IEC Power Quality Standards

Disturbance	IEEE standard	IEC standard
Harmonic environment	None	IEC 1000-2-1/2
Compatibility limits	IEEE 519	IEC 1000-3-2/4 (555)
Harmonic measurement	None	IEC 1000-4-7/13/15
Harmonic practices	IEEE 519A	IEC 1000-5-5
Component heating	ANSI/IEEE C57.110	IEC 1000-3-6
Under-Sag-environment	IEEE 1250	IEC 38, 1000-2-4
Compatibility limits	IEEE P1346	IEC 1000-3-3/5 (555)
Sag measurement	None	IEC 1000-4-1/11
Sag mitigation	IEEE 446, 1100, 1159	IEC 1000-5-X
Fuse blowing/upsets	ANSI C84.1	IEC 1000-2-5
Oversurge environment	ANSI/IEEE C62.41	IEC-1000-3-7
Compatibility levels	None	IEC 3000-3-X
Surge measurement	ANSI/IEEE C62.45	IEC 1000-4-1/2/4/5/12
Surge protection	C62 series, 1100	IEC 1000-5-X
Insulation breakdown	By product	IEC 664

SOURCE: EPRI's PEAC Corp. (*Courtesy of EPRI's "Signature."*)

electrical disturbances have on equipment. Finally, it discusses standards for mitigation solutions and measurement procedures categorized by electrical disturbances.

Voltage sag (dip) standards

Standards for voltage sags or dips use reliability indices to set voltage sag limits (IEEE uses the term *sag* or *momentary interruption,* but IEC uses the term *dip* or *short-time interruption* to refer to the same phenomenon). Voltage sags are typically the most important power

TABLE 3.5 Summary of United States Power Quality Problems

Example waveshape or RMS variation	Disturbance Description	Disturbance Sources	Effects Standards	Mitigation standards
	Impulsive transients Transient disturbance	ANSI/IEEE C62.41	ANSI/IEEE C62.45	UL 1449 IEEE 1100
	Oscillatory transients Transient disturbance	ANSI/IEEE C62.41	ANSI/IEEE C62.45	UL 1449 IEEE 1100
	Sags/swells RMS disturbance	IEEE P1346 IEEE 493	ANSI/IEEE C84.1 IEEE 446	IEEE 446 IEEE 1100 IEEE 1159
	Interruptions RMS disturbance	ANSI/IEEE C62.41	ANSI/IEEE C62.45	UL 1778 UL 1449 IEEE 1100
	Undervoltages/ overvoltages Steady-state variation	ANSI C84.1 Load variations Load dropping	Shorten lives of motors and lighting filaments	Voltage regulators Ferroresonant transformers
	Harmonic distortion Steady-state variation	ANSI/IEEE 519 ANSI/IEEE 929 ANSI/IEEE 1001 ANSI/IEEE 1035	ANSI/IEEE 519 ANSI/IEEE C57.110 ANSI/IEEE C84.1	ANSI/IEEE C57.110 ANSI/IEEE C37
	Steady-state variation Voltage flicker	ANSI/IEEE C84.1 IEEE 141 IEEE 141	Lights flicker Irritation	IEEE 519

quality variation affecting industrial and commercial customers. The *IEEE Gold Book* (Standard 493-1990, copyright © 1990, p. 38) already includes voltage sags in the definition of reliability:

> Economic evaluation of reliability begins with the establishment of an interruption definition. Such a definition specifies the magnitude of the voltage dip and the minimum duration of such a reduced-voltage period that results in a loss of production or other function of the plant process.

The most basic index for voltage sag performance is the *system average rms (variation) frequency index voltage* ($SARFI_x$). $SARFI_x$ quantifies three voltage sag parameters into one index. The three parameters are the number of voltage sags, the period of measurement, and the number of end users affected by the voltage sag. Consequently, $SARFI_x$ represents the average number of *specified* short-duration rms variation events per customer that occurred on a specific power system during a measurement time period. For $SARFI_x$, the specified disturbances are those rms variations with a voltage magnitude less than x for voltage drops or a magnitude greater than x for voltage increases. $SARFI_x$ is defined by Eq. (3.1):

$$SARFI_x = \frac{\Sigma N_i}{N_T} \tag{3.1}$$

where x = rms voltage threshold; possible values 140, 120, 110, 90, 80, 70, 50, and 10

N_i = number of customers experiencing voltage deviations with magnitudes above $Y\%$ for $x > 100$ or below $Y\%$ for $x < 100$ due to event i

N_T = number of customers served from the section of the system to be assessed

$SARFI_x$ is calculated in the same way as the system average interruption frequency index (SAIFI) value that many utilities have used for years (proposed IEEE Standard P1366). The two indices are, however, quite different. $SARFI_x$ assesses system performance with regard to short-duration rms variations, whereas SAIFI assesses only sustained interruptions. $SARFI_x$ can be used to assess the frequency of occurrence of sags, swells, and short-duration interruptions. Furthermore, the inclusion of the index threshold value x provides a means for assessing sags and swells of varying magnitudes. For example, $SARFI_{70}$ represents the average number of sags below 70 percent experienced by the average customer served from the assessed system.

Subindices of SARFI can be categorized by the causes of the events or by the duration of the events. For instance, a subindex of SARFI is the index related to voltage sags that are caused by lightning-induced faults. Other subcategories of indices include indices for instantaneous, momentary, and temporary voltage sags, as defined in IEEE 1159, 1995.

Indices have been developed for aggregated events. Examples of aggregated events are multiple voltage sags that often occur together because of reclosing operations of breakers and characteristics of distribution faults. Once a customer process is impacted by a voltage sag, the subsequent sags are often less important. To account for this effect, $SARFI_x$ uses an aggregate event method that results in only one count for multiple sags within a 1-minute period (aggregation period).

How do utilities estimate these indices? Utilities can use historical fault performance of transmission and distribution lines to estimate these indices. However, utilities have discovered that system monitoring at specific system locations provides a more accurate way to determine these indices. In order to obtain a more accurate measurement of these indices, many utilities (such as Consolidated Edison, United Illuminating, Northeast Utilities, San Diego Gas & Electric, TVA, Entergy, Baltimore Gas & Electric) have installed extensive monitoring systems that measure and record their systems' performance on a continuous basis. For example, Detroit Edison and Consumers Power installed monitoring systems to track performance at specific customers (automotive plants) as part of the contractual requirements associated with serving these customers.

Utilities and their customers find that the information they obtained by using these indices can be valuable for many different purposes. For instance, United Illuminating (UI) has installed power quality monitoring at all of its distribution substations. UI can use this power quality monitoring data to provide real-time system performance information to customer engineers, protection engineers, and operations engineers throughout the UI network. UI has used the data to calculate performance indices and has included this information in monthly and quarterly reports of the system performance. UI engineers use SARFI as one of UI's company performance drivers along with SAIFI, SAIDI, and CAIDI (interruption-based indices). They have used the SARFI-based ranking of substations to prioritize substation expansion and maintenance. For example, if $SARFI_{90}$ exceeds specified thresholds in any period, the UI engineers recommend a power quality investigation to determine the reason. UI engineers plan to include steady-state performance indices (voltage regulation, unbalance, harmonics) in their system performance reports.

Standards are needed for the effects of voltage sags on sensitive electronic equipment. A working group in the IEEE is developing such a standard. It is called IEEE Standard P1346, *Electric Power System Compatibility with Electronic Process Equipment.* This standard contains indices that will allow industrial engineers to evaluate how sensitive their industrial processes will be to voltage sags. In addition, IEEE has included the CBEMA curve described in IEEE Standard 446—1995 (*The Orange Book*) to show equipment susceptibility to voltage sags. In fact, the contract between Detroit Edison and its automotive customers used language from IEEE Standard 446—1995 to set limits as to acceptable voltage sags. Sometimes the CBEMA curve sets voltage sag limits that are not restrictive enough to protect some types of sensitive equipment. For example, Figure 3.2 shows an adjustable-speed drive that is more sensitive to voltage sags than indicated by the CBEMA curve.

Transients or surges

ANSI/IEEE C62.41-1991, *IEEE Guide for Surge Voltages in Low Voltage AC Power Circuits,* deals with transients in a building. This standard is concerned with the effect of transients on the load side of the meter. It categorizes the location of the transients and types of transient waveforms. The three locations are: category A, anything on the load side of a wall socket outlet; category B, distribution system of

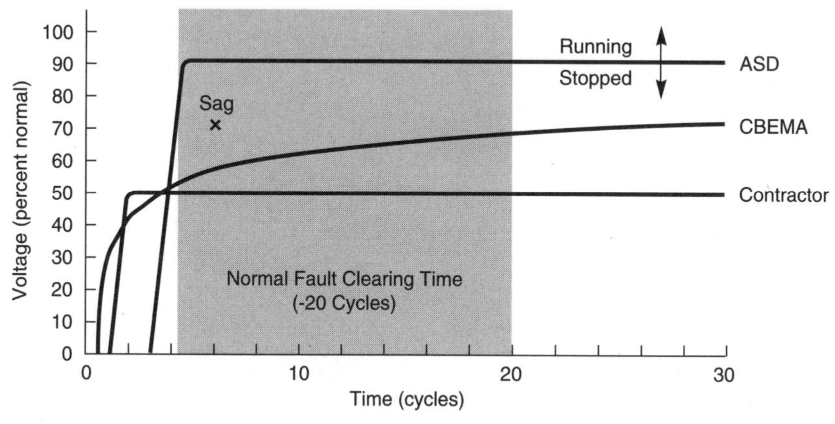

Figure 3.2 Example of equipment sensitivity to voltage sags.

the building; and category C, outside the building or on the supply side of the main distribution boards for the building. Their duration and frequency categorize the five types of transient waveforms:

1.2/50–8/20-μs combination wave

0.5-μs, 100-kHz ring wave

10/1000-μs unidirectional wave

5-kHz ring wave

Electrical fast transient

Transient voltage surge suppressor (TVSS) standards. The most popular way to protect your computer and other sensitive electronic equipment is to use a transient voltage surge suppressor or TVSS. This device "clamps" the voltage from a voltage transient or surge and keeps it from damaging your equipment. Clamping the voltage means reducing the voltage that enters the surge suppressor to a level that is safe for voltage-sensitive equipment. Once the voltage is clamped, it rises to what is called the "let-through" voltage. The let-through voltage is the voltage that the protected equipment receives. Figure 3.3 illustrates what is meant by clamp and let-through voltage. How surge suppressors work and how to evaluate their performance is discussed in the next chapter. The primary standard for the TVSS is Underwriters Laboratories Standard UL1449.

The original UL1449, developed in 1987, defines the requirements for TVSS devices based on the two classes identified in IEEE Standard C62.41: (1) permanently connected (category B) and (2) cord-connected (category B or A). Using the transient waves defined in C62.41, Underwriters Laboratories performs a series of tests. These tests are

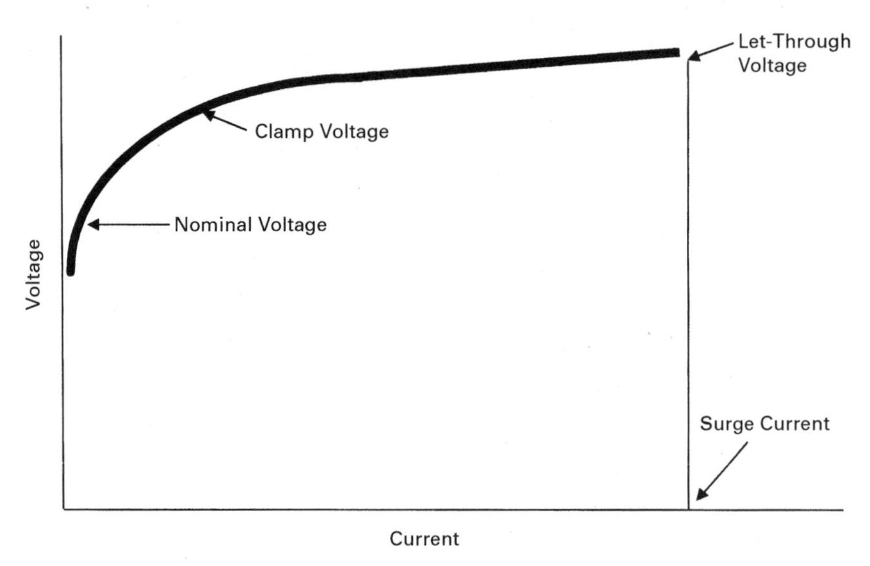

Figure 3.3 Clamp and let-through voltages.

designed to simulate the characteristics of voltage surges and the voltage that will pass through the TVSS. The tests included applying to the TVSS a voltage transient of 6000 V, 500 A, in 20-μs pulses. At the time, UL believed that this was the largest voltage transient that could be produced at a 120-V, 15-A outlet.

Since 1987, UL has revised UL 1449 for two reasons: open neutral-ground bonds cause 240 V to be applied to 120-V outlets and faults on the utility transmission system can bring 10,000 V into a 120-V outlet. Table 3.6 compares the requirements of the old UL 1449 with the new one.

Uninterruptible power supply (UPS) system. The purpose of an uninterruptible power supply system is to protect sensitive equipment from voltage surges and loss of power. A UPS often contains both a surge suppression device and some type of power supply, like a battery or a motor generator set. Most modern UPSs have a battery backup system and an inverter to convert dc to ac to operate equipment during an outage. Consequently, the same standards that are applicable to a surge protection device are applicable to UPSs, like UL 1449 and IEEE C62.41. For safety reasons, Underwriters Laboratories has developed a standard specifically for UPSs, UL 1778.

Voltage unbalance

The primary standard for voltage unbalance as well as steady-state voltage requirements is ANSI C84.1-1995. It specifies that equipment be designed to operate at voltages not to exceed +6 percent or less

TABLE 3.6 Comparison of Original and Revised UL 1449 Standard

Revised UL 1449	Original UL 1449
Protection against protector meltdown	Not required
Safe against catastrophic overvoltage	Not required
Safe against leakage/shocks after damage	Not required
Withstand two 3000-A and twenty 500-A surges	Withstand two 500-A and twenty-four 125-A surges
Specify protection modes	Not required
102pp	39pp

Source: Richard L. Cohen. 1998. "The New UL1449 Standard for Transient Voltage Surge Suppressors." *Power Quality Assurance*, vol. 9, no. 4, July/August, p. 37.

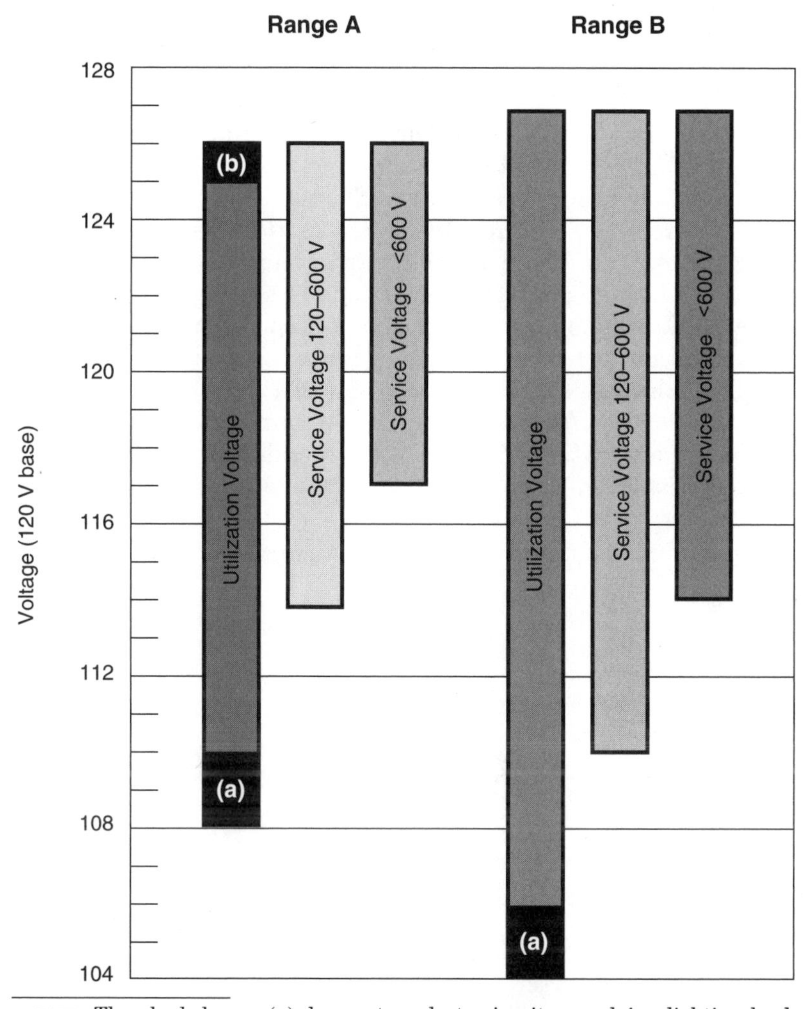

NOTE: The shaded area (a) does not apply to circuits supplying lighting loads. The shaded area (b) does not apply to 120- to 600-V systems.

Figure 3.4 ANSI C84.1-1995 steady-state voltage limits. (Reprinted from ANSI 84.1 by permission from National Electrical Manufacturers Association. Copyright © 1996 National Electrical Manufacturers Association.)

than −13 percent of the nominal 120/240 system voltage. Figure 3.4 summarizes these voltage standards. Range A applies to normal conditions. Range B applies to short-duration or unusual conditions.

ANSI C84.1-1995 defines voltage unbalance as the maximum deviation from the average of the three-phase voltages or currents, divided by the average of the three-phase voltages or currents, expressed in percent. Unbalance of voltage and current can damage motors. To prevent any damage to equipment, ANSI C84.1-1995 sets a maximum voltage unbalance at the meter under no-load conditions of 3 percent. Figure 3.5 provides an example of voltage unbalance statistics on a distribution feeder.

Voltage fluctuation or flicker standards

The primary United States standards for voltage fluctuation are contained in IEEE 519-1992, *Recommended Practices and Requirements for Harmonic Control in Electric Power Systems,* and IEEE 141-1995, *Recommended Practice for Electric Power Distribution for Industrial Plants* (*The Red Book*). There are no United States standards for measuring flicker at this time. The international standard for measuring flicker is IEC 1000-4-15, *Flickermeter—Functional and Design Specifications* (formerly IEC 868), and for setting flicker limits for individual appliances is IEC 1000-3-3, *Disturbances in Supply Systems Caused by Household Appliances and Similar Electrical Equipment* (formerly IEC 555-3). All these standards attempt to limit the lighting flicker so that it does not irritate a person seeing it.

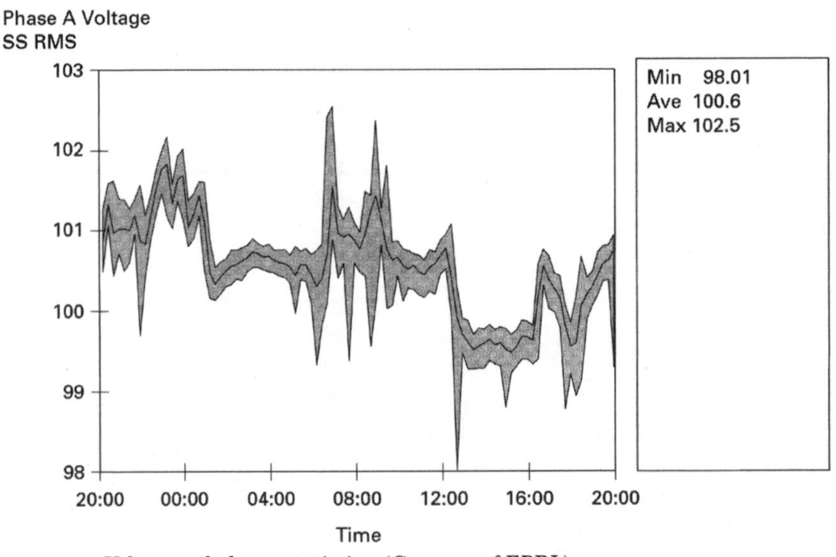

Figure 3.5 Voltage unbalance statistics. (*Courtesy of EPRI.*)

Because of the subjective nature of light flicker, standards organizations have had difficulty correlating voltage fluctuation standards to perceptible light flicker that is irritating to the observer. The two IEEE flicker standards attempt to solve this problem by using the GE flicker curve published in 1951, shown in Figure 3.6.

IEEE 519-1992 copyright © 1993, *Recommended Practices and Requirements for Harmonic Control in Electric Power Systems,* page 80, says "sources of flicker in industrial power distribution systems can be, for instance, the somewhat random variations of load typified by an arc furnace melting scrap steel or an elevator motor's starts and stops. A flicker source may be nearly periodic, as in the case of jogging or manual spot-welding. A source may also be periodic, as in the case of an automatic spot-welder." It mentions that static VAR (volt-amperes reactive) compensators at the flicker source keep the voltage steady under varying load conditions and therefore solve the flicker problem. A person experiencing irritating flicker can sometimes get rid of the flicker by changing the light bulb type. IEEE-519-1992 not only sets flicker standards but is the basis for setting harmonic standards in the United States.

Harmonic standards

The harmonic standard for the United States, IEEE 519-1992, *Recommended Practices and Requirements for Harmonic Control in Electric Power Systems,* recognizes that the primary source of har-

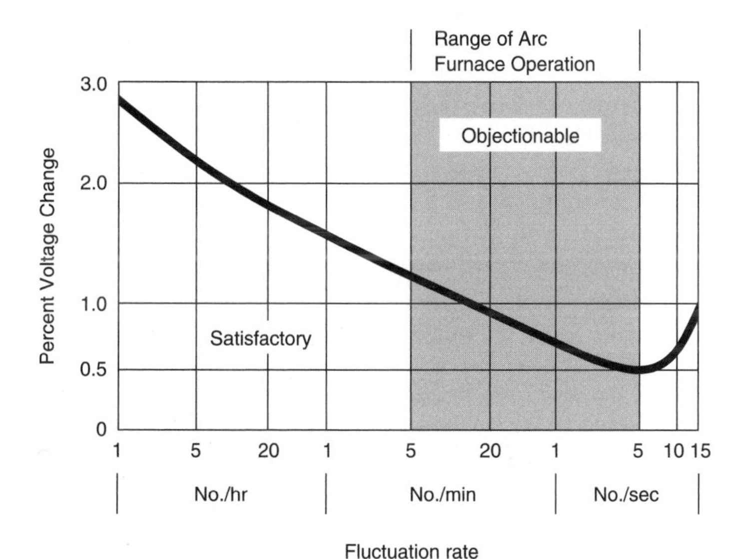

Figure 3.6 Flicker sensitivity curve. (*Source: GE.*)

monic currents is nonlinear loads located on the end-user (utility customer) side of the meter. However, the same standard indicates that capacitors located on the utility side of the meter can amplify the harmonic voltage. The utility can also transmit harmonic voltage distortion to other end users. IEEE 519-1992 sets current limits at the point of common coupling (PCC). Figure 3.7 shows that the PCC is where the utility connects to multiple end users.

IEEE 519-1992 defines harmonic limits on the utility side of the meter as the total harmonic distortion (THD) and on the end-user side of the meter as total distortion demand (TDD). This standard sets the voltage distortion limits or THD that the utility can supply to the end user at the point of common coupling. Table 3.7, from IEEE 519-1992, sets THD limits on the utility system at various voltages. The same IEEE standard sets limits on the harmonic current that the end user can inject into the utility's system at the point of common coupling. Table 3.8 provides the TDD limits in IEEE 519-1992.

In contrast to the IEEE setting harmonic limits at the point of common coupling, the IEC sets harmonic limits on individual loads, like adjustable-speed drives. These limits are contained in IEC 1000-3-2 (formerly IEC 555-2).

Harmonics cause electronic equipment to malfunction. Section 6 of IEEE-519-1992 discusses the effects of harmonics on electronic equipment. Electronic equipment cannot stand more than 5 percent harmonic voltage distortion factor, with a single harmonic being no more than 3 percent of the fundamental voltage. Higher levels of harmonics result in erratic malfunction of the electronic equipment. In addition to causing problems to sensitive electronic equipment, harmonics can cause relays and meters to malfunction. The magnitude and frequency of occurrence of harmonics needs to be considered as well as the duration of the harmonics.

Section 6 of IEEE-519-1992 also discusses the effects of harmonic currents on electrical equipment, like motors and transformers. Harmonics cause motors and transformers to overheat. IEEE 519-1992 sets harmonic current limits to prevent them from overheating motors and transformers. For instance, the upper current distortion limit of 5 percent is to prevent harmonic currents from overheating transformers.

Because of the potential damage to transformers caused by harmonics, IEEE developed a standard specifically to limit harmonics' effect on transformers. This is ANSI/IEEE C57.110-1996, *Recommended Practice for Establishing Transformer Capability When Supplying Nonsinusoidal Load Currents.*

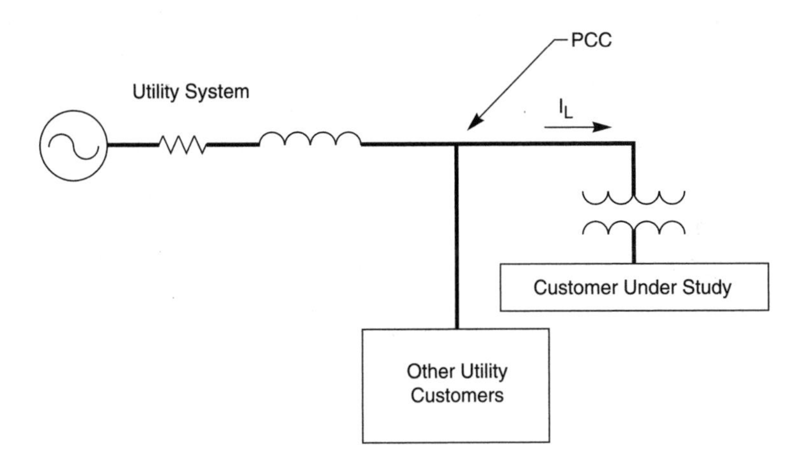

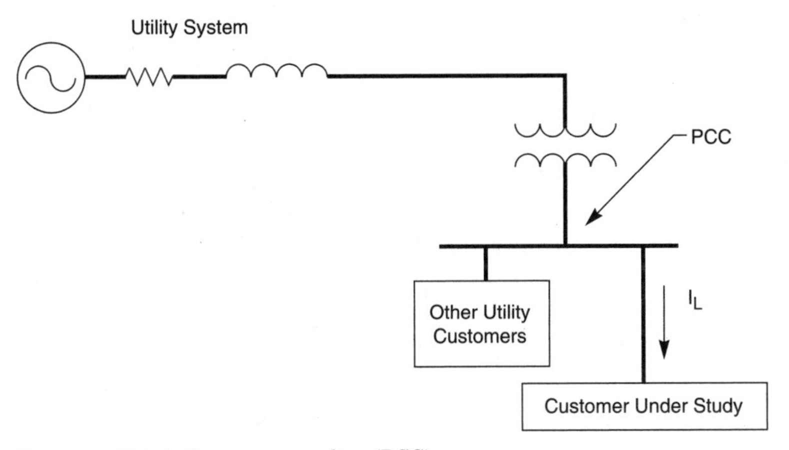

Figure 3.7 Point of common coupling (PCC).

TABLE 3.7 IEEE-519 Voltage Distortion Limits

Bus voltage	Maximum individual harmonic component, %	Maximum THD, %
69 kV and below	3.0	5.0
69 kV to 161 kV	1.5	2.5
Above 161 kV	1.0	1.5

NOTE: High-voltage systems can have up to 2.0% THD, where the cause is an HVDC terminal that will attentuate by the time it is trapped for a user.

SOURCE: IEEE Standard 519-1992, Copyright © 1993, IEEE. All rights reserved.

TABLE 3.8 IEEE-519 Harmonic Current Limits in Percent of I_L

$I_x < I_z$	$h < 11$	$11 \le h < 17$	$17 \le h < 23$	$23 \le h < 35$	$35 \le h$	TDD
			$v \le 69$ kV			
<20	4.0	2.0	1.5	0.6	0.3	5.0
20<50	7.0	3.5	2.5	1.0	0.5	8.0
50<100	10.0	4.5	4.0	1.5	0.7	12.0
100<1000	12.0	5.5	5.0	2.0	1.0	15.0
1000	15.0	7.0	6.0	2.5	1.4	20.0
			69 kV $< v \le 161$ kV			
<20*	2.0	1.0	0.75	0.3	0.15	2.5
20<50	3.5	1.75	1.25	0.5	0.25	4.0
50<100	5.0	2.25	2.0	0.75	0.35	6.0
100<1000	6.0	2.75	2.5	1.0	0.5	7.5
>1000	7.5	3.5	3.0	1.25	0.7	10.0
			$v > 161$ kV			
<50	2.0	1.0	0.75	0.3	0.15	2.5
≥50	3.5	1.75	1.25	0.45	0.22	3.75

NOTE: All power-generation equipment is limited to these values of current distortion, regardless of I_{SC}/I_L.

SOURCE: IEEE Standard 519-1992, Copyright © 1993, IEEE. All rights reserved.

Transformer overheating standards

ANSI/IEEE Standard C57 series addresses the problem of harmonics causing transformers to overheat. It does this by setting so-called *K*-factor ratings of transformers. Harmonics' major effect on transformers is to increase losses and heating in transformers. They increase both load and no-load losses. They increase load losses by causing skin effects, increasing eddy-current, I^2R, and stray losses. They increase no-load losses by increasing hysteresis losses. IEEE and UL have adopted standards to either derate regular transformers or to design special transformers that can withstand the effect of harmonics. These specially designed transformers are called *K*-factor transformers.

***K* factor.** Purchasers of transformers can use the *K*-factor value to pick a specially designed *K*-factor transformer or to derate a non-*K*-factor transformer. They need to first calculate the *K* factor. Then, they can decide whether to derate a standard transformer or purchase a specially designed *K*-factor transformer. If they decide to purchase a standard transformer, they use the *K*-factor to derate the standard transformer. If they decide to purchase a specially designed *K*-factor transformer, they use the *K* factor to pick the *K*-factor rating of the transformer.

A *K*-factor transformer has certain features that allow it to handle the extra heating of harmonic currents. It may have a static shield between the high- and low-voltage windings to reduce electrostatic noise caused by harmonics. It may use smaller-than-normal, transposed, and individually insulated conductors to reduce the skin-effect and eddy-current losses. It may also have a neutral conductor in the secondary winding large enough to carry the third-harmonic neutral currents. It may have core laminations that are individually insulated to reduce eddy currents in the core. It may have a larger core with special steel to reduce hysteresis losses and reduce the possibility of the transformer saturating because of high voltage peaks on the distorted bus voltage waveform. This special steel has less resistance to the changing magnetic fields. A larger core increases the area of steel and thus reduces the flux density and resistance to the changing magnetic fields. The transformer design engineer can also reduce the *K*-factor transformer flux density by increasing the number of turns in the winding. The *K*-factor transformer may have larger conductors than a standard transformer with the same nameplate rating to reduce heating caused by increased I^2R losses. Often it has added cooling ducts in the windings to reduce the increased heating effects of harmonics. Reducing the height of the conductor reduces eddy-current losses and decreases the flux density, as discussed previously.

Underwriters Laboratories, in UL 1561, *Standard for Safety for Dry-Type General Purpose and Power Transformers,* developed the *K*-factor constant to take into account the effect of harmonics on transformer loading and losses. IEEE 1100-1992 copyright © 1992, *Power and Grounding Sensitive Electronic Equipment,* (*The Emerald Book*), page 75, defines *K* factor as

$$K = \frac{\Sigma\,(I_h h)^2}{\Sigma\,(I_h)^2} \tag{3.2}$$

where I_h = harmonic current and h = harmonic value. This formula shows how to calculate *K* factor by summing the product of each harmonic current squared and the harmonic order squared and dividing by the summation of the harmonic current squared. Then, calculate the increased eddy current losses due to harmonics by multiplying the rated eddy current losses by the *K* factor.

The steps in calculating the *K* factor of a transformer are shown in the flowchart in Figure 3.8. This flowchart provides a step-by-step method for determining the *K* factor to be used either to derate a standard transformer or to specify the *K* rating of a *K*-factor transformer. Chapter 10 of *Energy Efficient Transformers* by Barry Kennedy explains how to calculate transformer *K* factor.

ANSI/IEEE C57.110-1986 does not mention *K* factor. However, ANSI/IEEE C57.110-1986 does provide the methods for calculating the losses and currents for a certain harmonic load that is the basis for determining *K*-factor values. This standard was intended for application

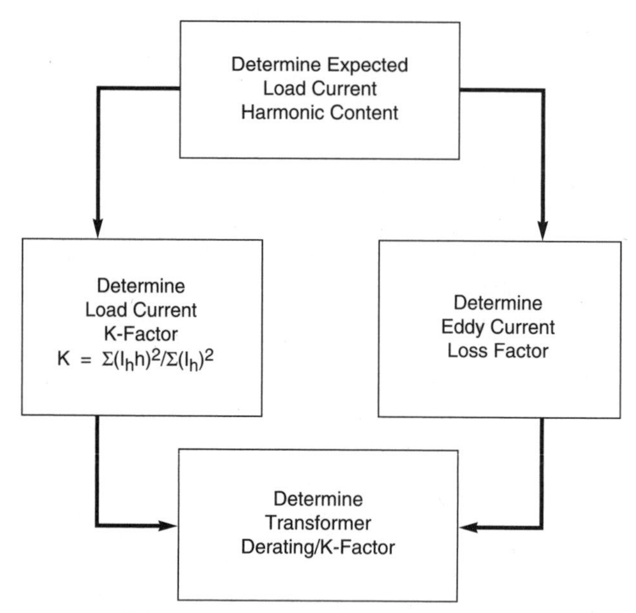

Figure 3.8 *K*-factor calculation steps.

with liquid-immersed transformers, yet K factor as derived from this standard is applied to dry-type transformers. This implies some inaccuracy in the calculation of K factor for dry-type transformers.

In summary, the purpose of the K-factor rating is to rank transformers for harmonics, reduce skin-effect losses, and reduce the possibility of core saturation. Transformer manufacturers design transformers with a special K-factor rating. Transformers with a K-factor rating have a note on their nameplates indicating that they are designed for nonsinusoidal current load with a certain K factor.

A K-factor transformer may cost approximately twice as much as a standard transformer and weigh 115 percent more than a standard transformer. It is recommended that purchasers of transformers use K-factor transformers rather than derate a standard transformer. This is to avoid unforeseen hot spots. A derated transformer may still contain hot spots due to harmonics that could result in overheating and transformer loss of life.

Harmonics not only can cause transformers and other equipment to overheat but also can cause cables to overheat as well. The neutral conductor in the cable is especially susceptible to overloading due to harmonics.

Neutral conductor loading standards

Why are standards needed to limit the excessive neutral current caused by harmonics and to size neutral conductors to carry them?

Single-phase nonlinear electronic loads will draw current only during the peak of the voltage waveform. These loads combined in a three-phase circuit produce triplen harmonics (multiples of third-order harmonics, like third, ninth, fifteenth). Triplen harmonics do not cancel one another but are additive and return exclusively through the neutral conductor. The resulting magnitude of the neutral current may increase to 173 percent of the rms phase current. Thus, the neutral current may exceed the capacity of the neutral conductor.

There are no United States standards for limiting harmonic currents from single-phase loads and no standards for sizing neutral conductors to accommodate them. The international community in IEC 1000 3-2, *Limits for Harmonic Current Emissions,* has set limits for triplen harmonics generated by various classes of single-phase equipment.

Static electricity

Static electricity can often be the hidden cause of poor power quality. A small discharge of 4000 V from a finger to ground is enough to damage sensitive electronic equipment. The National Institute of Standards and Technology recognized this concern when it published the Federal Information Processing Standards Publication (FIPS Pub.) 94, *Guideline on Electrical Power for ADP Installations,* in 1983. It defines static electricity as "electric discharges dislodged and trapped when insulating materials touch and are forcefully separated." Figure 3.9 from FIPS Pub. 94 shows a static discharge from a finger to a switch. A static discharge of 10,000 V can jump 0.5 inches, while a discharge of 20,000 V can jump 1 inch. FIPS Pub. 94 mentions that by keeping the humidity at least 50 percent, increasing the conductivity of carpeting, furniture, and upholstery, and treating shoes and clothing with antistatic preparations, you can reduce static discharges considerably.

IEEE Standard C62.47-1992 copyright © 1993, *Guide on Electrostatic Discharge (ESD): Characterization of the ESD Environment,* defines ESD as the peak current to the discharge voltage and sets limits for various types of objects. For example, it sets ESD limits of 36 to 38

Figure 3.9 Static discharge.

A/kV for various types of desks, cabinets, and telephones in the office environment. Telephones can be affected by poor power quality. Some standards have been developed to limit the noise of telephone lines.

Telephone power quality standards

The marriage of the telephone to computers and, in 1984, the divestiture of AT&T have increased the need for power quality standards in the telephone industry. Just as computers are sensitive to variations in power quality in other applications, they are sensitive to the same variations in their applications to telephones. Telephone standards were developed and implemented by AT&T prior to its breakup in 1984. Many of the standards developed after the divestiture of AT&T applicable to telephone equipment are the same standards that are designed to protect other sensitive electronic equipment. Table 3.9 lists some of those standards that are critical to protecting the smooth operation of telephone equipment.

In addition to these telephone standards, utilities have IEEE Standard 487, *Recommended Practice for the Protection of Wire Line Communications Facilities Serving Electric Power Stations.*

Grounding and wiring standards

The primary standards for wiring and grounding are IEEE Standard 446, *Emergency and Standby Power Systems for Industrial and Commercial Applications* (*The Orange Book*), IEEE Standard 141-1993, *Electric Power Distribution for Industrial Plants* (*The Red Book*), IEEE Standard 142-1991, *Grounding of Industrial and Commercial Power Systems* (*The Green Book*), IEEE Standard 1100, *Powering and Grounding Sensitive Electronic Equipment,* FIPS Pub. 94, and the **National Electrical Code® (NEC)®*, ANSI/NFPA 70. While the NEC is concerned with providing adequate grounding that protects the public from electrical shock, these other standards are concerned with setting grounding standards that protect sensitive equipment from damage or misoperation caused by extraneous ground current. They do this by specifying how to properly ground equipment to prevent ground loops, electrical noise, and static electricity from affecting sensitive electrical equipment. In addition to grounding standards, there are standards designed specifically for different types of sensitive electronic equipment.

Sensitive electronic equipment standards

Working through the various standards organizations, representatives from industries that use sensitive electronic equipment have developed

**National Electrical Code®* and *(NEC)®* are registered trademarks of the National Fire Protection Association, Inc., Quincy, Mass. 02269.

TABLE 3.9 IEEE Power Quality Standards by Title Applicable to Telecommunications

Title	Relevant standard
Recommended Practice on Surge Voltages in Low-Voltage AC Power Circuits	IEEE C62.41-1991
Guide for Application of Gas Tube Arrester Low-Voltage Surge Protective Devices	IEEE C62.42-1987
Surge Protectors Used in LV Data, Communications, and Signaling Circuits	IEEE C62.42
Recommended Practice for Grounding of Industrial and Commercial Power Systems	IEEE 142-1991
Recommended Practice for Electrical Power Systems in Commercial Buildings	IEEE 241-1990
Recommended Practice for Powering and Grounding Sensitive Electronic Equipment	IEEE 1100-1992
Recommended Practice on Monitoring Electric Power Quality	IEEE 1159-1995
Guide for Service to Equipment Sensitive to Momentary Voltage Disturbances	IEEE 1250-1995
Electric Power System Compatibility with Industrial Control Devices	IEEE 1346-1994
Electrostatic Discharge: ESD Withstand Capability Evaluation Methods	IEEE C62.38
Guide on ESD: Characterization of the ESD Voltage Environment	IEEE C62.467-1992

SOURCE: IEEE Standards, copyright © IEEE, IEEE. All rights reserved.

standards to protect their equipment from poor power quality. This is especially true of the semiconductor industries.

The semiconductor industry needs power quality standards for the design, operation, and maintenance of its facilities, like clean rooms. The Institute of Environmental Sciences and Technology (IEST) developed Recommended Practice-12, *Considerations in Clean Room Design.* The semiconductor industry's own Semiconductor Equipment and Materials International (SEMI) has recently developed three standards for electrical compatibility: SEMI E6-96, *Facilities Interface Specifications Guideline and Format,* SEMI E33-94, *Specification for Semiconductor Manufacturing Facility EMC,* and SEMI E51-95, *Guide for Typical Facilities Services and Termination Matrix.* Another industry that needs industry-specific power quality standards is the health care industry.

The health care industry's increased use of sensitive electronic equipment to monitor, diagnose, and sustain the vital functions of its patients has caused the IEEE, NFPA, and IEC to develop standards for the health care industry. These standards establish guidelines that prevent sensitive equipment from affecting each other through radiated electromagnetic interference or improper wiring and grounding and ensure the reliability of emergency backup systems. Some of these health care power quality standards are listed in Table 3.10.

Other industries, such as the pulp and paper industry, need industry-specific power quality standards. They have instead relied on general power quality standards for electrical systems, such as IEEE-519-1992. The IEEE doesn't provide power quality standards specific to equipment, while the IEC has established equipment-specific limits. For example, it has set harmonic current limits for lighting equipment in IEC 1000-3-2, *Harmonic Limits for Low Voltage Apparatus.*

Trends in Power Quality Standards

Over the years, various standards organizations have developed power quality standards whenever a particular power quality problem appeared. They started in the 1890s, setting limits for voltage and current. They have recently increased their activity. They will need to develop even more standards in the future as the use of sensitive electronic and computerized equipment proliferates and deregulation of the utility industry unfolds. Deregulation of the telecommunications industry drove the need for more standards to replace the uniform approach of monopolistic companies, like AT&T. Deregulation of the

TABLE 3.10 Health Care Facilities Power Quality Standards by Title

Title	Relevant standard
National Electric Code	NFPA 90
Healthcare Facilities	NFPA 99
Electric Systems in Healthcare Facilities	IEEE 602 (*White Book*)
Emergency and Standby Power Systems for Industrial and Commercial Applications	IEEE 446 (*Orange Book*)
Powering and Grounding Sensitive Electronic Equipment	IEEE 1100 (*Emerald Book*)
Electromagnetic Compatibility	IEC 1000-3,-4
Medical Electrical Equipment	IEC-601-01, Part 2
Electromagnetic Compatibility	ANSI C63.18-1997
Industrial, Scientific and Medical Equipment (ISM) Installed on User's Premises	ANSI/IEEE 139-1988
Industrial, Scientific, and Medical Equipment	FCC Part 18
Radio Frequency, ISM Equipment	CISPR 11
Guidance for Electromagnetic Compatibility of Medical Devices, Part 1	AAMI TIR 18
Medical Laser	IEC 825-1

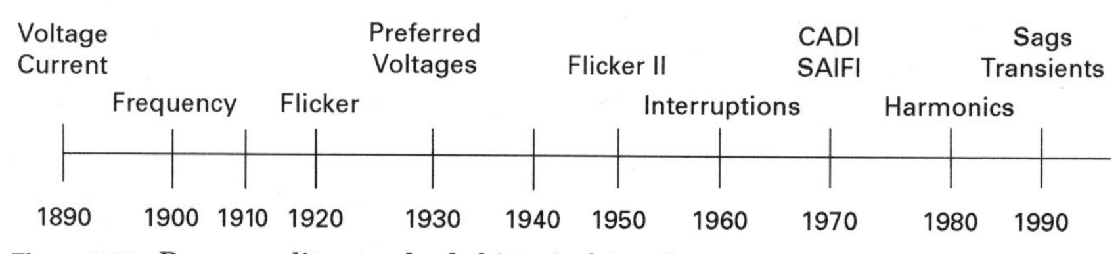

Figure 3.10 Power quality standards historical trend.

electric utility industry likewise will cause an increased need for standards, as electric utilities can no longer act as monopolies that are able to control the level of power quality for their customers. Figure 3.10 illustrates this historical trend.

One of the outcomes of utility deregulation will be the need to determine who is responsible for a particular power quality problem and what is expected of them to mitigate that problem. Contracts based on power quality standards will be essential to establish a satisfactory level of power quality between utilities and their customers. Utilities and their customers need contracts that describe how to resolve the situation when the power quality isn't satisfactory. They have already written power quality contracts to cover various power quality issues. The requirements of particular power quality contracts and the concerns that must be addressed will depend on the parties involved and the characteristics of the system. Chapter 8, "Future Trends," presents the various types of power quality contracts and the participants in those contracts when the utility industry is deregulated. Power quality contracts will require the party responsible for causing a power quality problem to solve it. The next chapter discusses how to solve power quality problems.

References

1. McGranaghan, Mark. 1998. "Part I—Overview of Power Quality Standards." URL address: *http://web.yuntech.edu.tw*~wangyj/PQ/PQS/paper2.htm:Electrotek Concepts, Inc.
2. Key, Tom. 1998. "Standards Update." *Signature Newsletter,* vol. 8, no. 1, winter/spring, p. 5.
3. O'Neill, Anne. 1996. "Why Develop Standards in IEEE If the Goal Is to Harmonize Standards with IEC?" *IEEE Standards Bearer,* p. 6.
4. McGranaghan, Mark. 1998. "Part II—Standards for Different Types of Power Quality Variations." URL address:*http://web.yuntech.edu.tw*~wangyj/PQ/PQS/paper2.htm: Electrotek Concepts, Inc.
5. Bush, William. 1994. "Understanding the Proliferation of Power, Grounding and Protection Standards." *Power Quality Assurance,* vol. 5, no. 1, January/February, pp. 53–60.
6. Ray, Larry. 1998. "Don't Let Sags and Interruptions Disturb You." *EC&M Electrical Construction & Maintenance,* vol. 97, no. 9, August, pp. 42–46.
7. Owen, Edward L. 1996. "Power Disturbance and Quality: Light Flicker Voltage Requirements." Paper presented at *IEEE 1994 IAS Annual Meeting,* Denver, Col.

8. Martzloff, Francois D., Arshad Mansoor, and Doni Nastasi. 1998. "Reality Checks for Surge Standards." *Power Quality Assurance.* URL address: http://www.power quality.com/art0040/art1.htm.
9. Cohen, Richard L. 1998. "The New UL 1449 Standard for Transient Voltage Surge Suppressors." *Power Quality Assurance,* vol. 9, no. 4, July/August, pp. 32–37.
10. McDonald, James N. 1998. "Joules vs. Peak Amps: TVSS Lock-In Specifications Can Be Misleading." *Power Quality Assurance,* vol. 9, no. 2, March/April, pp. 48–50.
11. Gilligan, Sidney. 1992. "Using Standard Voltages Produces Satisfied Customers." *Electric Light and Power,* vol. 70, no. 2, February, p. 14.
12. Halpin, Mark S., et al. 1999. "Voltage and Lamp Flicker Issues: Should the IEEE Adopt the IEC Approach?" URL address: *http://grouper.ieee.org/groups/1453/drpaper.html.* Available from the IEEE.
13. Kennedy, Barry W. 1999. "Application of IEEE 519 Standards in the Restructured Competitive Electricity Industry." *Power Systems World '99.* Chicago, Ill., November 9–11, 1999.
14. "IEEE Recommended Practice for Establishing Transformer Capability When Supplying Nonsinusoidal Load Currents." ANSI/IEEE Standard C57. 110-1986. Piscataway, N.J.
15. Kennedy, Barry W. 1998. *Energy Efficient Transformers,* New York: McGraw-Hill, pp. 147–159.
16. "Electromagnetic Compatibility (EMC)—Part 3: Limits—Section 2: Limits for Harmonic Current Emission (Equipment Input Current [≤] 16 A Per Phase)." IEC 1000-3-2. 1995.
17. "IEEE Guide on Electrostatic Discharge (ESD): Characterization of the ESD Environment." ANSI/IEEE Standard C62, 47-1992. Piscataway, NJ.
18. Clarke, Pat. 1991. "Telecom Power Quality Guidelines." *Power Quality,* vol. 2, no. 5, September/October, pp. 38–40.
19. Lewis, Warren. 1986. "Application of the National Electrical Code to the Installation of Sensitive Electronic Equipment." *IEEE Transactions on Industry Applications,* vol. IA-22, no. 3, May/June, pp. 400–415.

Chapter

4

Power Quality Solutions

There are four ways to solve and prevent power quality problems:

1. Design equipment and electrical systems to prevent electrical disturbances from causing equipment or systems to malfunction.
2. Analyze the symptoms of a power quality problem to determine its cause and solution.
3. Identify the medium that is transmitting the electrical disturbance and reduce or eliminate the effect of that medium.
4. Treat the symptoms of the power quality problem by the use of power conditioning equipment. Power conditioning equipment mitigates a power quality problem when it occurs.

This chapter will deal with all four of these approaches to solving and preventing power quality problems.

Reduce Effects on Sensitive Equipment

Manufacturers of sensitive equipment can reduce or eliminate the effects of power quality problems by designing their equipment to be less sensitive to voltage variations. For instance, they can simply adjust an undervoltage relay or add some device, like a capacitor, to provide temporary energy storage when the voltage sags too low. They can alter their equipment to desensitize it to power quality problems. For example, they can design special *K*-factor transformers that tolerate harmonics or use cables with neutrals large enough to carry triplen harmonics.

It is usually more cost effective to prevent a power quality problem before it occurs. Power quality problem prevention requires purchasers and manufacturers of electrical equipment to be aware of potential power quality problems and how to prevent them. They need to first examine the power quality of the existing power system before installing new equipment. They must perform a power quality investigation of the current, voltage, and frequency on both sides of the electrical meter. The steps in performing this investigation are called a power quality site survey. Chapter 7 describes in detail how to perform a power quality site survey. Most power quality surveys require the installation of power quality monitoring equipment to determine the status of the power quality inside a specific site. Chapter 6, "Power Quality Measurement Tools," explains how monitoring equipment works and how to use it. This chapter presents solutions to various power quality problems. Usually the most cost-effective solution is at the end-user level of the system, as illustrated in Figure 4.1.

A systematic approach to preventing a power quality problem is the best approach. A systematic approach requires procedures for designing and installing equipment that is sensitive to electrical disturbances as well as equipment that may cause electrical disturbances. *The Power Quality Workbook for Utility and Industrial Applications* developed by EPRI and the Bonneville Power Administration provides procedures for evaluating potential power quality problems caused by installation of new equipment or changes in the operation or wiring of existing equipment. Figure 4.2 shows the basic steps involved in a power quality problem evaluation. The workbook provides worksheets and flowcharts on how to prevent power quality problems caused by harmonic sources, voltage sags, interruption of electric service, flicker, voltage unbalance, transients, and poor wiring and grounding. It is designed to provide tools for advanced power quality evaluation. It also contains guidelines for writing power quality contracts. The flowchart in Figure 4.3 illustrates how the *Power Quality Workbook for Utility and Industrial Applications* helps prevent power quality problems.

Reduce or Eliminate Cause

Before a power quality engineer can reduce or eliminate the cause of a power quality problem, the engineer must diagnose the power quality problem to determine its source. The diagnostic procedure requires the power quality engineer to perform a power quality survey (see Chapter 7) and answer some basic questions. Is the problem's source located in the utility's transmission or distribution system? Or is the problem's source found inside the end user's facility? Who is responsible for causing the problem? Is it the utility or the end user?

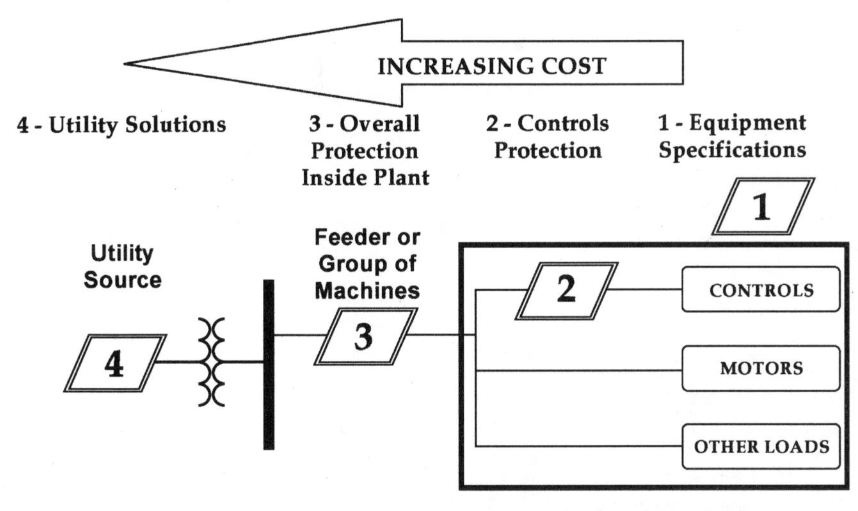

Figure 4.1 Increasing cost of solutions.

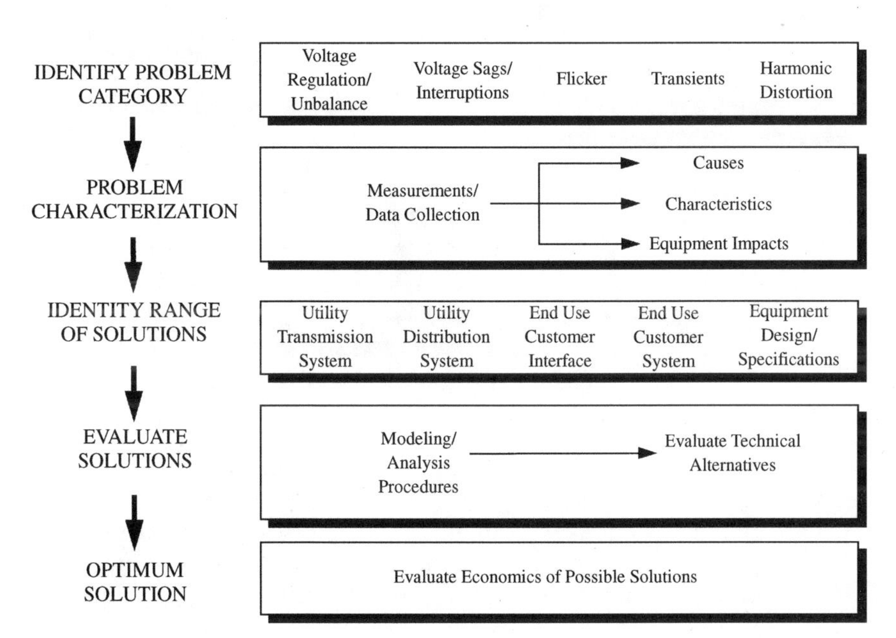

Figure 4.2 Basic steps involved in a power quality problem evaluation. (*Courtesy of EPRI.*)

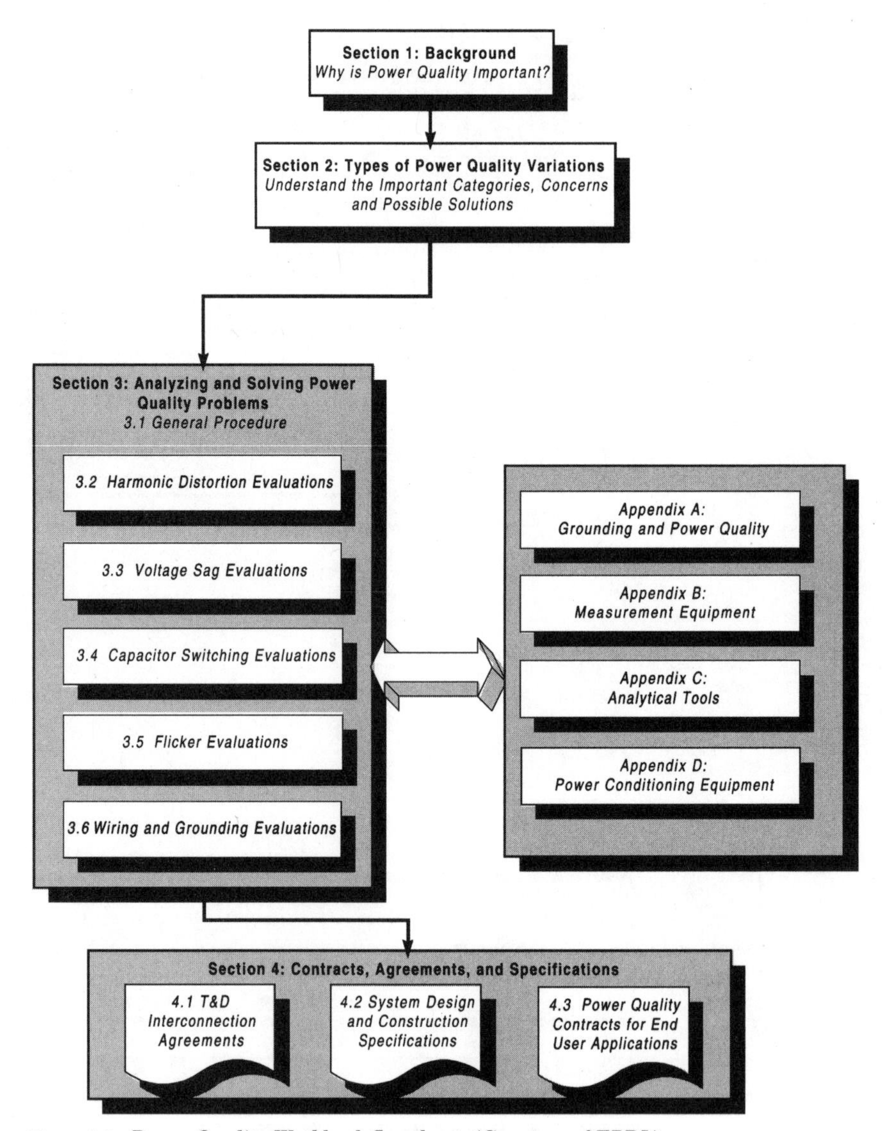

Figure 4.3 *Power Quality Workbook* flowchart. (*Courtesy of EPRI.*)

The location of the disturbance usually determines who is responsible for solving the power quality problem, as illustrated in Figure 4.4.

The type of power quality problem and its cause often determine the solution. Changing the medium transmitting the power quality problem, whether wire or air, may be the best solution to a power quality problem.

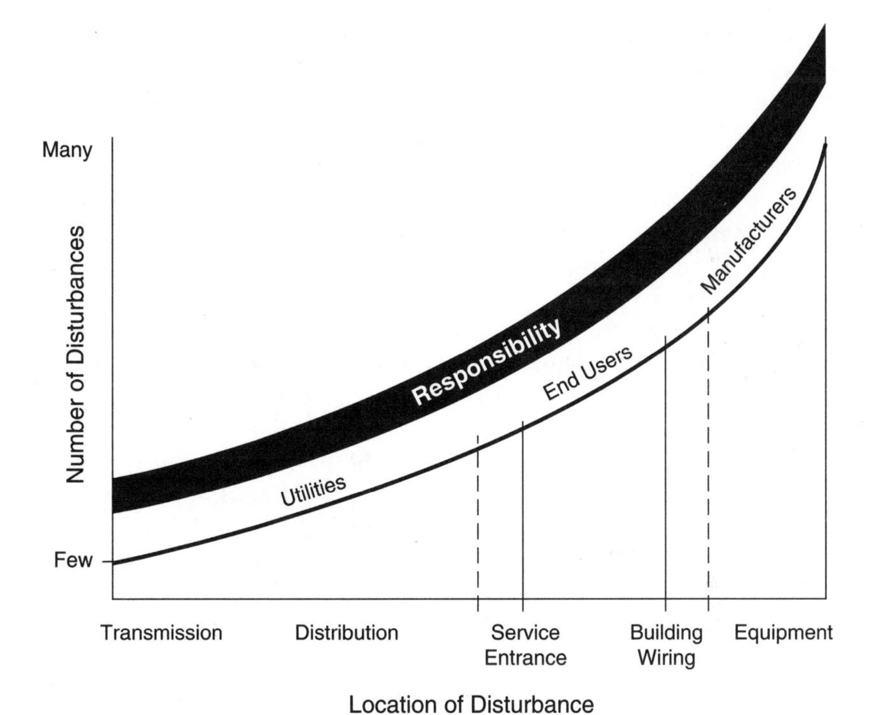

Figure 4.4 Power quality solution responsibility. (*Courtesy EPRI-PEAC System Compatibility Research.*)

Reduce or Eliminate Transfer Medium

Often the transmission and distribution system act as a conduit for transmitting harmonics, transients, voltage sag, or flicker from one end user to another. In that case, it is not practical to move the sensitive equipment to a location that is not connected to the system transmitting the power quality problem. For example, a transformer located in Idaho was generating third harmonics that were transmitted by the utility's interconnected transmission and distribution system to digital clocks in Washington and causing them to blink off and on. The best solution was to eliminate the source of the third harmonics by replacing the transformers or installing filters on the transformers.

Inside a facility, a sensitive load and an electrical disturbance load may be connected to the same circuit. Simply reconnecting the sensitive equipment to another circuit may be the solution to that type of power quality problem. This solution requires a survey of the site, including power quality monitoring, to determine the source of the power quality problem and how it is being transmitted to the offended equipment. For example, a hospital used a radiation machine to treat cancer patients. The radiation machine overradiated a patient whose arm was cancerous. The arm had to be amputated. The patient sued the hospital for several million dollars. A power quality engineer performed a power quality survey and discovered a laser copier in a room next to the radiation machine. The engineer knew that the laser copier heated the toner each time it started. The heating process required a large amount of current from the power supply. The large amount of current caused the service voltage to dip. The dips in service voltage caused the radiation machine to misoperate and overradiate. The engineer concluded that the source of the power quality problem was the copier and decided that the best solution was to disconnect the radiation machine from the circuit connected to the copier and connect it to its own dedicated circuit. This type of problem is called a wiring and grounding problem. Chapter 5, "Wiring and Grounding," discusses these types of problems and solutions.

Also inside a facility, a sensitive load may be located in close proximity to a piece of equipment causing a disturbance to the sensitive equipment. The disturbance may be transmitted through the air. Simply moving the sensitive equipment to another location may be the solution to this type of power quality problem. This solution requires a survey of the site, including power quality monitoring, to determine the source of the power quality problem and how it is being transmitted to the offended equipment. For example, a person complains that a computer monitor wiggles and calls a power quality engineer to solve the problem. During a visual survey, the power quality engineer locates a microwave oven in the room adjacent to the computer monitor. The engineer measures the electromagnetic field (EMF) radiating from the microwave oven. The engineer concludes from the measurements that the microwave oven is the source of the monitor wiggles. Figure 4.5 shows that the solution to this problem is to increase the distance between the monitor and the microwave enough to reduce the EMF to a level where it does not interfere with the monitor's output. Another very common way to solve power quality problems is through the purchase and installation of power conditioning equipment.

Install Power Conditioning Equipment

Power conditioning equipment provides essential protection against power quality problems. What is power conditioning equipment? Is it the same as power quality mitigation equipment? Yes. Is it similar to hair conditioners that soften hair after shampooing and keep static electricity from causing hair to stand up? Or is it similar to the water conditioners that soften water and make it "cleaner"? There are some

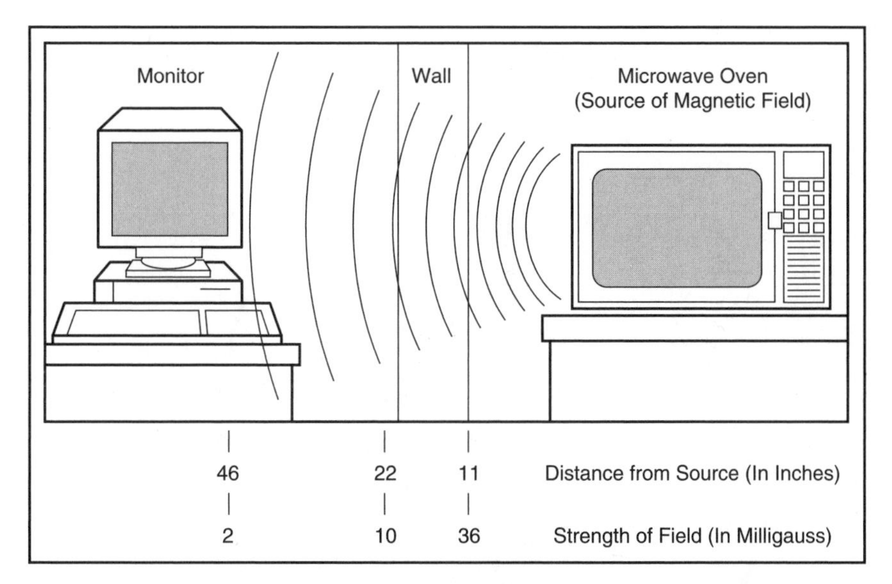

This diagram shows how the strength of a magnetic field rapidly decreases as the distance from its source increases. As shown here, moving the monitor another 2 ft away from the source eliminates the jitters caused by the oven on the other side of the party wall. (Data source is from the 1991 EPRI EMF Science & Communication Seminar, "Magnetic Field Source Characterization," EPRI.)

Figure 4.5 Distance effect on EMF strength. (*Courtesy of EPRI.*)

similarities. Power conditioning equipment does refer to devices that are supposed to make "dirty" power "clean." Power conditioning equipment improves the power quality just as water conditioners improve the water quality. Technically, power conditioning equipment includes devices that reduce or eliminate the effect of a power quality disturbance. Depending on the type of equipment, it conditions (modifies) the power by improving the quality and reliability of the power at any part of the power system. It can be used to condition the source, the transmitter, or the receiver of the power quality problem. In other words, utilities as well as residential, commercial, and industrial end users use it. It often provides a barrier between electrical disturbances and sensitive electronic equipment, as illustrated in Figure 4.6.

The most common types of power conditioning equipment include uninterruptible power supplies (UPSs), line conditioners, and surge suppressors. Other types of power conditioning equipment include isolation transformers, passive and active filters, superconducting magnetic energy storage (SMES), dynamic voltage restorers (DVRs), constant-voltage transformers (CVTs), and various types of motor-generator sets. What are the theory and applications of power conditioning equipment?

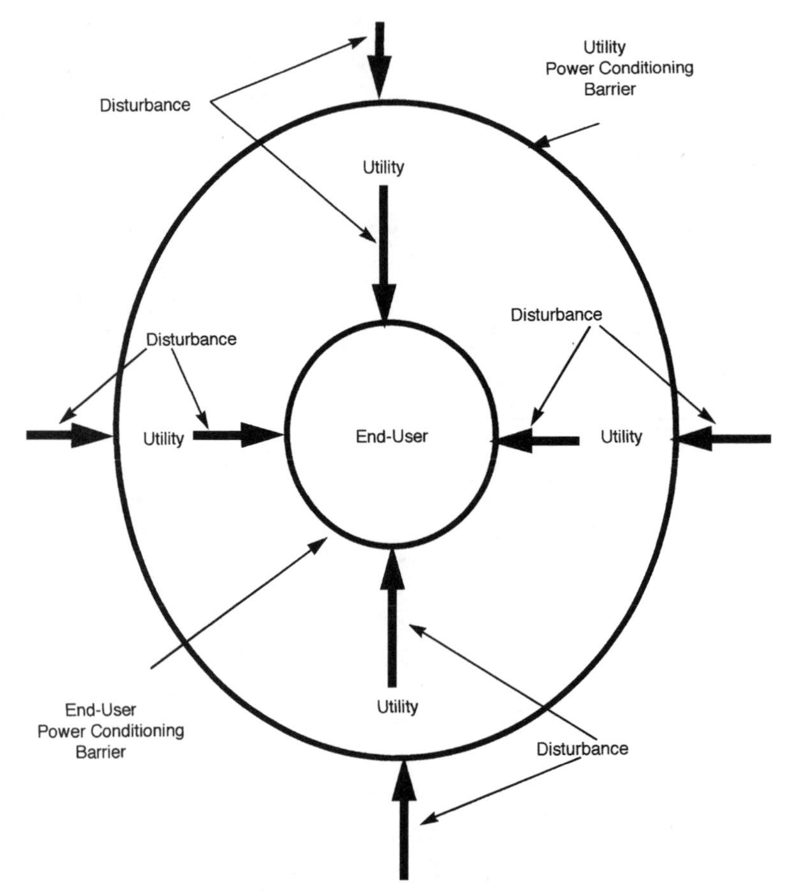

Figure 4.6 Power conditioning barrier.

How does it work?

Power conditioning usually involves voltage conditioning because most power quality problems are voltage quality problems. Most devices condition or modify the voltage magnitude or frequency. They employ technology to reduce the effect of transient and steady-state voltage changes or isolate the sensitive equipment from the disturbance. For example, surge suppressors clamp, i.e., limit, the transient voltage amplitude, and regulators keep the steady-state voltage from deviating from the specified nominal voltage. Isolation transformers keep power quality disturbances from reaching sensitive electronic equipment. Filters reduce or eliminate voltages and currents that have frequencies other than 60 Hz. The main types of voltage conditioners include voltage regulators and tap changers of various types, ferromagnetic devices, harmonic filters, solid-state surge suppressors, and static VAR compensators.

Other types of power conditioners include devices that provide alternative sources of energy. These types of power conditioners include devices for energy storage or for switching to alternative sources. Solid-state switches provide an alternative energy source by quickly switching from a power supply feeder to an alternative feeder during a disturbance. Energy storage systems include batteries, capacitors, superconducting magnets, motor-generator sets, and flywheels. Each one of these technologies provides energy that can be accessed during an electrical disturbance such as a temporary interruption or a voltage sag. They all have the advantage of providing isolation from the disturbance. They have the disadvantage of providing only a limited amount of energy for a limited amount of time.

Power conditioning equipment is sometimes referred to as *mitigation equipment.* The equipment can be divided into nine categories:

Surge suppressors

Noise filters

Isolation transformers

Low-voltage line reactors

Various line-voltage regulators

Motor-generator sets

Dual feeders with static transfer

Uninterruptible power supplies (UPSs)

Harmonic filters

The most popular power conditioning equipment is surge suppressors.

Surge suppressors

The most common devices for preventing power quality problems from damaging equipment are surge suppressors. Surge suppressors protect sensitive equipment from being zapped by voltage surges or lightning strokes on the power system. They are the shock absorbers or safety valves of electrical power systems. If they are located on the utility side of the meter, they are called *surge* or *lightning arresters.* If they are located on the end-user side of the meter, they are called transient voltage surge suppressors (TVSSs). They divert to ground or limit the transient voltage caused by lightning or switching surges to a level that will not harm the equipment they are protecting. They are connected so that the transient "sees" the surge suppressor before it reaches the protected equipment.

Utilities specify and locate arresters near equipment they wish to protect, like transformers, distribution lines, and substation equipment. They install arresters on the high-voltage side of distribution transformers. As shown in Figure 4.7, they use surge suppressors on the high-voltage and low-voltage side of substation transformers.

Figure 4.7 Transformer surge arresters. (*Courtesy of Bonneville Power Administration.*)

End users locate surge suppressors or TVSSs inside their facilities, between the power outlet and sensitive electronic equipment, such as computers, adjustable-speed drives, and communication devices, or at the main power supply panelboard. Figure 4.8 shows TVSSs located at the power outlet and panelboard. There are two basic types of surge suppressors: crowbar and voltage-clamping devices.

Crowbar devices. The term *crowbar* comes from the idea of putting a crowbar across a line to short-circuit the current to ground. Surge arresters short-circuit voltage transients to ground. How do they work? They have a gap filled with a material that acts like a short circuit to voltage transients. These materials include air, special ionization gas, or a ceramic-type material like silicon carbide for low voltages or zinc oxide for medium and high voltages. The gap acts as an nonconducting insulator when the voltage is normal. The gap becomes a conductor when the transient voltage exceeds the breakdown voltage of the material in the gap. The high transient voltage arcs across the gap. The surge's energy is then dissipated harmlessly to ground. Figure 4.9 shows the operation of a crowbar arrester.

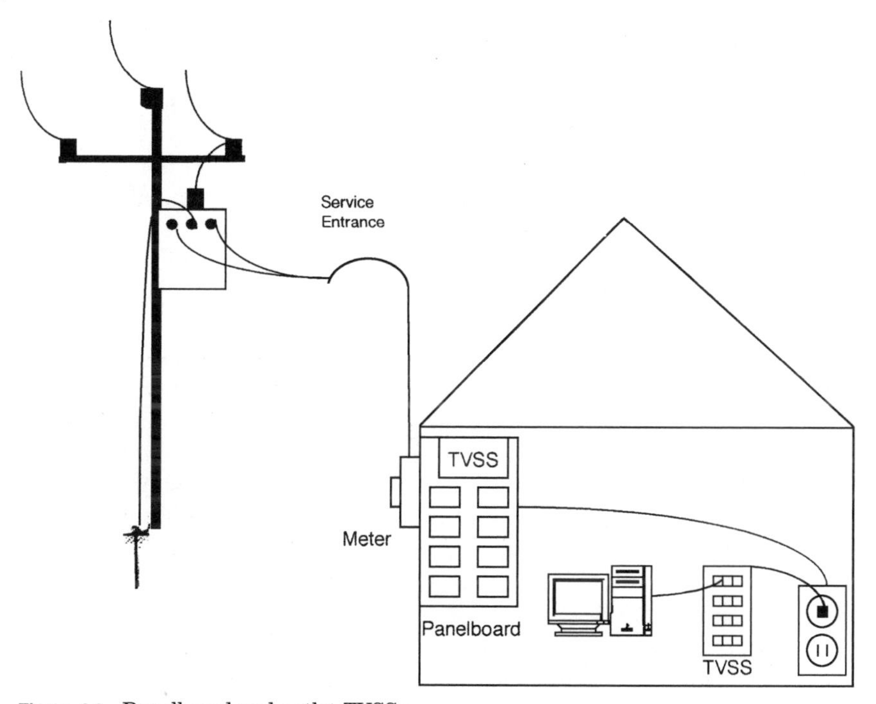

Figure 4.8 Panelboard and outlet TVSSs.

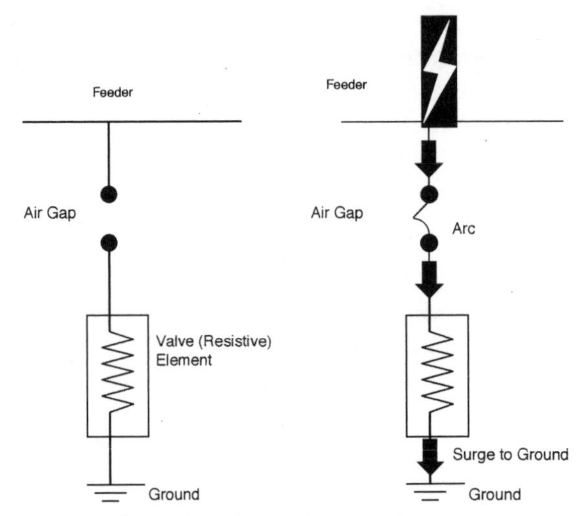

Figure 4.9 Operation of a crowbar arrester.

Voltage-clamping devices. Voltage-clamping surge suppressors usually contain a material that clamps the voltage of a transient. This material is a nonlinear resistor (varistor) whose resistance decreases as the voltage across it increases. They usually contain metal oxide varistors (MOVs) or silicon avalanche (zener) diodes that clamp, i.e., limit, excessive line voltage and conduct any excess impulse energy to ground. Clamping a voltage means that the top of the transient voltage will be looped off so that the protected equipment is not damaged by the excessive voltage of the transient, as shown in Figure 4.10.

UL 1449 rates a TVSS according to its clamping voltage and energy suppression in joules. A *joule* is a metric measurement of energy equal to 0.7376 ft-lb. The joule rating depends on three variables: let-through voltage, current, and pulse duration. A new and improved TVSS can have a reduced let-through voltage, given the same peak pulse current. Therefore, a TVSS with reduced let-through voltage will have a reduced joule rating. Consequently, a reduced joule rating is not an indication that a TVSS has reduced capabilities. This confusion over joule rating has caused the power quality industry to move toward specifying the peak pulse current rather than joule rating for a TVSS performance. For example, the National Electrical Manufacturers Association (NEMA) standards rate a TVSS not by joules but surge current.

Often when a transient occurs, it destroys the MOV in the TVSS. The green indicator light on the TVSS is supposed to go off when the MOV has been destroyed. This means it is time to replace the old TVSS with a new one. Many TVSSs contain a fuse or circuit breaker

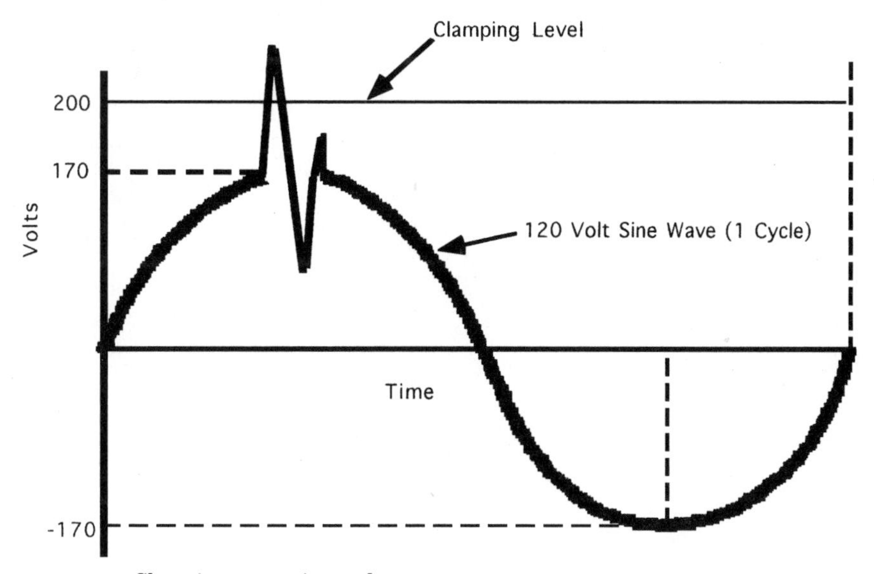

Figure 4.10 Clamping a transient voltage.

that detects current overloads and automatically trips the unit off. They also contain a master switch that controls power to all the receptacles. Figure 4.11 shows the various features of a TVSS.

TVSS manufacturers design plug-in type TVSSs to protect telephones as well as computers. High-voltage transients can come down the telephone line and fry telephones as well as personal computers and can even be transmitted in the air when computers are unplugged. Figure 4.12 shows how the TVSS needs to be connected to the telephone line as well as the power circuit.

TVSSs are installed either at the site of the equipment being protected or at the main power panel. Many utilities are offering programs that allow their customers to pay a monthly charge added to their power bill to lease panelboard-type TVSSs. TVSSs can cost as little as $20 and as much as $200. Be careful when buying surge suppressors. Many of the cheap ones are fancy extension cords. Look at the specifications marked on the suppressors. Has Underwriters Laboratories tested it to meet UL 1449 or IEEE 587? What is the magnitude of the transient voltage that the TVSS can clamp? It should be able to clamp transient voltages up to 6000 V. Does it clamp the voltage down to an acceptable level? It should clamp the voltage at 300 to 400 V for a 120-V outlet. The specifications marked on the outside of the TVSS reflect the components inside the TVSS. Figure 4.13 shows the inside of a TVSS.

Manufacturers design surge suppressors to prevent the damage caused by the excessive voltage from a transient. They do not normally

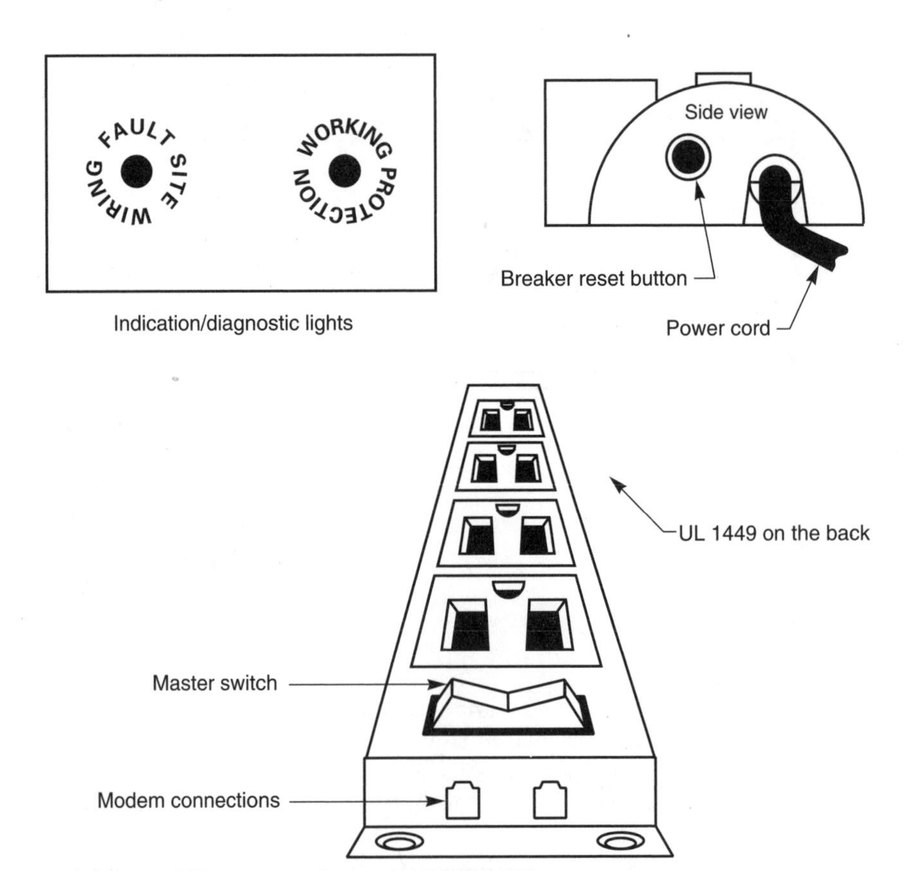

Figure 4.11 TVSS features. (*Courtesy of NWPQSC.*)

protect equipment from the damage caused by the high-frequency noise in a transient. They usually add filters to surge suppressors to provide protection from the noise. Filters made of capacitors and inductors (chokes) keep high-frequency voltages from reaching sensitive electronic equipment.

The current diverted by a TVSS can cause common-mode noise problems that can damage sensitive electronic equipment. Noise filters are needed along with the TVSS. Noise filters condition the frequency of the voltage or regulate steady-state voltage rather than suppress excessive voltages from reaching protected equipment.

Noise filters

All filters, including noise filters, prevent unwanted frequencies from entering sensitive equipment. They do this by using various combinations of inductors and capacitors. Inductors produce impedances that

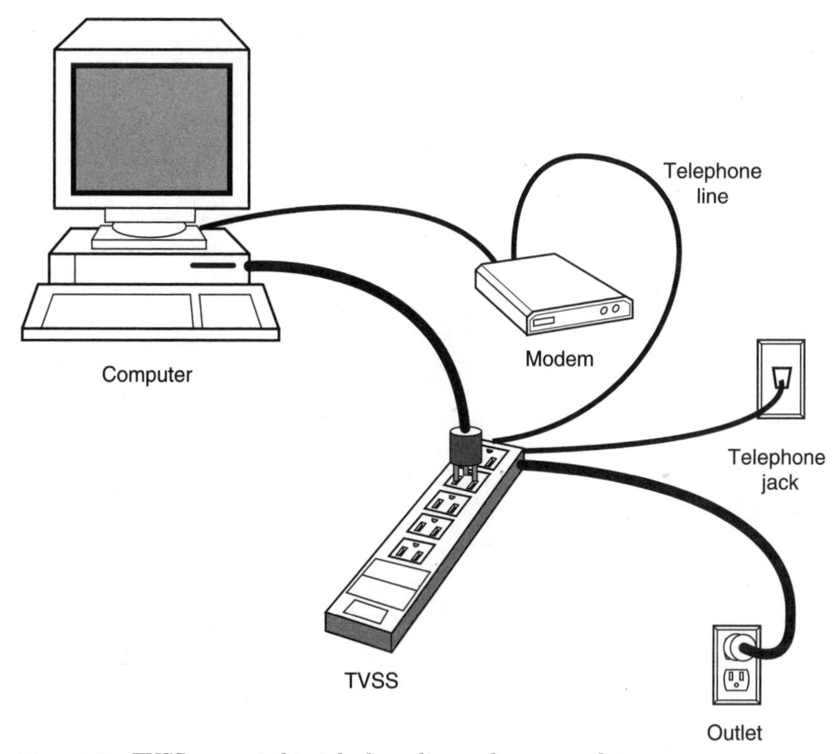

Figure 4.12 TVSS connected to telephone line and power outlet.

increase proportionately to the magnitude of the frequency. Capacitors produce impedances that reduce proportionately to the magnitude of the frequency. Connecting inductors and capacitors in various configurations reduces and diverts voltages and currents of various frequencies. Noise filters are low-pass filters. Inductors in noise filters allow the low-frequency fundamental signal of 60-Hz power to pass through. The capacitors in parallel with the inductors divert the high frequencies of common-mode and normal-mode noise to ground. Figure 4.14 shows the components of a low-pass noise filter.

Noise is any signal that comes through the electrical wiring that is not the primary 60-Hz signal. Normal-mode noise refers to the noise between the hot wire and neutral. Common-mode noise refers to the noise that occurs between the black hot wire and ground or the white neutral wire and ground in a three-wire 120-V wall socket. Noise comes from transient surges caused by lightning or switching on the utility power system. It also can come from motors, laser printers, resistive heating elements, transformers, and loose connections. Figure 4.15 shows normal-mode noise and Figure 4.16 shows common-mode noise.

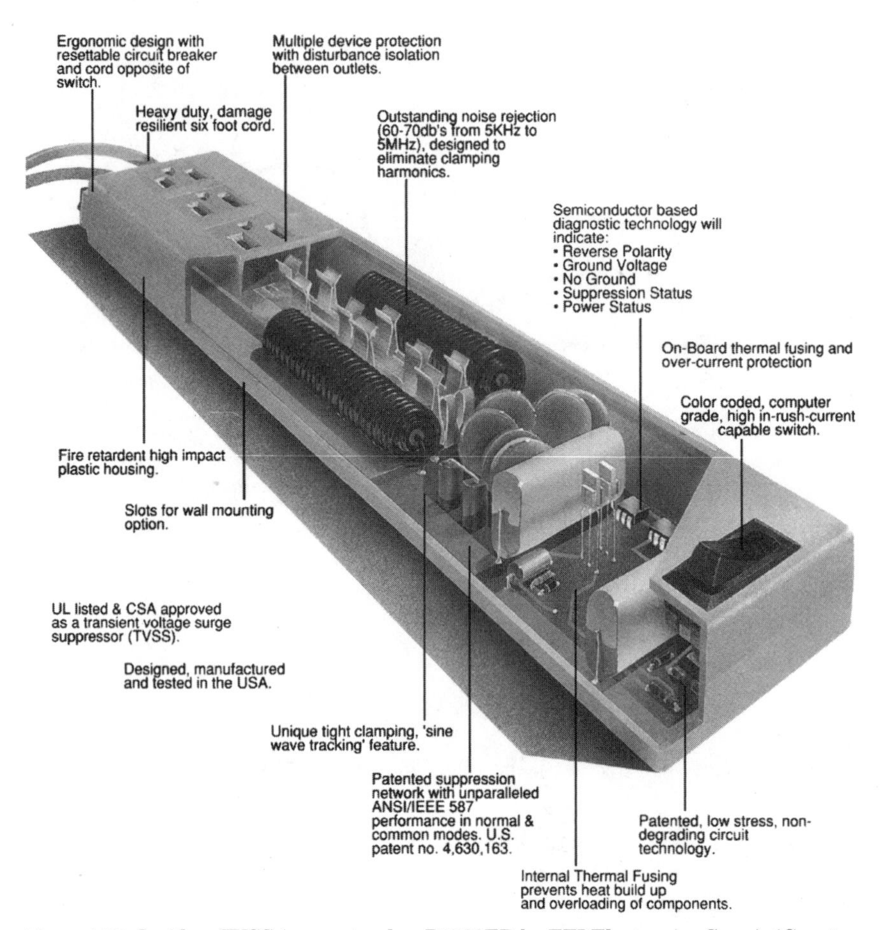

Figure 4.13 Inside a TVSS (power tracker P1500ED by EFI Electronics Corp.). (*Courtesy of EFI.*)

Noise filters are either stand-alone or are part of a TVSS. They should be able to reduce noise by a factor of 100 over a frequency range of 400 kHz to 30 MHz. They are not always effective in reducing noise. They do not eliminate common-mode noise but only control it. A more effective method for protecting equipment from high-frequency noise, especially common-mode noise, is the use of isolation transformers.

Isolation transformers

Shielded isolation transformers are very popular power-conditioning devices. They isolate sensitive loads from transients and noise caused by the utility. They can also keep harmonics produced by end-user nonlinear equipment from getting onto the utility's system. They especially

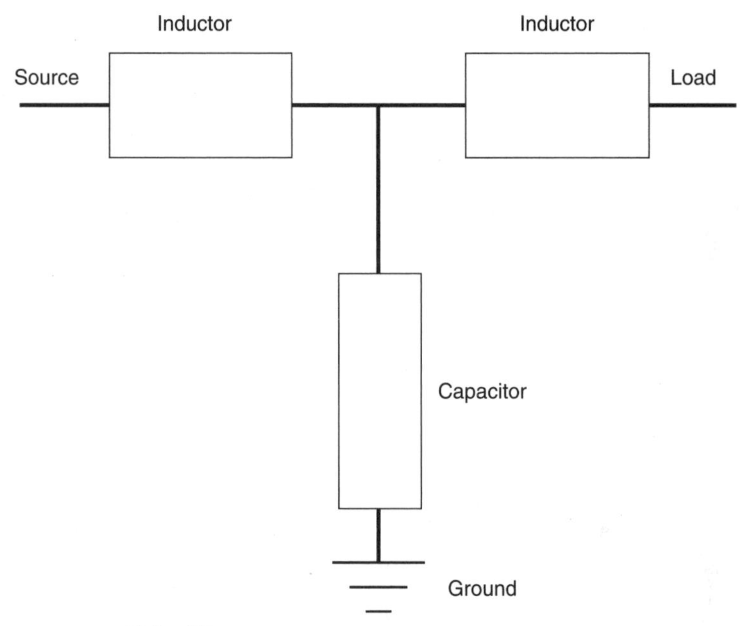

Figure 4.14 Noise filter components.

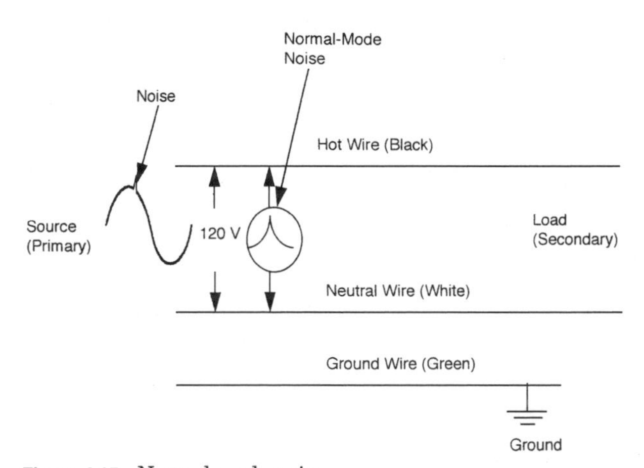

Figure 4.15 Normal-mode noise.

eliminate common-mode noise. How do they protect end-user equipment from utility-caused power quality problems?

The isolation transformer, as its name implies, isolates sensitive equipment from transients and noise produced by the utility. How does the isolation transformer isolate sensitive equipment? The components of the isolation transformer provide a path for transients and

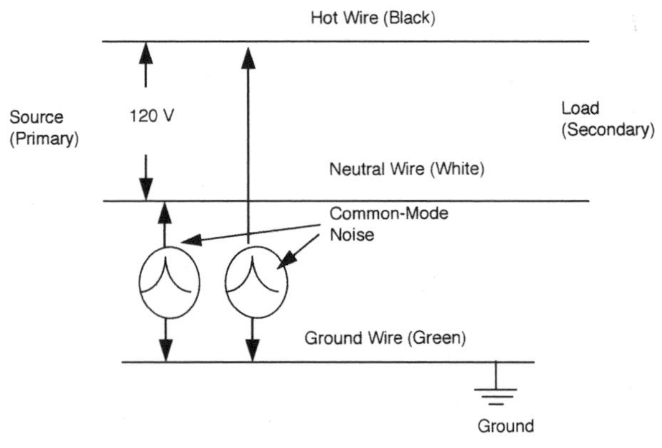

Figure 4.16 Common-mode noise.

noise. Isolation transformer components include a primary and secondary winding with a magnetic core and a grounded shield made of nonmagnetic foil located between the primary and secondary winding. Any noise or transient that comes from the utility is transmitted through the capacitance between the primary and the shield and on to the ground and does not reach any sensitive equipment. Figure 4.17 shows the configuration of a single-phase isolation transformer.

Isolation transformers prevent common-mode noise from reaching and damaging sensitive electronic equipment. If the secondary of the isolation transformer is a grounded wye, then no common-mode noise can reach the protected sensitive equipment. The NEC requires that the secondary neutral be bonded to ground. This bond eliminates any voltage that the load may see between the neutral and ground. With the isolation transformer, the NEC allows the secondary to be grounded.

In addition to protecting the end user from transients caused by the utility, the delta-wye isolation transformer protects the utility from triplen harmonics (third, ninth, fifteenth, etc.). How does the delta-wye isolation transformer keep triplen harmonics from the utility's system?

The isolation transformer transfers the triplen harmonics from the wye secondary to the delta primary of the transformer. The triplen harmonics remain in the delta primary circulating around and generating heat in the transformer but not getting on the utility's system. Figure 4.18 shows a diagram of a shielded delta-wye isolation transformer.

Shielded isolation transformers are often used in conjunction with surge suppressors. They do not regulate the voltage or protect equipment from voltage sags. Various types of line-voltage regulators provide that type of protection.

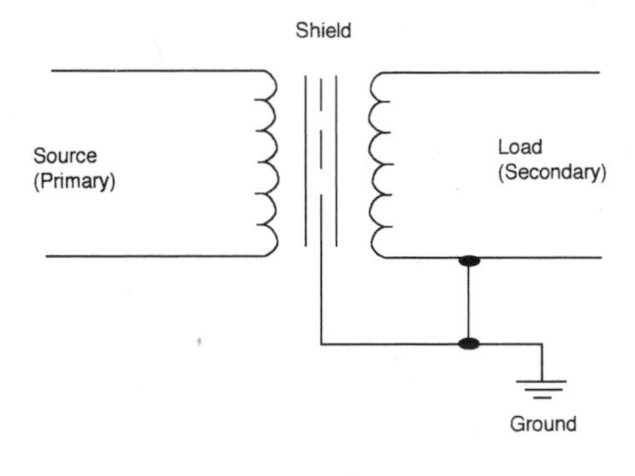

Figure 4.17 Single-phase shielded isolation transformer.

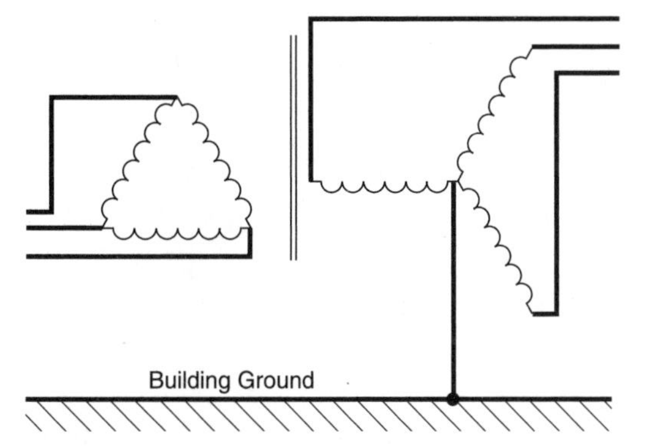

Figure 4.18 Three-phase delta-wye isolation transformer.

Line-voltage regulators

Line-voltage regulators are transformers specially designed to regulate, control, or hold the output voltage constant when the input voltage changes. They are based on the transformer principle that the input voltage E_1, when applied to the primary coil, induces a corresponding voltage E_2 in the secondary coil with a magnitude that is directly proportional to the ratio of the turns in the two coils. The current I_1 flowing in the primary coil or conductor causes a corresponding current I_2 to flow in the secondary coil or adjacent conductor with a magnitude that is inversely proportional to the ratio of the turns (N_1/N_2) in the two coils or conductors. Equation (4.1) and Figure 4.19 illustrate this principle:

$$\frac{E_1}{E_2} = \frac{I_2}{I_1} = \frac{N_1}{N_2} \tag{4.1}$$

Voltage regulators are used on the utility's transmission and distribution system and inside an end user's facility to prevent long-duration

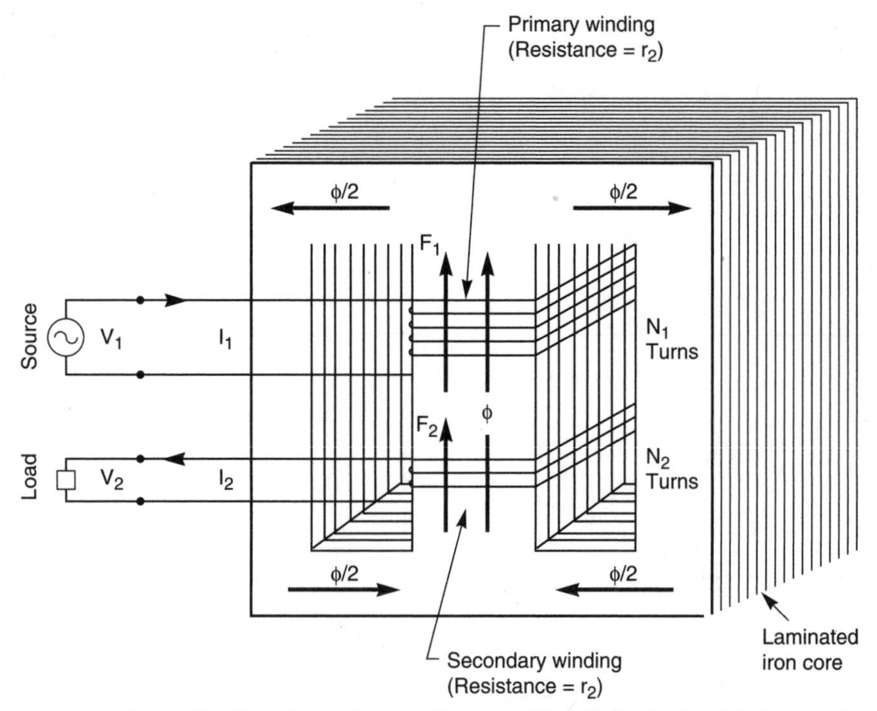

Figure 4.19 Magnetic effect of transformer. (*Courtesy of Oak Ridge National Laboratory.*)

voltage sags, dips, and surges. Power transformers make big changes in voltages by increasing the voltage to a more economical level to allow power to be transmitted over long distances. Voltage regulators make some small changes in voltage to keep the voltage relatively constant. They come in various sizes and shapes. They use tap changers, buck-boost regulators, or constant-voltage transformers (CVTs) to keep the voltage constant. What are tap changers and how do they help keep the voltage constant?

Tap changers. Tap changers on transformers allow utilities to regulate the voltage on their transmission and distribution systems. They also allow end users to regulate the voltage that is being supplied to their sensitive electronic equipment. How do tap changers regulate voltage?

Tap changers regulate the voltage by changing the ratio of turns between the primary and secondary of the transformer. When the input voltage drops, the tap changer changes to a tap that increases the ratio. The increased ratio results in the transformer output voltage not changing. Figure 4.20 illustrates the configuration of a tap-changing transformer.

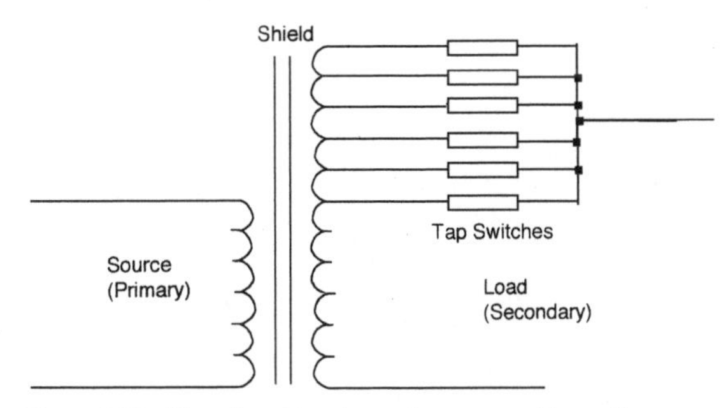

Figure 4.20 Tap-changing transformer configuration.

The two basic types of tap changers on utilities' systems are load and no-load tap changers. The load tap changer automatically senses the need to change taps and usually makes the change when the alternating voltage reaches zero. A no-load tap changer requires the utility operator to disconnect the transformer before engaging the tap changer. The utility operator can operate no-load tap changers manually or by a motor. Load tap changers are often under oil to prevent arcing. Tap changer manufacturers design load tap changers to change the voltage at ±10 percent of the nominal voltage. Tap changer manufacturers design no-load tap changers to change the voltage at ±5 percent of the nominal voltage.

End users can use tap-changing regulators to regulate voltage applied to their sensitive electronic equipment. These tap changers sense a need to switch taps on the voltage regulator and use thyristors (solid-state switches) to switch the taps automatically. They are referred to as *electronic tap-switching regulators.* They are often included with the computer power supply. They can keep the output voltage within ±3 percent for a 20 to 40 percent change in input voltage. Another type of regulator that performs like a tap-switching regulator is a buck-boost regulator.

Buck-boost regulators. Buck-boost regulators regulate a voltage by adding transformer windings that either reduce (buck) or increase (boost) the voltage. They compare the output voltage to the input voltage and use electronic solid-state switches, like thyristors, to switch the windings from the buck to the boost state or from the boost to buck state to keep the output voltage constant. They can maintain the output constant within ±1 percent for a 15 to 20 percent change in input voltage. They can provide isolation and common-mode noise reduction like an isolation transformer if electrostatic shielding is added in the buck-boost regulator. Figure 4.21 shows configuration of a buck-boost regulator. Another type of transformer that keeps the

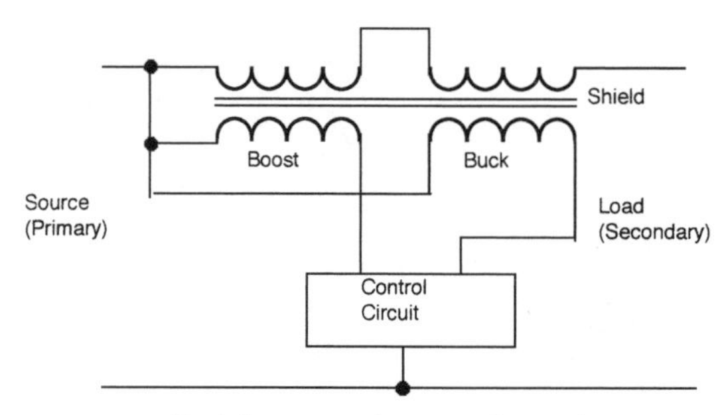

Figure 4.21 Buck-boost regulator configuration.

output voltage constant is a ferroresonant transformer or a constant-voltage transformer (CVT).

Constant-voltage transformer (CVT). The constant voltage transformer does what its name says. It provides a constant output voltage when the input voltage increases above or decreases below the nominal voltage. EPRI's Power Electronics Applications Center (PEAC) performed tests verifying that a CVT maintains a voltage output of 16 percent to 213 percent when the input voltage swings from 120 percent to 220 percent. It is the end user protection from voltage sags and transients. It prevents increased currents from reaching sensitive equipment. It an old and established technology. How does it keep the output voltage constant?

It uses two basic electrical principles that transformer designers usually try to avoid: resonance and core saturation. Resonance occurs when the impedance of the capacitor equals the impedance of the inductor. In this case a capacitor is in series with the induction of the CVT coil. This causes the current to increase to a point where it saturates the steel core of the CVT. Transformer saturation means the magnetic core (steel) cannot take any more magnetic field. Like a waterlogged sponge, it stops absorbing current and produces a constant output voltage. In a transformer, a current in the primary winding produces a magnetic flux that induces a current and voltage in the secondary winding. There is a point where increased current in the primary saturates the core with too much magnetic flux. This is the saturation point. At this point the transformer no longer transforms the voltage or current according to the ratio of the primary and secondary turns. Figure 4.22 shows the configuration of a constant voltage transformer.

If not used properly, however, the CVT can cause its own power quality problems. It does not like harmonics and will overheat, like any other transformer, in the presence of harmonics. In fact, the CVT can

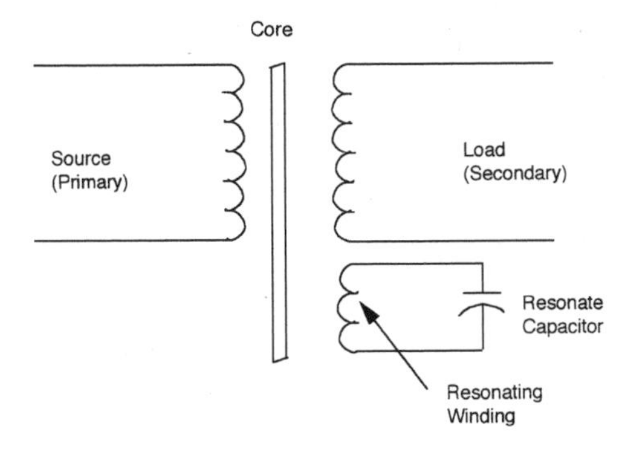

Figure 4.22 Constant-voltage transformer configuration.

generate harmonics. It produces a square wave. This is a sine wave that is clipped on the top and sides. So it is a good idea to include a harmonic filter when a CVT is used. The CVT will also generate transients. Thus, a TVSS should be used in conjunction with a CVT. A CVT can be inefficient: 80 percent efficiency at heavy load and 50 percent at light load. A CVT can be hard on your ears when it goes into resonance. Saturation causes the core to vibrate noisily. It doesn't like high inrush current. It needs to be sized to handle inrush currents. Even with all these drawbacks, the CVT has been used extensively as protection for advanced data processing equipment. Some claim that this technology is becoming obsolete. Another type of power conditioning device that has been used for many years is the motor-generator set.

Motor-generator sets

Motor-generator (M-G) sets provide an old (30 to 40 years) but reliable and economic way to solve power quality problems. They isolate the sensitive load from disturbances and provide backup during power outages. The motor connects to the utility supply power and runs the generator through a shaft or belt. The generator provides clean power to critical equipment. The conversion from electrical energy to mechanical energy and back to electrical energy isolates sensitive electronic equipment from voltage sags, harmonics, transients, overvoltage, and undervoltage disturbances. If power is interrupted, the generator keeps supplying power to critical loads by using diesel or natural gas as the fuel. Figure 4.23 shows a typical motor-generator set. Why are they not used as extensively as they were used in the past?

One of the major drawbacks to M-G sets is their inability to provide power during the initial loss of utility power. There is a delay between the time the utility loses power and the M-G comes on line. This 5-second delay before the M-G set kicks in is called "ride-through." There are two basic ways of providing power during the ride-through period.

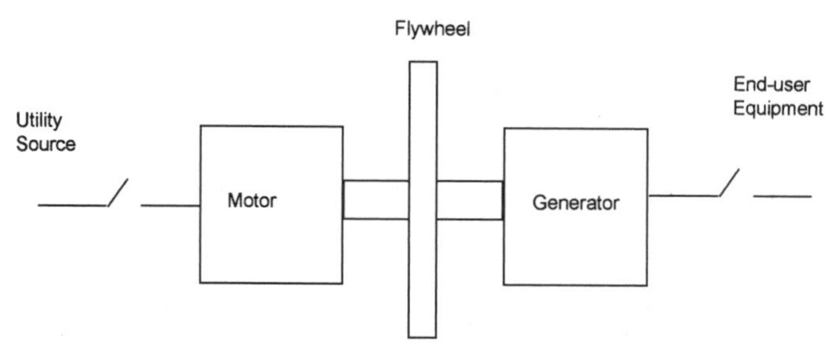

Figure 4.23 Typical motor-generator set.

One way is to use a mechanical rotational device called a *flywheel.* A flywheel is a rotating wheel located between the motor and the generator. It stores rotational energy and keeps the generator operating during the critical ride-through time. Commercial flywheels can add 10 to 20 seconds duration time for loads up to 500 kW.

Another way is to use a rechargeable battery pack. A battery pack provides the power to critical equipment during the 5-second delay before the M-G set comes on line. Battery packs can be expensive and large.

EPRI and Precision Power have developed a new type of M-G set. It is called the "written-pole" M-G set. It provides ride-through by writing the poles on the stator (the stationary part of the motor) and using a large rotor located on the outside of the stator. The rotor acts as a flywheel and stores enough energy to keep the M-G set running during the ride-through time. Another type of power conditioner that provides some energy storage is the magnetic synthesizer.

Magnetic synthesizers

Magnetic synthesizers combine power conditioning devices previously discussed. They use resonant circuits made of nonlinear inductors and capacitors to store energy, pulsating saturation transformers to modify the voltage waveform, and filters to filter out harmonic distortion. They supply power through a zigzag transformer. The zigzag name comes from the way the transformer changes the phase angle between voltage and current. The zigzag transformer traps triplen harmonic currents and prevents them from reaching the power source.

Applications of magnetic synthesizers include protection of large computer installations, computerized medical imaging equipment, and industrial processes, like plastic extruders, especially from voltage sags. They protect sensitive loads not only from voltage sags but also from transients, overvoltage, undervoltage, and voltage surges. However, they can be bulky and noisy. The block diagram in Figure 4.24 illustrates the main components of a magnetic synthesizer.

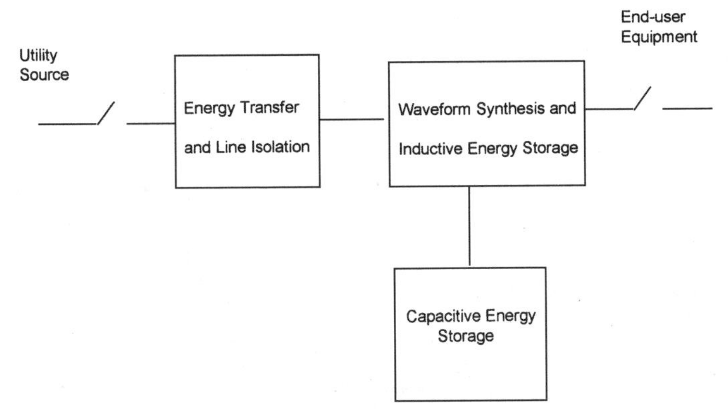

Figure 4.24 Magnetic synthesizer components.

Static VAR compensators (SVCs)

Static VAR compensators use a combination of capacitors and reactors to regulate the voltage quickly. They replaced old-style synchronous condensers. Synchronous condensers supplied continuous reactive regulation but were too expensive to buy, operate, and maintain. SVCs are less expensive to operate and maintain. They use solid-state switches that insert the capacitors and reactors at the right magnitude to keep the voltage from fluctuating. Utilities use SVCs on their high-voltage power systems.

Utilities use SVCs to keep the voltage from sagging during a fault on a transmission line. They are quite large and expensive. For example, the Bonneville Power Administration (BPA), a federal bulk power marketing agency located in the U.S. Pacific Northwest, installed in 1994 two $25 million 350 million VAR capacitive reactance and 300 million VAR inductive reactance SVCs to counter voltage instabilities. BPA installed them to provide voltage support during loss of transmission capacity to the highly industrialized Portland, Oregon, and Seattle, Washington, areas. Another major SVC application is to prevent flicker caused by arc furnaces.

Industrial plants using electric arc furnaces to melt metal use SVCs to reduce voltage flicker. Electric arc furnaces are notorious for causing voltage flicker. How do they cause flicker? The electric arc furnace's energized electrodes cause flicker when they melt the scrap. During scrap melting, the electrodes produce electric arcs that vary in length and move around in the furnace. The variation in characteristics of the electric arcs cause the power line voltage to fluctuate. This voltage fluctuation is called *flicker* and is very annoying.

Voltage flicker from arc furnaces affects lighting not only inside the plant but outside the plant as well. Any end user's lighting connected to the same line feeding the arc furnace will flicker. Even though an SVC installed at the arc furnace can cost as much as $1.5 million, it usually is the lowest-cost alternative to reducing flicker to an acceptable level. Figure 4.25 shows a typical SVC at an arc furnace plant.

Distribution utilities use SVCs designed for reducing flicker on their distribution systems. Westinghouse Electric Corp. and EPRI have developed a distribution compensator called DSTATCOM (distribution static compensator). It is an SVC installed on a distribution system to reduce flicker originating from several sources. For example, as shown in Figure 4.26, American Electric Power installed a DSTATCOM near Swayzee, Indiana, on its 12.47/7.2-kV distribution circuit to mitigate flicker caused by the starting up of large motors at three rock crusher plants nearby.

Uninterruptible power supply (UPS)

What is a UPS? How does it work? A UPS conditions the voltage and power. It conditions the voltage by providing a constant voltage even during a voltage dip (sag). It conditions the power by providing a source of power during an outage. It provides a constant voltage and power source from a static or rotary source. The static source is usually a battery but can be a magnetic source, like a ferroresonant transformer or a superconducting magnetic, while the rotary source is usually a diesel-fueled motor-generator set. Often motor-generators use some technology, like a flywheel or "written-pole" motor, to provide power during the time it takes to bring the motor-generator on line. Typical UPS units have the battery charged continually by the main source of power.

A UPS contains basic components or building blocks that can be connected in various configurations. The basic building blocks of a UPS system include the battery, an inverter, and a rectifier. The battery is usually lead acid with a 1- to 5-year usable life and 5- to 60-minute backup capability, depending on the battery size. The inverter is a sol-

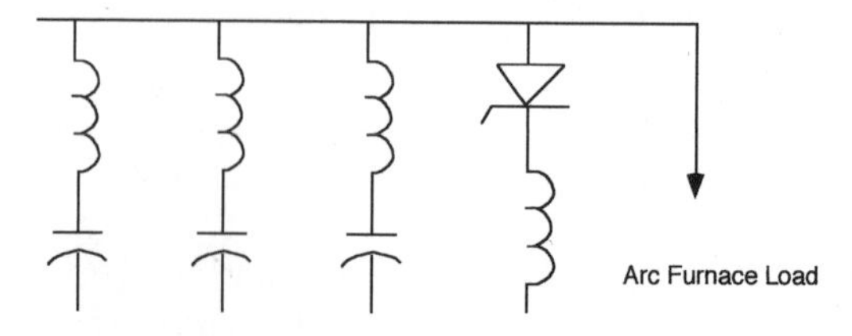

Figure 4.25 Typical SVC configuration for an arc furnace plant.

Figure 4.26 DSTATCOM installation. (*Courtesy of EPRI.*)

id-state device containing thyristors that convert dc to ac, usually with a modified square wave. The rectifier or battery charger is similar to the adapter that connects your Walkman to an ac circuit. It consists of diodes or thyristors connected in such a way that they convert ac to dc. These UPS building blocks or modules are usually connected together with static switches that protect the UPS from overloads and the sensitive equipment from failures in the UPS. They are connected in three different configurations: on line, off line, and line interactive.

An on-line UPS, as shown in Figure 4.27, provides a fully charged battery backup available all the time. It has the advantage of conditioning the power from surges, sags, or outages continuously. It has the disadvantage of shorter battery life because the continuous charging and discharging of the battery wears the battery out.

An off-line or standby UPS, as shown in Figure 4.28, turns off the inverter connected to the battery during normal operation. The UPS turns the inverter on to convert dc power to ac only during an outage. Consequently, it saves battery life by not continuously charging and recharging the battery. However, there is a time delay of 4 to 10 milliseconds to engage the UPS during an interruption.

Finally, the line-interactive UPS, as shown in Figure 4.29, is a hybrid of the on-line and off-line configurations. It charges the battery during normal operation. When there is an outage it reverses operation and converts the dc power from the battery to ac power to be used

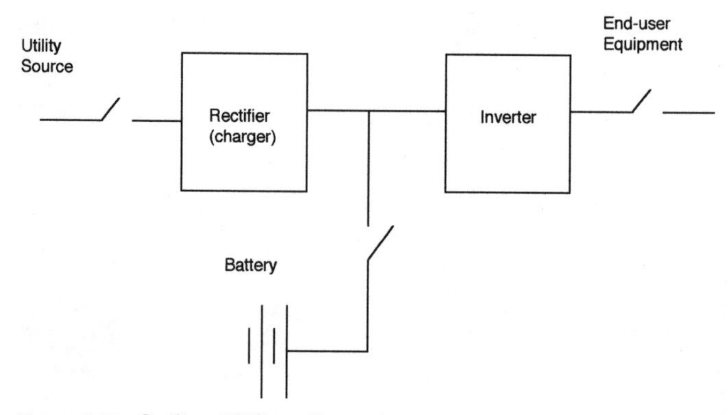

Figure 4.27 On-line UPS configuration.

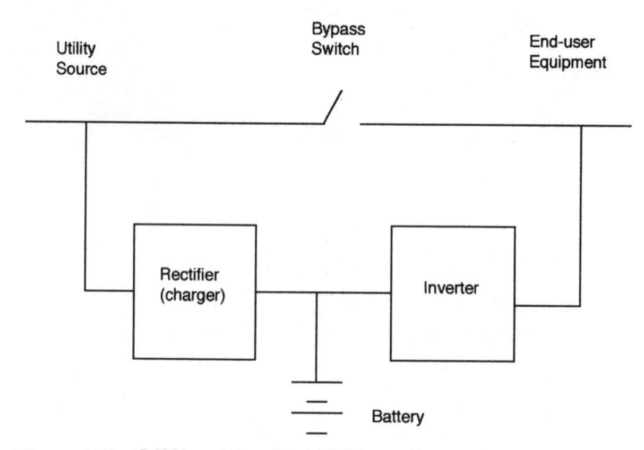

Figure 4.28 Off-line (standby) UPS configuration.

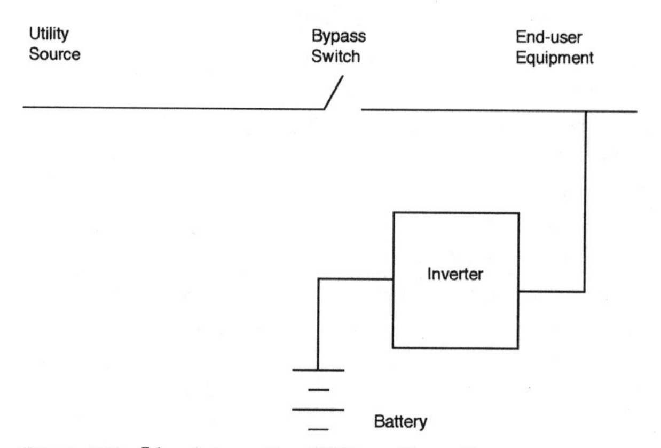

Figure 4.29 Line-interactive UPS configuration.

by the sensitive electronic equipment. It has the advantages of requiring a minimal time to switch from ac to dc power and avoiding continuously discharging and charging cycles on the battery.

In highly critical areas, such as silicon fabrication plants and hospitals, an end user can add a rotary UPS module to the static UPS module, as shown in Figure 4.30, or can use a separate motor-generator set in place of a battery-operated static UPS. The rotary module consists of a motor-generator (M-G) set that provides ride-through and isolation from voltage surges, impulses, and sags. The motor turns the generator during normal operation. Consequently, power is always conditioned by the M-G set, because the generator produces a voltage waveform independent of the incoming voltage from the utility. During an outage, the generator provides electrical power to the sensitive load.

Proper sizing and selection of a UPS system can be critical to its successful operation. Such things as size and type of load, whether the load is single phase or three phase, installation location, and cooling and lighting requirements of the load all need to be taken into consideration in selecting and purchasing a UPS system. The Northwest Power Quality Service Center (NWPQSC) has developed a brochure for selecting single-phase UPSs that are smaller than 2000-VA UPS systems. The title of the brochure is *Uninterruptible Power Supply Specifications and Installation Guide.* Information about this and similar brochures can be obtained from the NWPQSC Web site at *www.nwpq.com.* Figure 4.31 shows a small UPS system for a personal computer and a large UPS system for a commercial application.

Two new UPS technologies that have become commercially available are the written-pole motor-generator set and the superconducting magnetic energy storage (SMES).

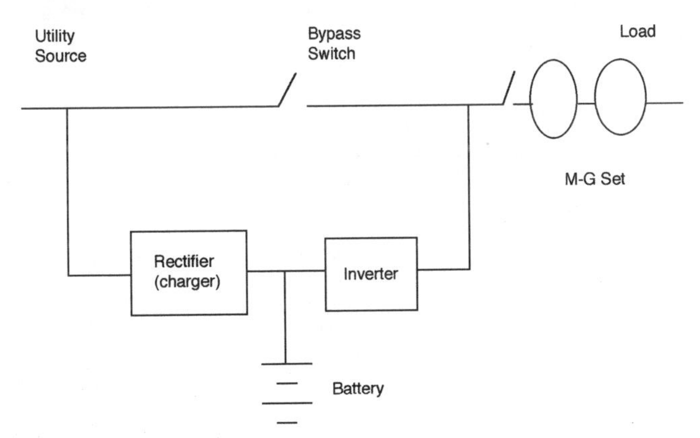

Figure 4.30 Static and rotary UPS.

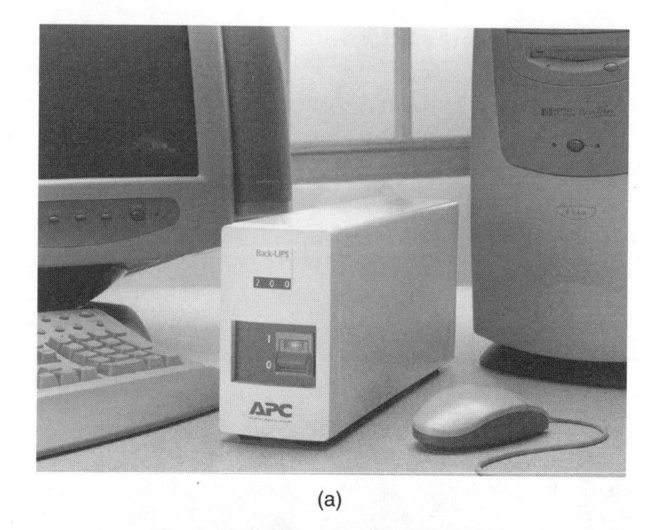

(a)

(b)

Figure 4.31 Small and large UPSs. (*Courtesy of American Power Conversion.*)

Superconducting magnetic energy storage (SMES). How do you store electrical energy that is available in milliseconds? Superconducting magnetic energy storage is one of the advanced technologies for storing electrical energy that has been commercially available for several years. As the name implies, SMES uses superconductors, which virtually eliminate losses in electrical equipment. The media have recently reported that high-temperature superconductors can function at 140 kelvins (K), or −133°C, rather than the normal low-temperature superconductor requirement of 0 K (−273°C). SMES uses traditional low-temperature superconductors. SMES stores electrical energy within a magnet that contains superconducting coils. Figure 4.32 shows how the SMES works.

SMES provides a large amount of energy (750 kVA to 500 MVA) for a short time (2 seconds) very quickly (within 2 milliseconds). Depending on its size, utilities can use SMES to provide large bulk energy storage in remote areas, emergency standby for loss of transmission and distribution capacity, and ride-through electrical energy to critical loads during a voltage sag or outage. Figure 4.33 shows the cryostat that contains a superconducting storage device and the trailer containing the cryostat, refrigerating equipment, and the inverter for converting dc to ac. For example, Carolina Power & Light Co. installed in 1994 a SMES at a 500-kW angleboard plant in Hartsville, South Carolina. For 3 years the SMES protected the angleboard plant from 283 of the 289 voltage sags and outages.

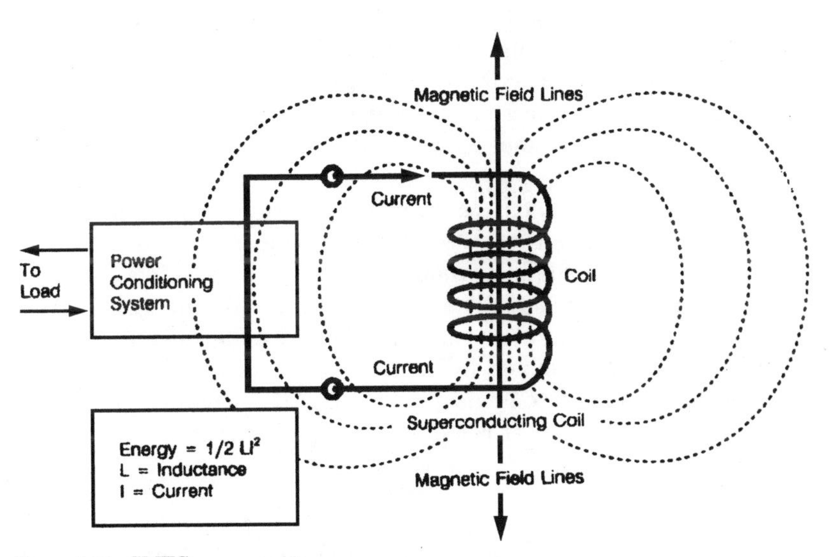

Figure 4.32 SMES components.

Figure 4.33 SMES cryostat and trailer. (*Courtesy of American Superconductor.*)

Written-pole, or Roesel, motor-generator sets. One of the major problems with M-G sets is their inability to maintain a stable frequency during a disturbance. This occurs because the speed of the induction motor is somewhat slower than the speed of the synchronous generator. One solution to this problem is the use of a synchronous motor instead of an induction motor. One disadvantage of the synchronous motor is that it takes at least 100 milliseconds to start. If the disturbance exceeds 100 milliseconds, then the synchronous motor has to be restarted. Another solution to the problem of M-G sets riding through a disturbance is the addition of a flywheel. The flywheel stores energy that is available during the disturbance. A new technology that provides constant frequency during a disturbance and causes the M-G set to ride through a disturbance is the written-pole motor developed by John Roesel.

The written-pole technology rewrites the position of the poles on the motor when the motor slows down. This provides a stable frequency output for 15 to 20 seconds without battery backup. It does this by using the rotor as a flywheel. The rotor spins outside the stator windings. This rotation of the rotor continues for 15 seconds even when the motor is shut down. As shown in Figure 4.34, the writing poles in the stator magnetize the ferrite layer in the motor. This technology is being used in hospital UPSs and at remote sites, like radar installations.

Solid-state switches

Solid-state switches have become a cost-effective alternative to UPS systems. They require two power sources from the utility. The pri-

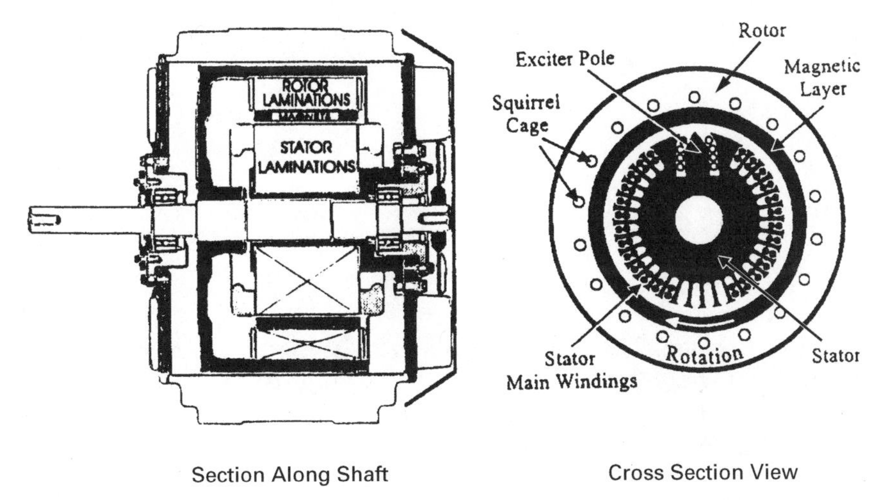

Figure 4.34 Written-pole motor with external rotor. (*Courtesy of EPRI.*)

mary power line, or feeder, has a solid-state (silicon) closed switch connected between the sensitive load and the primary utility source. The secondary feeder has a solid-state open switch connected between the sensitive load and the secondary utility source. If the primary feeder fails or voltage drops to an unacceptable level, the primary switch opens and the secondary switch closes. This transfers power from the failed feeder to the backup feeder. This transfer takes 2 to 10 seconds. Some utilities, like Baltimore Gas and Electric and Consolidated Edison, offer a premium power quality program using the solid-state switch scheme. This offering has been quite successful in providing continuous service to critical loads, like hospitals, semiconductor factories, and financial institutions' computer centers. IEEE Standard 446-1987 presents the requirements for transfer switches. Figure 4.35 provides a layout of the dual solid-state switch transfer scheme.

Harmonic filters

Utilities use harmonic filters on their distribution systems, while end users use harmonic filters in their facilities to keep harmonic currents from causing their electrical equipment to overheat and to detune resonating circuits. Harmonic filters are the "shock absorbers" of electricity and work on the principle that inductors and capacitors connected together will either block harmonic currents or shunt them to ground. Filters containing inductors and capacitors block or pass certain frequencies, because an increase in frequency increases an inductor's impedance while reducing a capacitor's impedance. There are many

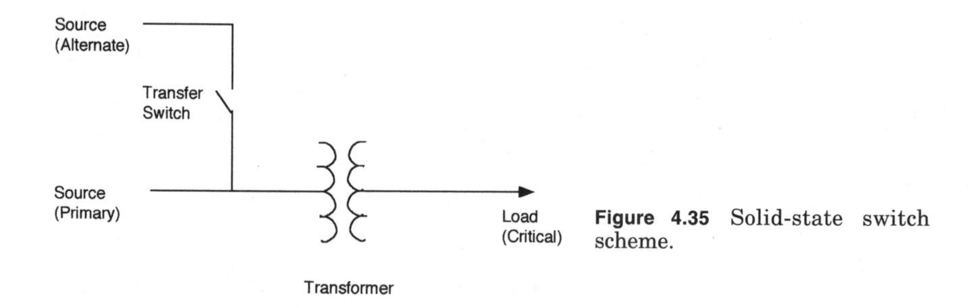

Figure 4.35 Solid-state switch scheme.

types of harmonic filter configurations. The two basic types of harmonic filter configurations are series and shunt filters.

The series filter refers to the filter made of a capacitor and inductor connected in parallel with each other but in series with the load. This type of filter provides a high-impedance path for harmonic currents and blocks them from reaching the power supply but allows the fundamental 60-Hz current to pass through. This type of configuration has the drawback of having to carry the full load current.

The other type harmonic filter configuration is a shunt filter that consists of a capacitor and inductor connected in series with each other but in parallel or shunt with the load. This type of filter configuration provides a low-impedance path for harmonic currents and diverts them harmlessly to ground. The shunt filter is more common and less expensive, because it doesn't have to carry the full load current. However, if shunt filters are not selected carefully, they can resonate with existing electrical components and cause additional harmonic currents. Both the series and shunt filters are shown in Figure 4.36.

Passive filters. Passive harmonic filters use static inductors and capacitors. Static inductors and capacitors do not change their inductance (henries) and capacitance (farads) values. They are designed to handle specific harmonics. They are called passive because they do not respond to changes in frequency. They include small plug-in devices and large hard-wired devices. They are often connected to electrical devices that cause harmonics, such as variable-speed drives and fluorescent lights. Harmonic filters sometimes are referred to as *traps* or *chokes*. They may become ineffective if the harmonics change because the load changes. Active filters may be the answer to changing harmonic currents.

Active filters. Active harmonic filters are sometimes referred to as *active power line conditioners* (APLCs). They differ from passive filters in that they condition the harmonic currents rather than block or divert them. Active harmonic filters use electronic means (bridge

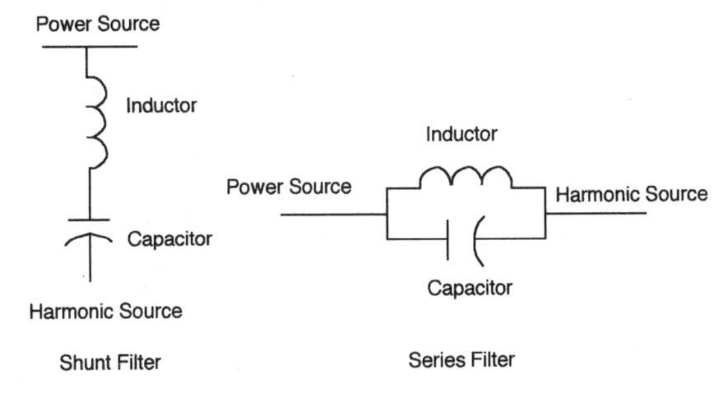

Figure 4.36 Shunt and series harmonic filter schematics.

inverters and rectifiers) to monitor and sense the harmonic currents and create counterharmonic currents. They then inject the counter-harmonic current to cancel out the harmonic current generated by the load. They also regulate sags and swells by eliminating the source voltage harmonics. While expensive in the past, they are becoming more cost effective. They are most effective in compensating for unknown or changing harmonics.

Other Harmonic Solutions

The first way to prevent harmonic problems is to design equipment so that it is not affected by the harmonics. Equipment can be designed to withstand the heating effects of harmonics. For example, the engineer can design neutral conductors large enough to carry large neutral currents caused by the additive effects of triplen (third, sixth, ninth, twelfth, etc.) harmonics. Transformer engineers can design special *K* factor transformers to withstand the effects of harmonics. Before purchasing a transformer, transformer buyers will need to calculate and specify the *K* factor using the procedure presented earlier in this chapter.

The second way to prevent harmonic problems is to properly design and specify equipment that is the source of harmonics or the cause of amplifying harmonics. For example, adjustable-speed drives are the most common nonlinear source of harmonics. There are normally two types of adjustable-speed drive: 6 pulse and twelve pulse. IEEE-519 allows higher harmonic levels for 12-pulse adjustable-speed drives than for 6-pulse adjustable-speed drives. The purchaser of adjustable-speed drives can take advantage of the higher harmonic levels by specifying 12-pulse adjustable-speed drives or by paralleling a delta-wye transformer with a delta-delta transformer to convert a 6-pulse drive into a 12-pulse arrangement. Figure 4.37 illustrates how paralleling a delta-wye transformer with a delta-delta transformer converts a 6-pulse adjustable-speed drive to a 12-pulse adjustable-speed drive.

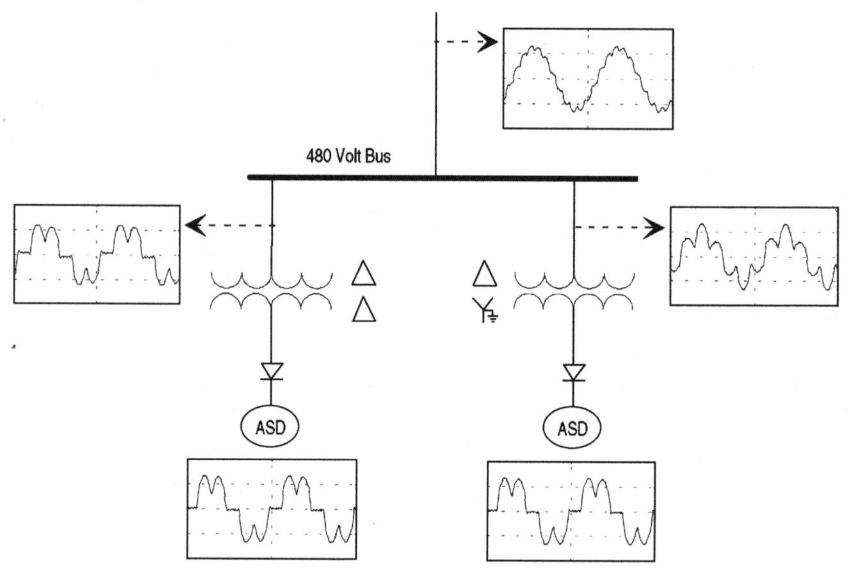

Figure 4.37 Converting a 6-pulse ASD to a 12-pulse ASD. (*Courtesy of EPRI.*)

Utilities and end users concerned about avoiding power quality problems specify capacitors that do not resonate with the existing harmonics. They know that they can avoid having their capacitors amplify harmonics. They know how to calculate the resonant point and pick a capacitor with a kVAR value that is smaller or larger than that resonant point. They know that the utility's capacitors can interact with the end user's capacitors and try therefore to avoid switching them on at the same time. Figure 4.38 illustrates how utilities and end users can specify capacitors that avoid resonating with existing harmonics.

Selection of Appropriate Power Conditioning Equipment

End users should implement the following seven steps before selecting the appropriate power conditioning equipment to mitigate their problem:

1. Determine the power quality problem.
2. Correct wiring and grounding and faulty equipment problems before purchasing power conditioning equipment.
3. Evaluate alternative power conditioning solutions.

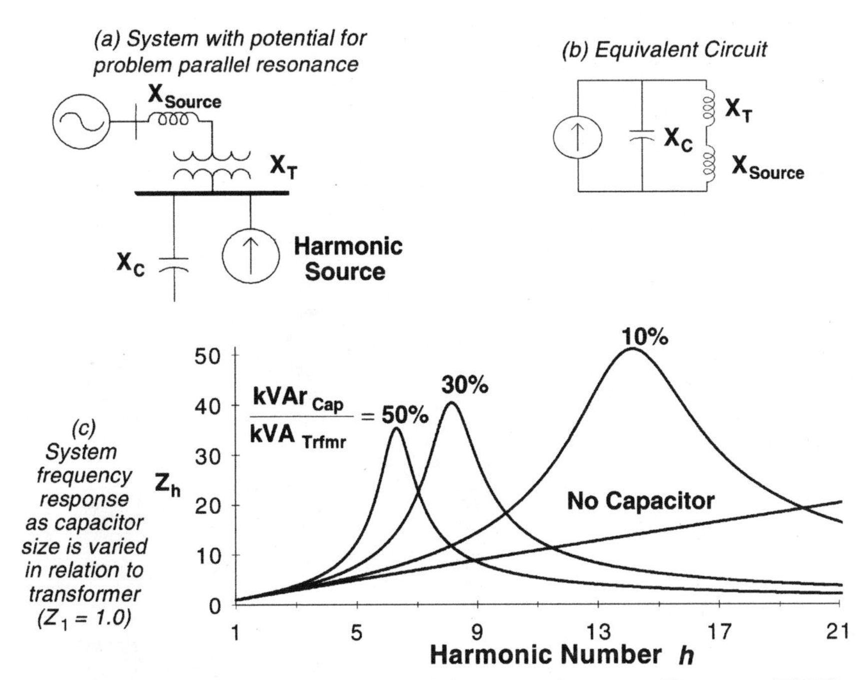

Figure 4.38 Effect of capacitor size on parallel resonant frequency. (*Courtesy of EPRI.*)

4. Develop a power conditioning plan.
5. Determine if the utility source is compatible with the load.
6. Select and install power conditioning equipment.
7. Operate and maintain power conditioning equipment properly.

Finally, utilize Table 4.1, taken from page 162 of IEEE 1100-1992, *Power and Grounding Sensitive Electronic Equipment* (*The Emerald Book*), to select the appropriate power conditioning technology to match the power quality problem.

Grounding and Wiring Solutions

Recent surveys by EPRI and others indicate that improper grounding and wiring cause 80 to 90 percent of the power quality problems. However, many end users overlook improper grounding and wiring in their facilities. They should always investigate the wiring and grounding in their facilities before purchasing and installing expensive power conditioning equipment. The next chapter will show how to identify and solve power quality problems caused by improper wiring and grounding.

TABLE 4.1 Matching Power Conditioning Technology to Power Quality Condition*

Power Quality Condition		Power Conditioning Technology: Transient Voltage Surge Supressor	EMI/RFI Filter	Isolation Transformer	Voltage Regulator (Electronic)	Voltage Regulator (Ferroresonant)	Motor Generator	Standby Power System	Uninterruptible Power Source	Standby Engine Generator
Transient Voltage Surge	Common Mode	▒		■	▒	■	■	▒	▒	
Transient Voltage Surge	Normal Mode	▒			▒	■	■	▒	■	
Noise	Common Mode		▒	■	▒	■	■	▒	▒	
Noise	Normal Mode		▒	▒	▒	■	■	▒	■	
Notches				▒		■	■		■	
Voltage Distortion						▒	■		▒	
Sag					▒	▒	▒	▒	■	
Swell					▒	■	■	▒	■	
Undervoltage					■	■	■	▒	▒	
Overvoltage					■	■	■	▒	▒	
Momentary interruption							▒	■	■	
Long-term interruption										■
Frequency Variation								▒	■	▒

■ It is reasonable to expect that the indicated condition will be corrected by the indicated power conditioning technology.

▒ There is a significant variation in power conditioning product performance. The indicated condition may or may not be fully correctable by the indicated technology.

*Output power quality varies, depending on the power conditioning technology and its interaction with the load.

SOURCE: IEEE Standard 1100-1992.

References

1. Lonie, Bruce. 1993. "Things to Consider Before Buying Mitigation Equipment." *Power Quality Assurance,* vol. 4, no. 6, November/December, pp. 6–15.
2. Freund, Arthur. 1987. "Protecting Computers from Transients." *EC&M Electrical Construction & Maintenance,* vol. 86, April, pp. 65–70.
3. Martzloff, Francois D. 1998. "What Are the Lights on Your Surge Protector Telling You?" *Power Quality Assurance,* vol. 9, no. 4, July/August, pp. 68–72.
4. Kowalczyk, Stan W. 1992. "Root Out the Silent Effects of Electrical Noise." *Chemical Engineering,* vol. 99, no. 6, June, pp. 145–148.
5. Rensi, Randolph. 1995. "Why Transformer Buyers Must Understand Load-Tap Changers." *Electrical World,* vol. 209, no. 6, June, pp. 21–28.
6. Mosman, Mike, and Grett Korn. 1998. "UPS Systems and Engine-Generator Compatibility." *Power Quality Assurance,* vol. 9, no. 2, March/April, pp. 14–21.
7. Gruzs, Thomas M. 1991. "An Overview of Power Conditioning Technologies: Part II—Power Synthesizers." *Power Quality,* vol. 2, no. 1, January/February, pp. 24–28.
8. Reason, John. 1995. "Live-Fire Fault Test of SVC: A Lesson in Power Quality." *Electrical World,* vol. 209, no. 8, August, pp. 34–37.
9. Volkommer, Harry T. 1998. "DSTATCOM Stamps Out Voltage Flicker." *Transmission & Distribution World,* vol. 50, no. 13, December, pp. 11–17.
10. Pfendler, Tom. 1995. "At the Heart of UPS." *Cellular Business,* March, pp. 62–70.
11. Baumann, Philip D. 1993. "Superconducting Magnetic Energy Storage: A Key Technology for the 21st Century." *Public Utilities Fortnightly,* vol. 131, no. 6, March 15, pp. 33–34.
12. Hoffman, Steve. 1997. "Written-Pole™ Revolution." *EPRI Journal,* vol. 22, no. 3, May/June, pp. 27–34.
13. Reason, John. 1995. "Solid-State Transfer Will Eliminate Voltage Sags. *Electrical World,* vol. 209, no. 10, October, pp. 64–63.
14. Jakwani, Asif, and Paul Jeffires. 1998. "Actively Eliminate the Harmonics in Your Facility." *Power Quality Assurance,* vol. 9, no. 1, January/February, pp. 48–53.
15. EPRI. 1998. *Active Harmonic Filter Technology and Market Assessment.* TR-111088: Palo Alto, Calif. EPRI.
16. Woodley, Neil H. "Tomorrow's Custom Energy Center Using Emerging Power Electronics." *Power Quality Assurance.* URL address: *http://www.powerquality.com/art0038/art1.htm.*

Chapter

5

Wiring and Grounding

Often the least expensive solution to a power quality problem is proper wiring and grounding, and the more expensive solution is the purchase and installation of the power conditioning equipment described in Chapter 4. However, many electricians wire and ground facilities according to the National Electric Code (NEC) but ignore the power quality aspects of wiring and grounding. Therefore, improper wiring and grounding practices cause most power quality problems (80 to 90 percent). A more cost-effective solution to these types of problems would be to correct the cause of the problem instead of just the symptom. Many times a power quality problem is caused by a loose connection, too small a neutral conductor, incorrect grounding, or a damaged conductor. This chapter explains why correctly wiring and grounding for power quality as well as safety is important and how to solve power quality problems caused by poor wiring and grounding practices. Before discussing wiring and grounding power quality problems and how to solve them, this section presents basic wiring and grounding principles.

Wiring Principles

The three basic principles of wiring:

1. Keep the length of the wire to a minimum to avoid an unnecessary voltage drop on the conductor.
2. Connect wires solidly to panels and switchboards.
3. Size and select the type of wires to match the current-carrying requirements of the load.

All three of these principles try to minimize the conductor and connector resistance. These principles apply as much to the inside of an end user's home as to a utility's high-voltage transmission line. For example, a residential or small commercial facility has a main panelboard to distribute power to the various circuits inside, as shown in Figure 5.1, while an industrial or large commercial facility has not only a main panelboard but also several branch panelboards, as shown in Figure 5.2.

End users and utilities select conductors that are the most cost-effective material. Most industrial, commercial, and residential end users use copper conductors in their electrical distribution systems, while most utilities use aluminum conductors in their high-voltage transmission and low-voltage distribution lines. The larger conductors have steel reinforcement. In the past, many utilities used copper conductors on their power lines. Since World War II, utilities have stopped using copper on their power lines, because it is too expensive. Many utilities use aluminum conductor steel-reinforced (ACSR) cable on their power transmission lines. Figure 5.3 shows a typical ACSR conductor.

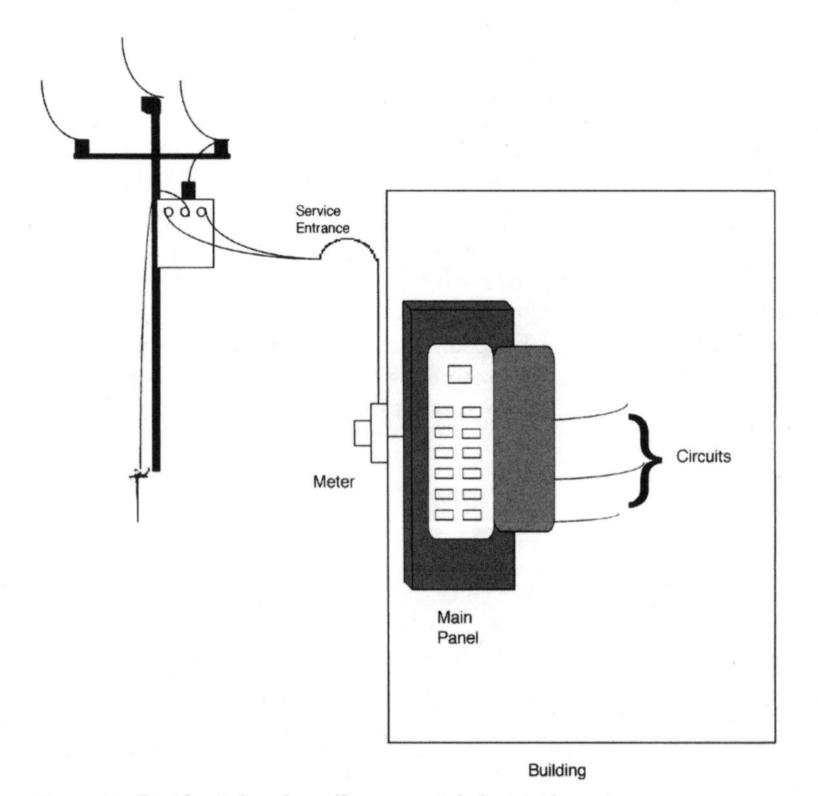

Figure 5.1 Residential and small commercial electrical service.

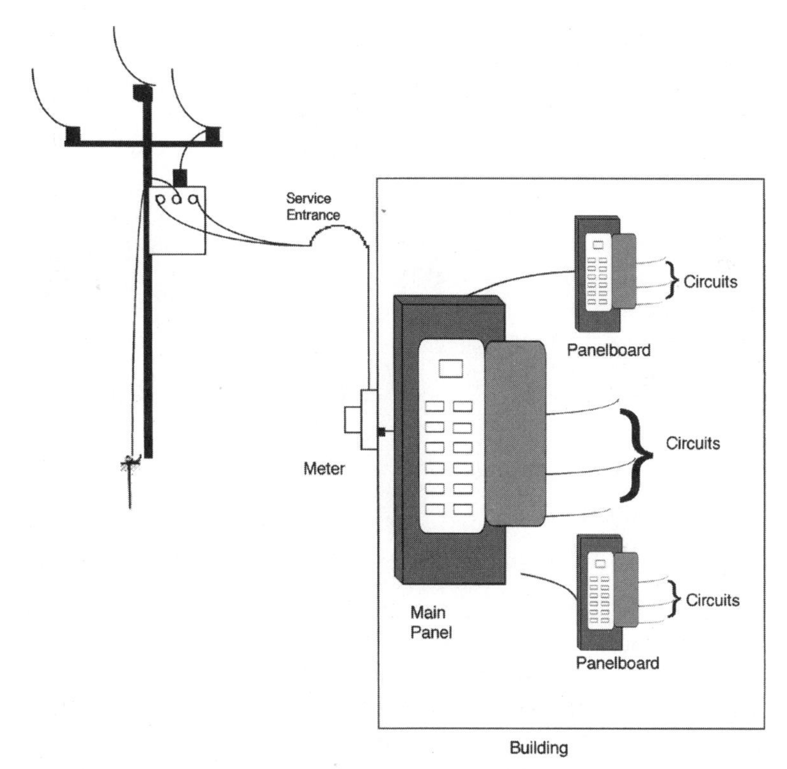

Figure 5.2 Large commercial and industrial electrical service.

Figure 5.3 Power transmission line ACSR conductor. (*Courtesy of Western Area Power Administration.*)

Grounding Principles

Proper grounding is critical to the safe and effective operation of all electrical equipment. The term *grounding* means connecting an object or electric circuit to ground or earth with an electrical conductor. The *object* is usually a piece of equipment. The *electric circuit* usually refers to the utility's transmission and distribution system or the end user's power distribution system. However, circuit grounding can include grounding of a telephone utility's and end user's telecommunication system. The symbol for grounding is ⏚.

Grounding has four basic purposes:

1. Protect people from electrical shock and equipment from a short-circuit fault
2. Provide a zero reference point
3. Provide noise control
4. Provide a path for lightning and switching surge faults

The *National Electrical Code* is the "bible" for proper grounding in the United States. It deals only with protecting the public from electrical shock and electrical fire hazards. It does not deal with power quality. Chapter 3 discussed three good guides for grounding requirements to prevent power quality problems. They are IEEE Standard 1100-1992, "IEEE Recommended Practice for Powering and Grounding Sensitive Electronic Equipment" (*The Emerald Book*); IEEE Standard 142-1991, "IEEE Recommended Practice for Grounding of Industrial and Commercial Power Systems" (*The Green Book*); and the "Federal Information Processing Standards (FIPS) Publication 94."

The *NEC* describes the two basic types of safety grounding as system grounding and equipment grounding. *System grounding* includes grounding for electrical power and telecommunication systems. Section 250-2(a)* of the *NEC* explains that grounding electrical systems to earth is done to "limit voltages imposed by lightning, line surges, or unintentional contact with higher voltage lines, and...stabilize the voltage to earth during normal operation." Section 250-2(b) of the *NEC* explains that "conductive materials enclosing electrical conductors or equipment, or forming part of such equipment, shall be connected to earth so as to limit the voltage to ground on these materials." The *NEC* requires equipment and equipment enclosure grounding to prevent people from receiving a shock when they touch the equipment or equipment enclosure.

Grounding equipment for power quality purposes requires connecting microprocessor-controlled equipment to ground to provide a zero reference point. Microprocessor-controlled equipment needs a zero reference point to operate properly and control noise. It is usually grounded to a grounding ring or electrode driven into the ground.

Grounding power systems includes grounding the utility's and end user's power systems. What are the components of the electrical power system involved in system grounding?

Power System

The entire electrical power system from the generator to the load can be divided into five levels: generation, transmission, subtransmission, distribution, and secondary systems. Each one of these systems is distinguished by the nominal operating voltage level. The generation voltage is usually at 13.8 kV, transmission at 230 kV and above, subtransmission at 115 to 230 kV, distribution at 34.5 to 69 kV, and secondary at 120 to 600 V. While all transmission and subtransmission systems are three phase, most distribution systems are three phase but can be single phase. Most secondary systems are single phase but can be three phase. Figure 5.4 shows a simplified utility power system at the transmission (includes subtransmission), distribution, and secondary voltage levels.

Each one of these power system levels has its own grounding requirements. What are the grounding requirements for the utility power system?

Utility power system grounding

The utility power grounding system includes the generation, transmission, subtransmission, and distribution grounding systems. Throughout the world, all utilities ground their generators. However, in different parts of the world, utilities ground their transmission and distribution power systems according to the IT, TT, or TN grounding systems. In many parts of Europe, utilities use the IT grounding system. In the IT grounding system, they either do not bond their power system's neutral to the generator ground or end user's ground or instead ground it through an impedance, as shown in Figure 5.5. In Asian countries, utilities use the TT grounding system. In the TT grounding system, as shown in Figure 5.6, they connect their power system's neutral to the generator ground but not to the end user's ground. Finally, U.S. utilities primarily use the TN grounding system. In the TN grounding system, utilities bond their power system's neutral to the generator's ground and end user's ground and at individual transmission and distribution towers, as shown in Figure 5.7.

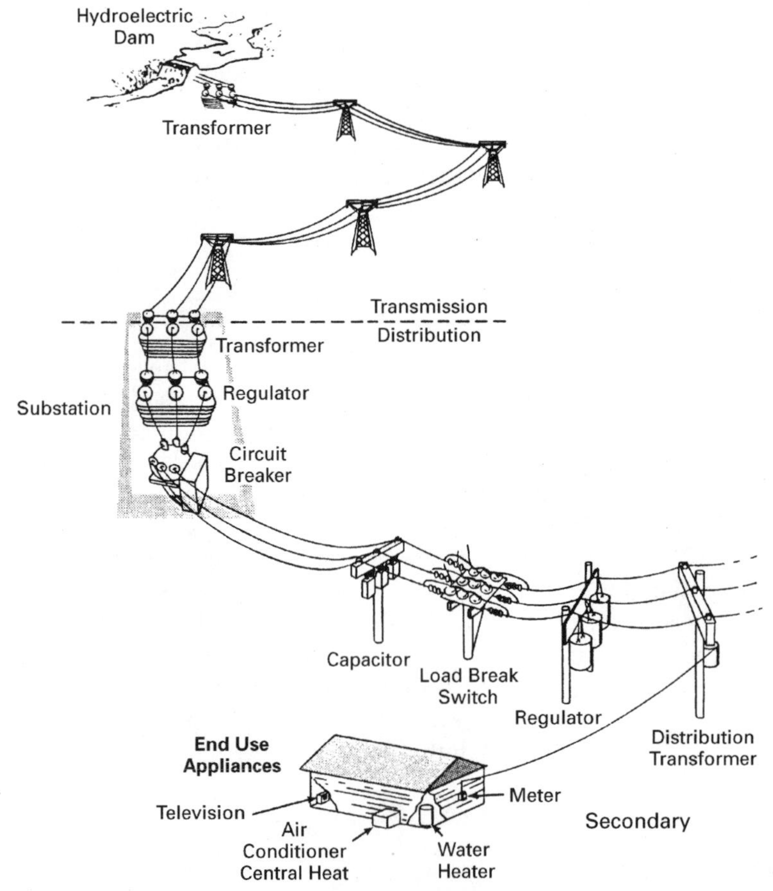

Figure 5.4 Simplified utility power system. (*Courtesy of Bonneville Power Administration.*)

U.S. utilities ground their power systems in order to provide a path for lightning and ground-fault currents. Utilities set relays to detect ground-fault currents and isolate the source of the fault by sending a signal to open appropriate breakers. Utility engineers also use the power system ground as a reference for insulation coordination. They also design ground wires to be strung above the power-line conductors and connected to the ground to shield the phase conductors from lightning strokes, as shown in Figure 5.8.

The transformer secondary circuit neutral is usually grounded. For example, the distribution transformer used to step the distribution voltage of 7200 V down to 240/120 V for use in a home has a neutral connected to the ground conductor. Figure 5.9 shows the neutral of the service entrance distribution transformer ground wire.

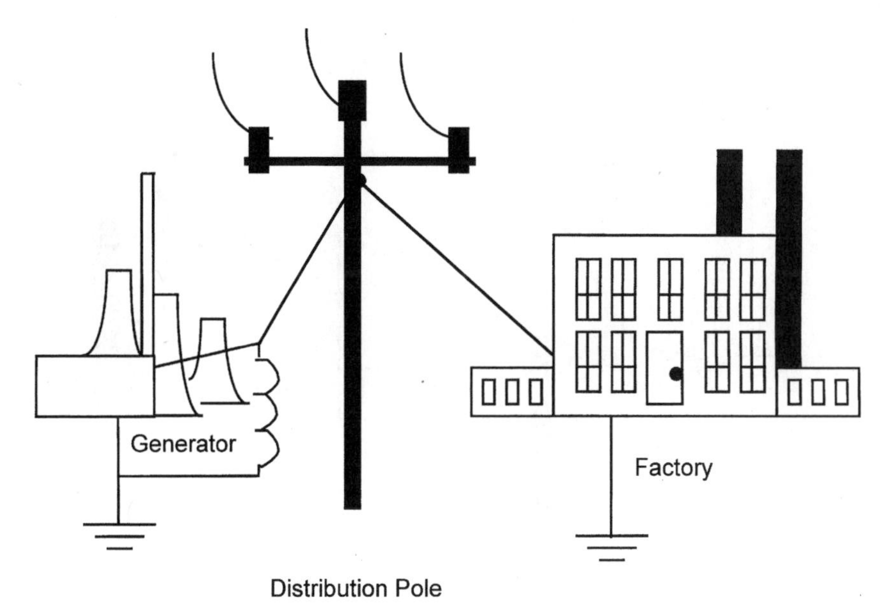

Figure 5.5 IT grounding system.

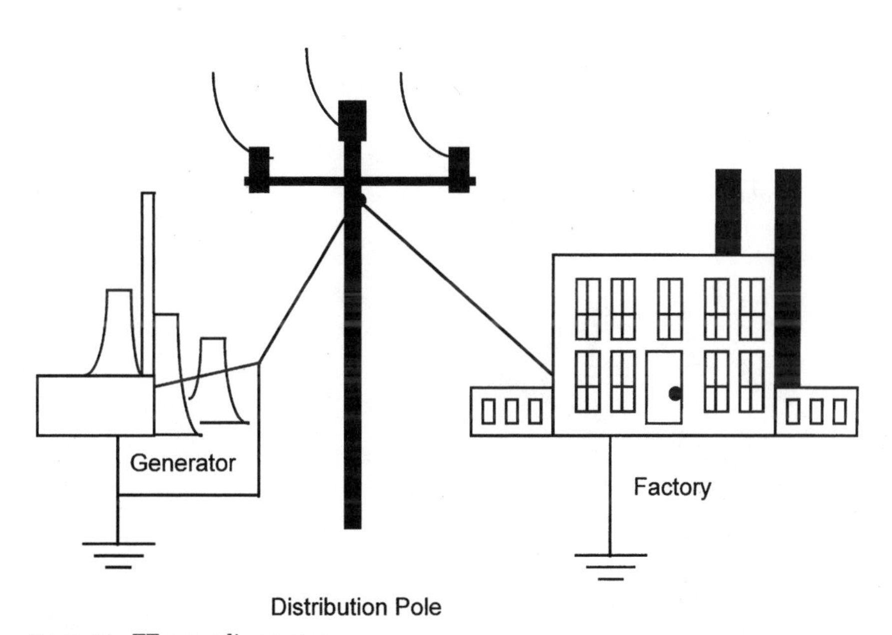

Figure 5.6 TT grounding system.

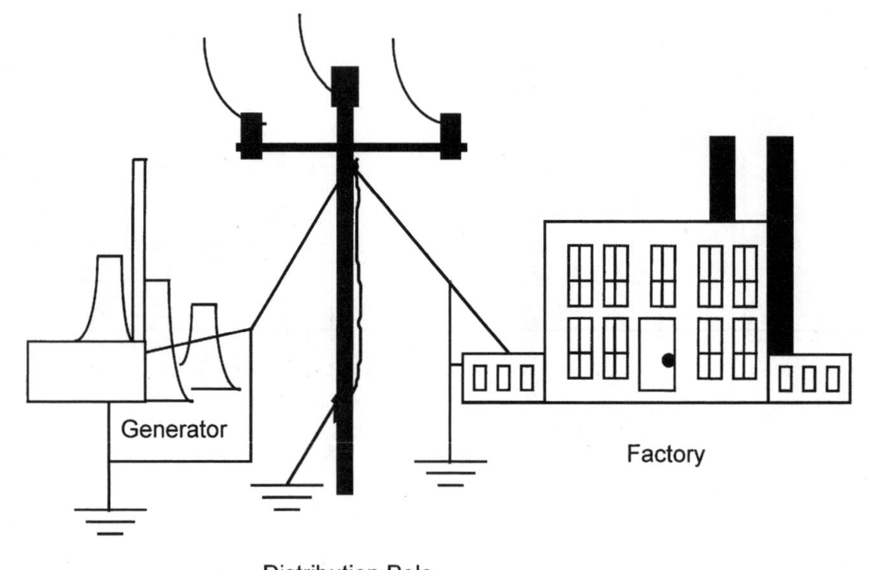

Figure 5.7 TN grounding system.

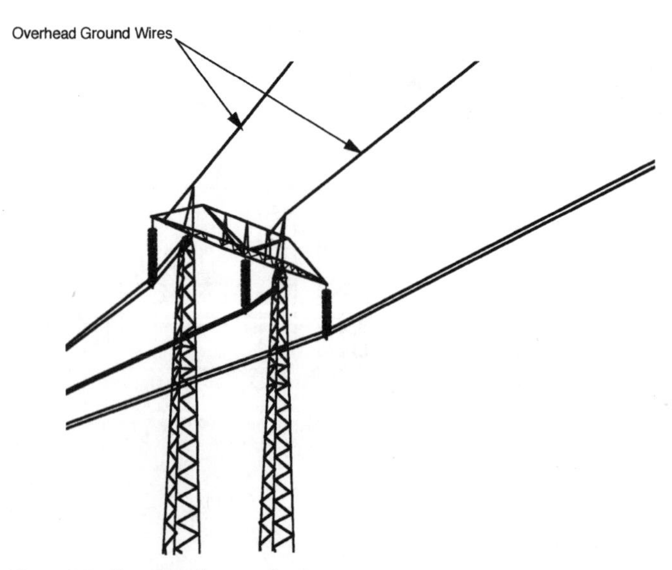

Figure 5.8 Overhead ground wires.

U.S. utilities have various practices for grounding underground and overhead lines. For example, utilities should but often do not ground the shield of underground cables. Figure 5.10 shows that lightning can strike a tree and produce electrical transients in the ground. Unground buried cables can transmit the electrical transients in the ground to sensitive electronic equipment, resulting in damaged equipment.

In the case of overhead lines, many U.S. electric utilities connect the neutral to ground at the foot of the transmission tower, as shown in Figure 5.11. Utilities connect a wire to the neutral and run it along the tower to, usually, an 8-ft ground rod (electrode) or to a counterpoise wire encased in concrete. Grounding the transmission tower neutral allows linemen to climb towers safely and diverts lightning strokes away from the transmission line.

U.S. utilities have a standard practice to ground the neutral of a four-wire distribution system to meet NEC requirements. The NEC requires that a distribution line must have a minimum of four ground rods per mile and ground rods should be 25 ohms (Ω) or less. The

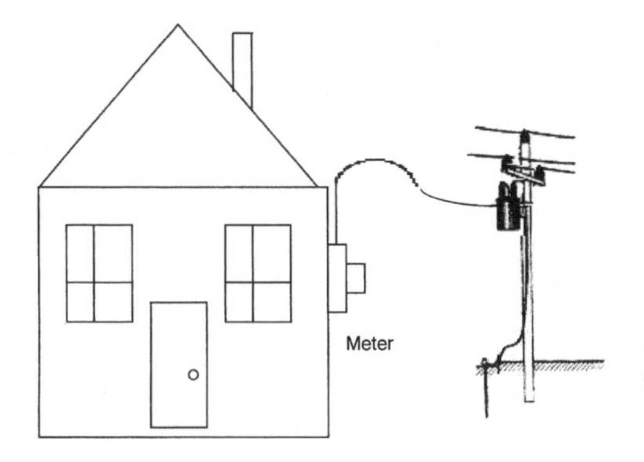

Figure 5.9 Distribution transformer grounding at distribution pole.

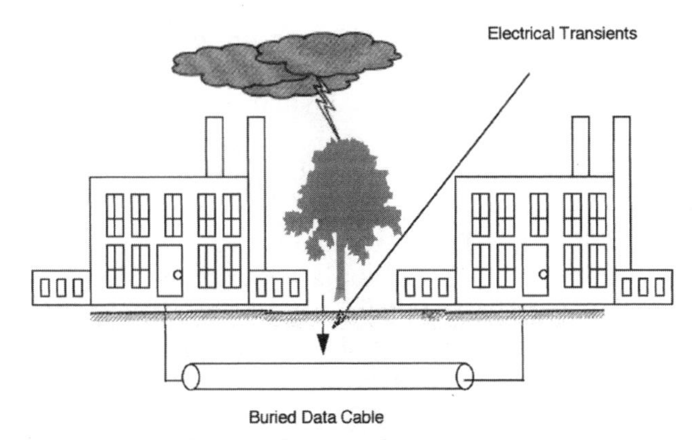

Figure 5.10 Underground electrical transients.

Figure 5.11 Transmission tower grounding.

grounded neutral provides a low-resistance path to earth for fault current caused by a lightning stroke or a fault on the system. In addition, a well-grounded neutral reduces induced voltages from radio transmitters that can interfere with the power signal on the distribution line. They not only ground the distribution line neutral to ground rods but, as shown in Figure 5.12, also connect substation equipment, like substation transformers, to a grounding grid.

Utilities build high-capacity, high-voltage direct-current lines to transmit large amounts of power long distances or for submarine cables. As shown in Figure 5.13, they design direct-current lines to use the ground as a path for both monopolar and bipolar dc transmission. In the case of monopolar dc transmission, the ground path carries load current during normal conditions. In the case of bipolar dc transmission, the ground provides a path for load current during an outage of one of the power-line conductors. They use monopolar (i.e., one conductor between the converter stations) transmission primarily for submarine cables. They use bipolar (i.e., two conductors of opposite polarity) for long-distance overhead transmission lines. The ground includes a pad for sending the load current through the ground at one end of the direct-current line and a pad for receiving the load current from the ground at the other end of the direct-current line.

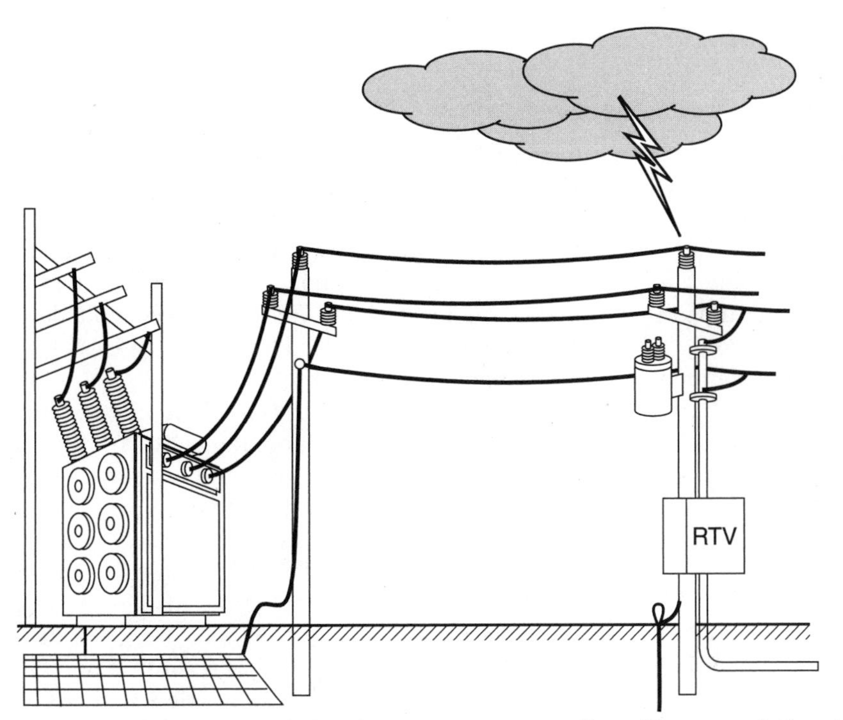

Figure 5.12 Substation and distribution system grounding. (*Courtesy of Georgia Institute of Technology.*)

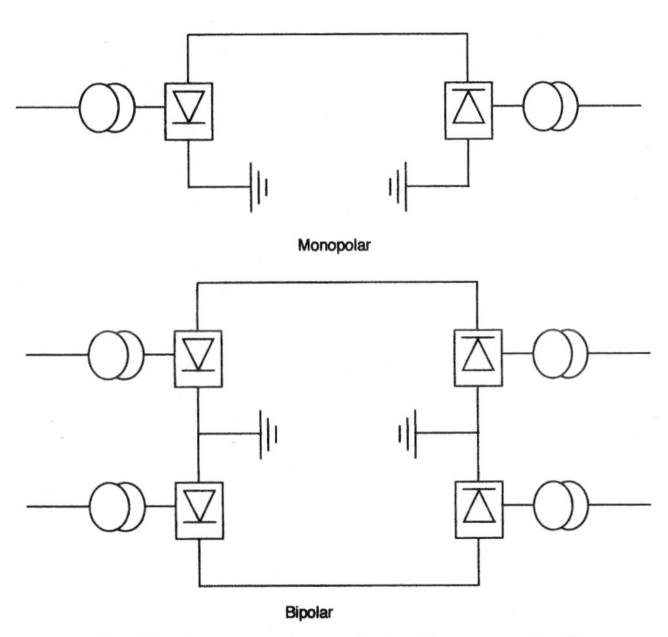

Figure 5.13 Direct current transmission line grounds *(Courtesy of Bonneville Power Administration.)*

Telecommunication system grounding

Telecommunication systems include telephone lines, local-area networks (LANs) for connecting computers, and cable TV. Telecommunication system grounding has become more important in the office, factory, and home with the advent of LANs and multiple phone lines in the office and factory, and cable TV and several telephone lines in the home. Telecommunication systems are used today for both voice and data transmission. They include computers connected together in a LAN system and computer-controlled telephone systems.

They have become just as critical to the operation of an office or factory as the power system. Most offices and many homes today rely on e-mail and telephone communication for their day-to-day operation.

Just as the *NEC* has developed rules for grounding power systems, it has developed grounding requirements for telecommunication systems as well. Where a coaxial cable coming into a building is exposed to power conductors, *NEC* Article 820-22 requires the cable's metal sheath to be grounded. The TV cable entering a home should also be grounded. Computers too must be grounded for 60-Hz operation according to *NEC* requirements. But often grounding for telecommunication systems is not done correctly, with consequent power quality problems.

The breakup of AT&T into smaller companies in the 1980s has compounded power quality problems with telephone systems. When AT&T was a monopoly, it set the standards for installing and operating telephone systems. Today, there no longer is one company providing a standard way of installing and operating telephone systems. Thus, there is a greater need for the telecommunication industry to set standards. Is this a precursor to the power quality problems that will possibly increase after the deregulation of the electrical utility industry? Chapter 9, "Future Trends," discusses the possible effects of deregulation of the electric utility industry on power quality.

One of the major power quality problems with telecommunication systems is noise caused by ground loops. The noise affects the clarity of voice transmission and the accuracy of data transmission. The signal ground connected to the power ground at various points usually causes ground loops in telecommunication systems. The ground loop can then couple with the signal grounding system and cause noise and interference in the telecommunication system. One way to reduce noise caused by ground loops is to connect data and voice (phone) equipment signal ground to the power system ground at a single entry point, as shown in Figure 5.14.

End-user power system grounding

The end-user power system includes the industrial, commercial, and residential secondary power systems. End-user power system grounds

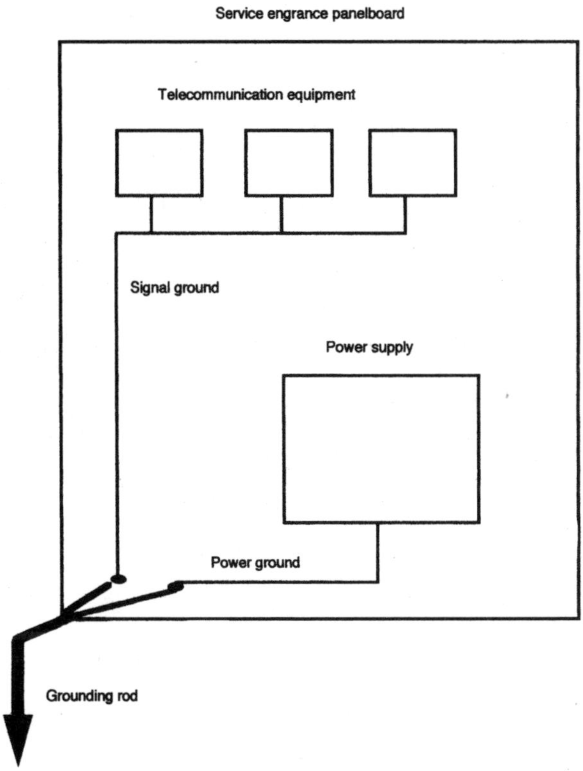

Service engrance panelboard

Telecommunication equipment

Signal ground

Power supply

Power ground

Grounding rod

Figure 5.14 Telecommunications equipment grounding.

have a different purpose from the utility power system ground. Rather than provide a path for lightning and faults, their main purpose is to prevent shock to personnel and provide a power and signal reference. The *NEC* allows end-user ground connections to be smaller than the equipment ground conductors because they do not have to carry fault currents. Normal grounding practice for residences is to ground incoming power to a ground rod and a water pipe and to ground telephone circuits and TV cable to a water pipe, as shown in Figure 5.15.

Grounding practice for commercial facilities with computer equipment is to ground all power equipment to a grounding grid, as shown in Figure 5.16.

The *NEC* requires ground connections to be less than 25 Ω, while power quality experts recommend that, to minimize power quality problems, grounds for sensitive electronic equipment should not exceed 5Ω. *NEC* Article 250-51 requires that an effective grounding path meet the following requirements:

1. Continuous and permanent
2. Capacity capable of handling fault current

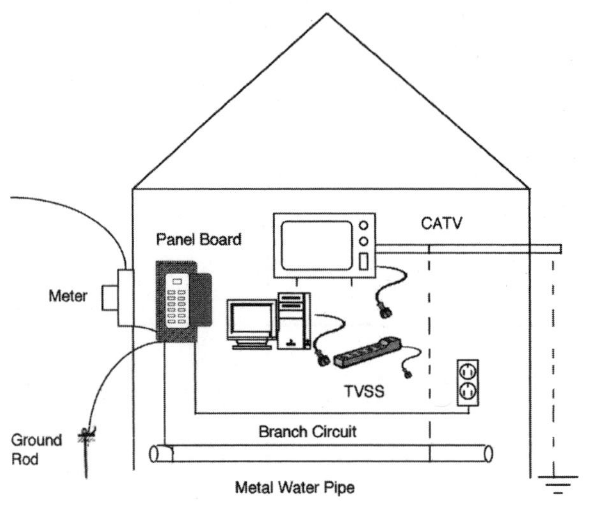

Figure 5.15 Residential grounding.

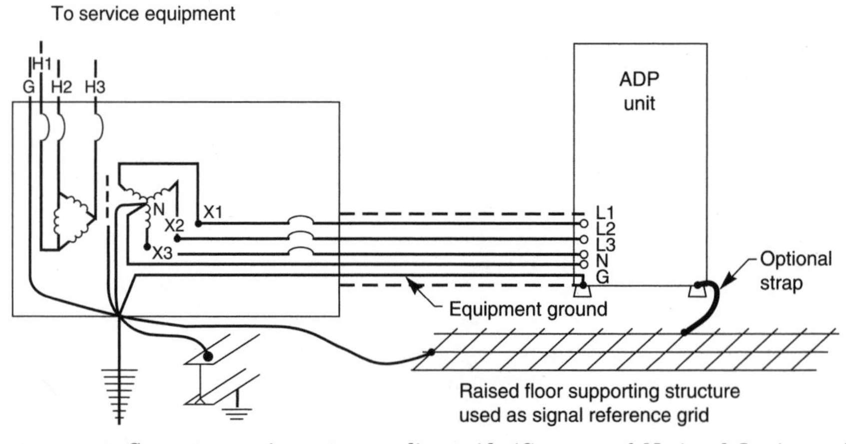

Figure 5.16 Computer equipment grounding grid. (*Courtesy of National Institute of Standards and Technology.*)

3. Low enough impedance to limit voltage drop to ground and allow relays to be tripped
4. Earth should not be the only ground

Grounding to *NEC* requirements does not prevent wiring and grounding problems from occurring. Improper wiring and grounding of sensitive electronic equipment cannot only cause power quality problems but can also prevent power conditioning devices from protecting sensitive electronic equipment.

The purpose of grounding for power quality is to control noise and transients. Grounding for power quality requires an equipotential ground system or plane. The IEEE Standard 1100-1992, "IEEE Recommended Practice for Powering and Grounding Sensitive Electronic Equipment" (*The Emerald Book*), page 92, describes an equipotential ground plane as "a mass (or masses) of conducting material that, when bonded together, provide a low impedance to current flow over a large range of frequencies." It is, therefore, critical to avoid voltage variations between different ground locations to prevent stressing insulation and causing ground currents that cause power quality problems.

Wiring and Grounding Problems

An experienced power quality inspector can identify many wiring and grounding power quality problems by sight while performing a power quality survey as described in Chapter 7, "Power Quality Surveys." Nevertheless, a power quality inspector often has to identify power quality problems by measurements of voltage, current, and impedance, using the power quality measuring instruments described in Chapter 6, "Power Quality Measurement Tools." When a power quality inspector diagnoses a power quality problem, the inspector needs to know the causes of wiring and grounding problems. What are some of the causes of power quality problems related to improper wiring and grounding?

One common cause of wiring and grounding power quality problems is the confusion and conflicts between the application of *NEC* safety grounding standards and power quality standards of IEEE and other standards organizations. Mixing power and telecommunications systems and 60-Hz electrical power with high-frequency communication signals in a facility adds to this confusion. The latter issue is primarily caused by the fact that electricity acts differently at 60 Hz and 10,000,000 Hz. At high frequencies, long inductive wires increase in impedance and act like open circuits, blocking the flow of electricity. Therefore, long inductive wires should be avoided, if possible. There are many types of power quality problems caused by improper wiring and grounding. They include ground loops, electromagnetic interference (EMI) noise, loose connections, poor grounds, lightning, insufficient neutral conductor, missing safety ground, and superfluous ground rods. Probably the most important power quality problem caused by improper wiring and grounding in commercial and industrial facilities is ground loops.

Ground loops

What are ground loops? What causes them? How do you detect ground loops? What kind of power quality problems do they cause? And how do you get rid of them? These are the questions we will answer in this section on ground loops.

IEEE Standard 1100-1992, "IEEE Recommended Practice for Powering and Grounding Sensitive Electronic Equipment" (*The Emerald Book*), page 28, defines ground loops as occurring "when two or more points in an electrical system that are nominally at ground potential are connected by a conducting path such that either or both points are not at the same ground potential." In other words, ground loops are electric current flowing in the ground connection between two or more pieces of equipment. Magnetic coupling between the "hot" wire and the ground wire can cause small ground currents. These are usually too small to cause any problems. However, large and damaging ground currents can be created when two pieces of equipment are grounded at two different points. If a communication path is present, it can also provide a path for ground loops. Figure 5.17 illustrates a ground loop between two computers grounded at two different points.

Long wires from computers to power outlets several feet away can cause ground loops. The impedance of a long power cable will increase with the frequency of the signal. Therefore, high-frequency noise signals will avoid the long power cable and flow in the neutral conductor and cause ground loops. They will also flow inside the computer equipment and cause voltage drops that damage sensitive microchips.

There are a few ways to detect ground loops. One way is to install a current transformer around the ground wire and measure the amount of current flowing in the ground wire. Another way of detecting ground loops is to measure the voltage between the neutral and ground to determine if it is greater than zero. A voltage greater than zero between the neutral and ground indicate that ground loops are present.

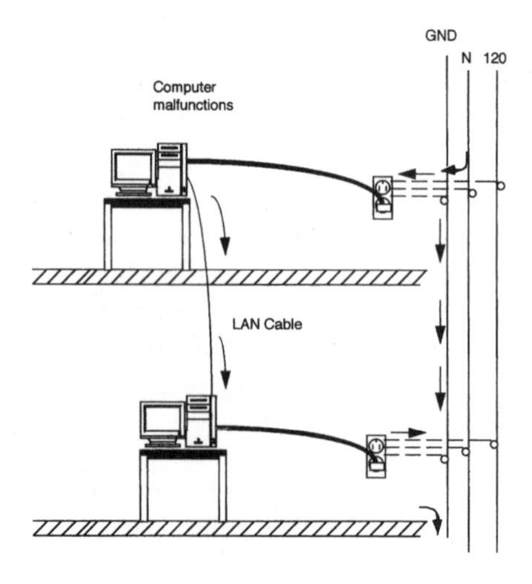

Figure 5.17 Ground loop between computers. (*Courtesy of National Institute of Standards and Technology.*)

Ground loops not only damage sensitive electronic components but can cause problems with communication equipment. They can cause ground wires to act as a loop antenna and transmit a humming-type noise that interferes with communication signals.

There are several ways to solve ground loop problems. The easiest solution to ground loop problems is to ground equipment and the service panel to a common point. The use of fiber-optic communications cable is an expensive but effective solution to eliminating noise in communications circuits caused by ground loops. If fiber-optic cable isn't feasible, a signal isolator inserted in the communications circuit will stop the flow of ground currents and allow communications signals to pass. Shortening the cable can reduce ground loops caused by long power cables. Locating power outlets or power strips near the computers is an effective way to reduce the length of power cables.

Electromagnetic interference (EMI) noise

Electromagnetic interference noise is a high-frequency signal on power lines and circuits. It is transmitted three ways: through the air, over a power line, or through the ground. Its source can be any common electrical appliance such as a fan, microwave oven, or fluorescent lights with magnetic ballasts. Other sources include transformers, electrical switchboards, and some uninterruptible power supplies. EMI causes computer monitors to wiggle and sensitive electronic devices to misoperate, such as a computer that controls gasoline flow in a gas station or an automatic teller machine in a bank. As mentioned in Chapter 4, EMI noise's effect can be removed by moving the source a safe distance from the device being affected. If the EMI noise is being transmitted via the ground, shielding the affected equipment or data cables can be an effective way of solving this power quality problem.

In the past, harmonic currents from a distribution power line induced harmonic currents in the open telephone lines built under the distribution line on the same pole, as illustrated in Figure 5.18. The harmonic currents often were at the same frequency as the telephone signal and caused noise and interfered with the telephone signal. This problem has been eliminated by the advent of shielded coax cables for telephone lines. However, high currents in the distribution line can be induced in the telephone and TV cables and cause a voltage drop across the reference ground.

Recently, fiber-optic cables built under high-voltage power lines have been experiencing failures from currents induced in the fiber-optic insulation by the high-voltage line. The arcing from the induced currents is called "dry band" tracking. Research is being conducted to find a cost-effective insulation that does not experience this problem.

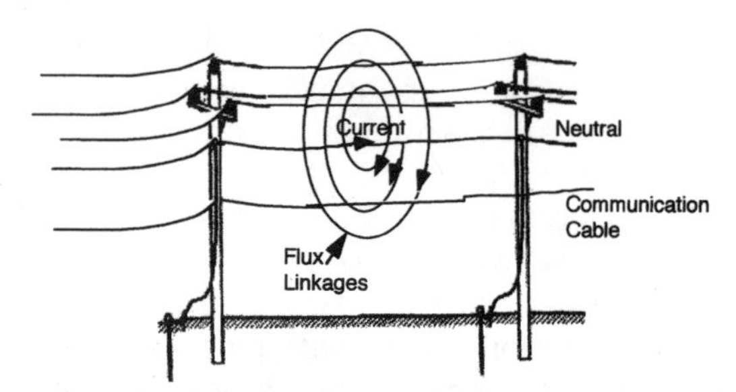

Figure 5.18 Inductive coupling between distribution and communication lines.

Loose connections

Loose or bad connections often cause power quality problems. They can cause noise that damages electronic equipment. They can cause heating and burning of insulation. The burnt insulation allows bare wires to touch and causes a short circuit.

One of the first places to look for a loose connections is in the service panel. An ozone smell or an arcing noise is one tip-off of a bad connection. Loose connections to ground can cause high-resistance grounds that do not divert lightning faults to ground.

Grounding for lightning and static electricity

Grounding for lightning is an effective way to protect computers and telecommunication systems and other sensitive electronic equipment from being damaged by lightning. Lightning is simply a large static discharge of current from a cloud to ground or from one cloud to another. This static discharge contains a large amount of electrical current that averages about 20,000 A but can be as high as 270,000 A. Lightning strokes cause extreme temperatures (as much as 60,000°F) and burn everything in their path. Like all electric current, lightning follows Ohm's law and seeks to follow the lowest-impedance path. If the lowest-impedance path is a human body, it can cause cardiac arrest and death. It does the same thing to sensitive electronic equipment. If the lowest-impedance path is through computer-related equipment, lightning can fry the heart of the computer, the microchip. It is important to have good ground paths for lightning strokes. Otherwise, lightning will damage computers and other sensitive equipment.

The best way to prevent lightning from causing damage to equipment is by diverting to ground the high current flow of a lightning stroke. Ben Franklin developed the lightning rod in the second half of the eighteenth century to divert lightning strokes away from buildings. Today's lightning grounding systems protect sensitive electronic equipment from lightning strokes by diverting them to ground and away from sensitive equipment. The National Fire Protection Association's Code 780, *Code for Protection Against Lightning,* gives the detailed requirements of a lightning grounding system. Figure 5.19 shows a building grounding system made of grounding rods and ring to protect computer-related equipment inside the building from damage caused by lightning.

Static electricity inside the building can be just as damaging as lightning outside the building.

Shoes rubbing a carpet and building up an electric charge on a person often cause static electricity. When a person touches a grounded object, electric current is discharged from the person to the object. If the object contains microchips, the electric discharge can destroy the microchips. Grounding the person before he or she touches sensitive equipment is an effective way to prevent this event from happening. People working on electronic equipment, as shown in Figure 5.20, commonly use grounding straps on wrists. Another grounding method is to provide a static drain path by grounding floor tiles or mats to the nearest grounded metal.

Attack of the triplens

The neutral conductor often becomes a path for the feared odd triplen harmonic currents. Chapter 4 briefly mentioned odd triplen

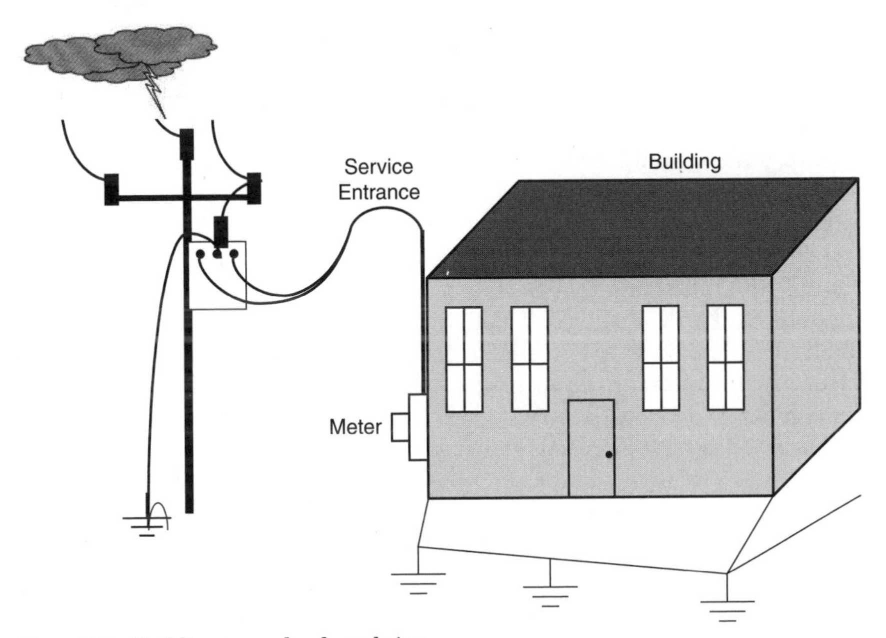

Figure 5.19 Building ground rods and ring.

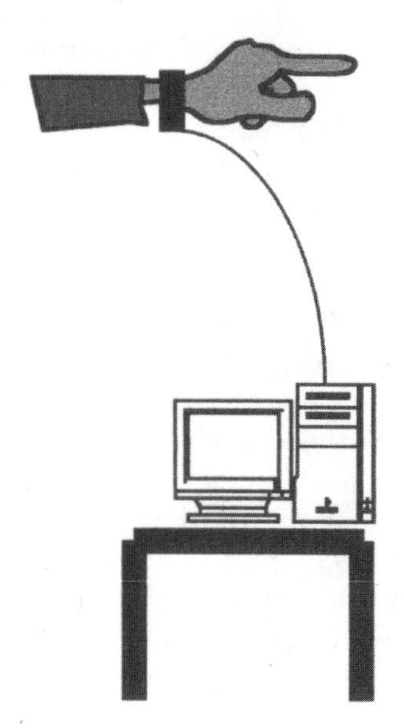

Figure 5.20 Grounding wrist straps.

harmonics. What are odd triplen harmonics? They aren't "odd" because they are strange but odd because they have odd numbers. They not only have odd numbers but also are odd multiples of the third harmonic. The third harmonic of the fundamental current frequency of 60 Hz is 180 Hz. Therefore, the third, ninth, fifteenth, and twenty-first harmonics are odd triplen harmonic orders at 180 Hz, 560 Hz, 900 Hz, and 1260 Hz, respectively. Why are they feared? They are feared because they have one nasty characteristic that the neutral conductor doesn't like. They are zero sequence currents that are in phase. This means they add to each other as well as to the 60-Hz neutral conductor normal current. This can result in the neutral current increasing to 2 to 3 times the phase current value. This can be devastating to a small neutral conductor that wasn't designed to handle such large currents. The neutral conductor can become overheated and cause a fire. Even if the neutral current is not large enough to cause a fire, a neutral current can cause a large voltage drop in the neutral, according to Ohm's law, and induce noise into nearby signal circuits.

How do these triplen harmonics get into the neutral conductor? They get into the neutral conductor from three single-phase nonlinear loads connected to the neutral. If these nonlinear loads have switched-mode power supplies, they will contain triplen harmonic currents. Switched-mode power supplies are notorious for generating triplen harmonic currents because they demand current at the peak of the voltage waveform. All of today's computer-controlled equipment contains switched-mode power supplies. Also, static power converters in adjustable-speed drives and uninterruptible power supplies change the fundamental waveform and create triplen harmonics. Figure 5.21 shows how each nonlinear load contributes to triplen harmonic currents in the neutral that are 3 times the triplen harmonic phase currents.

How can triplen harmonic currents be kept out of the neutral conductor? What can be done about them? The former Computer Business Equipment Manufacturers Association (CBEMA)—now the Information Technology Industry Council (ITIC)—recognized this problem. They suggested doubling the size of neutral conductors and providing separate neutrals for each phase conductor feeding a nonlinear load. Other solutions to triplen and other types of harmonic problems include the use of filters and delta-wye and zigzag transformers. However, some power quality solutions cause problems of their own.

Solutions that cause problems

Many electrical equipment installers think they are solving power quality problems but instead are causing new problems. This is because they do not have a clear understanding of how electricity operates under different conditions. This happens when people misuse power quality solutions, like isolated grounds, additional ground rods, and multiple neutral-to-ground connections. They think they have solved power quality grounding problems. They incorrectly apply these solutions and compound the very power quality problems they are trying to solve. Suffice

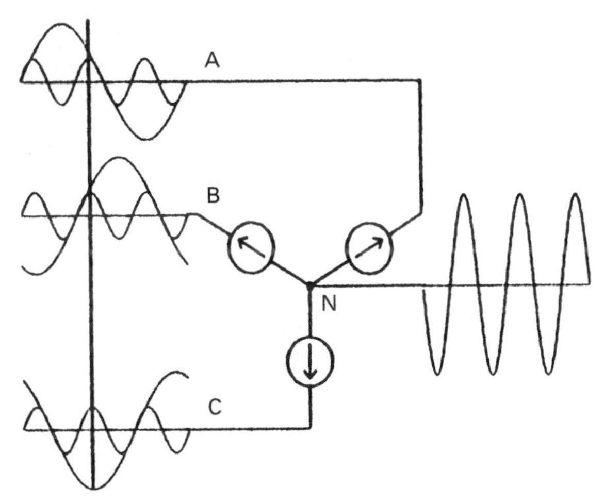

Figure 5.21 Nonlinear loads contribute to triplen harmonics in the neutral. (*Courtesy of Electrotek Concepts, Inc.*)

it to say that they can cause new power quality and safety problems when they misunderstand how grounding works.

Ground is not a drain for "dirty" power like a bathroom drain that gets rid of dirty water. Unlike dirty water, dirty power has a tendency to return to an electrical system and foul it up again. Electricity is not absorbed into the ground never to be seen again. It often finds a new path to return to the power system. This is especially true with electric currents at high frequencies. High-frequency electric currents have a tendency to follow new paths. Often adding grounds to get rid of dirty power instead results in it returning from a new direction ready to "gum up" electronic equipment. Installation of more grounds may provide a new path for noise and lightning and cause additional power quality problems. Yes, grounding does not always solve power quality problems but instead can be the very cause of more problems. Therefore, it is important to understand how electricity behaves under differing conditions and how to correctly apply the *NEC* grounding requirements for safety as well as IEEE standards for power quality.

Sometimes electricians try to solve power quality problems while causing a safety problem. They should never violate *NEC* safety standards in order to solve a power quality problem. Safety always comes first. Figure 5.22 illustrates how an electrician, in trying to solve a power quality problem by using a dedicated isolated ground (thinking it is supposed to be kept from the dirty utility ground by replacing a

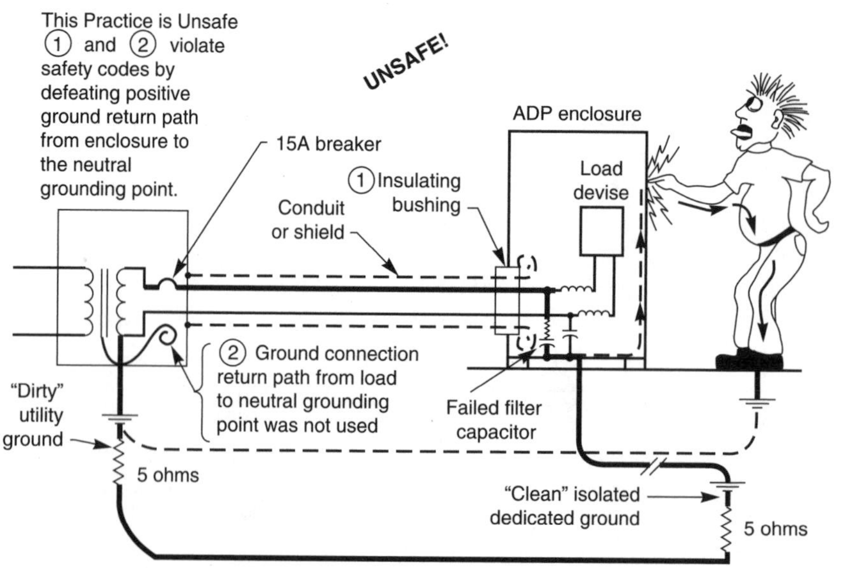

Figure 5.22 Electric shock hazard of isolated grounds. (*Courtesy of National Institute of Standards and Technology.*)

connection to the utility ground with an insulating bushing), caused an unsafe condition that violated the *NEC* safety requirements.

What wiring and grounding principles can be followed so as to avoid compounding power quality and safety problems? What solutions to wiring and grounding problems don't cause more problems?

Wiring Solutions

Power quality experts know that wiring solutions to power quality problems involve three basic electrical "S" principles:

1. Separation
2. Selection
3. Shielding

Let's examine each one of these principles and see how they prevent power quality problems.

Separation

Sometimes separation is the easiest and lowest-cost solution. Forcing sensitive and nonsensitive equipment to work together is not a good idea. One drives the other crazy. For example, when a large motor load starts up, the resulting large inrush current causes a voltage sag on the circuit serving the motor. If there are computers connected on the same circuit, the voltage sag can cause the computers to lose data and freeze. Separation is the answer.

Separation means supplying electricity to computers and other sensitive electronic equipment from a separate dedicated circuit all the way back to the panelboard. It simply means dedicating a separate circuit to serve sensitive loads. This includes a separate ground, neutral, and power conductor. Separation also keeps harmonics generated by nonlinear equipment, like adjustable-speed drives, from affecting equipment, like computers and telephones, that don't like harmonics.

Separation in the office environment means providing separate circuits for laser printers. A laser printer requires a large inrush current to heat up the toner each time it prints a document. This large inrush current causes the voltage to drop and lights to dim. This happens in offices where the lights and the laser printer are on the same circuit. Every time someone starts the laser printer, the fluorescent lights flicker. A dedicated separate circuit for the laser printer, as shown in Figure 5.23, is the best solution to this problem.

Separation of the neutral is a good strategy for dividing and conquering the attack of the evil triplens. Remember triplens march in unison,

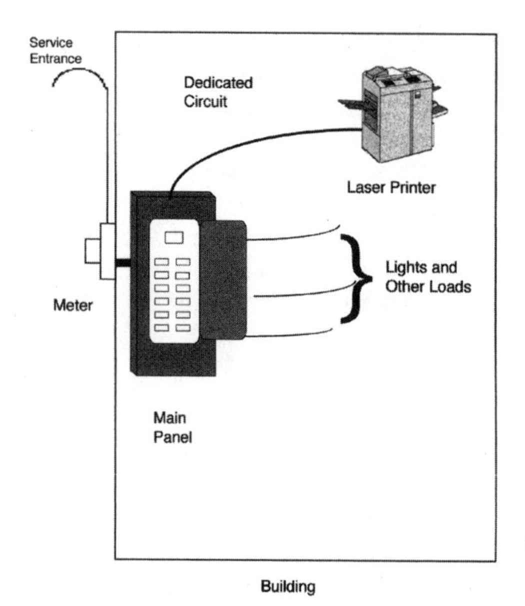

Figure 5.23 Dedicated circuit.

working together to overload neutral conductors. Separate neutrals distribute the triplens and keep them from overloading one neutral.

Separation of incompatible appliances in the home is another important strategy to keep electrical appliances from interacting with one another. For example, one residential customer rewired her home and discovered strange things happening with the dishwasher. It would start up on its own. Was it haunted? The residential customer called the local utility for help. The utility sent a power quality expert to investigate. The power quality expert found that the customer installed a fireplace insert with a thermostatically controlled fan to distribute the heat from the fireplace. Each time the fan came on the dishwasher started up too. This seemed strange until the power quality expert showed the customer that the fan control was wired to the dishwasher switch.

An orange plug on the outlet often identifies a separate ground. "Isolated grounds" in the section "Grounding Solutions" discusses the purpose and configuration of orange plugs.

If installing a separate dedicated circuit is not practical, another alternative is to avoid plugging too many sensitive loads on one circuit. This will minimize the interaction between different types of equipment. A good rule of thumb is to limit the number of outlets used per circuit to 6 rather than to the 13 allowed by the *NEC*.

Selection of wire and cables

Selection of the size and type of wire and cables for both power and data transmission is important in avoiding and solving power quality problems. In the case of power service, the electrician usually selects the wire and cable size to match the capacity of the circuit breaker at the panelboard, according to the requirements set by the *NEC*. But there are some power quality concerns that affect the selection of the power wire or cable. They include voltage drop and noise.

Keeping the voltage drop caused by the fundamental 60-Hz current flowing in a circuit to a minimum is critical to preventing power quality problems. Voltage drop follows Ohm's law. Ohm's law says that a voltage drop will occur across an impedance when current is flowing in it. The *NEC* limits the voltage drop to 3 percent in a branch circuit. Most manufacturers recommend the voltage drop should not exceed 1 percent in order to avoid power quality problems on branch circuits feeding sensitive electronic equipment.

According to Ohm's law, the voltage drop is reduced by reducing the impedance of the circuit. Increasing the size or gauge of the wire reduces the impedance of the circuit. In order to avoid power quality problems, the size of the wire should be above what the *NEC* requires.

In addition to the concern about voltage drop caused by the fundamental 60-Hz current, there is the concern about the voltage drop caused by harmonic currents. Harmonic currents cause a voltage drop with a crest factor that is much larger than the voltage drop caused by the fundamental current. The crest factor is the ratio of the crest value of the waveform to the RMS value. The crest factor caused by harmonic currents may be 3 or 4 times higher than the fundamental, while the crest factor for the fundamental is only 1.414 ($\sqrt{2}$) times higher than the fundamental. Consequently, the wire selection tables based on fundamental waveforms will provide erroneous voltage drops for harmonic currents. This is why it is important to select wire sizes possibly 2 to 3 times larger than required by the *NEC* to avoid power quality problems. In addition to choosing wire sizes to reduce voltage drop, wire and cable sizes should be selected so as to minimize the effect of noise on data and voice circuits.

In the case of voice and data service, the technician has a choice of three main types of cables: (1) twisted pair, (2) coaxial, and (3) fiber optic. Figure 5.24 shows the configuration of each of these types of communication cable. Each of these types of cable has its advantages and disadvantages.

The twisted pair has the advantage of being lowest in cost but the disadvantage of being the most noisy. It contains pairs of insulated small-size (24-gauge) wires in an insulated tube twisted together to reduce noise. Used more than 80 percent of the time, it is the most common cable of the three types. The large impedance of the small-gauge wire limits the distance a signal can be effectively transmitted over a twisted pair to approximately 330 ft. It is also susceptible to cross talk—the leakage of signals from one pair to another. Shielding

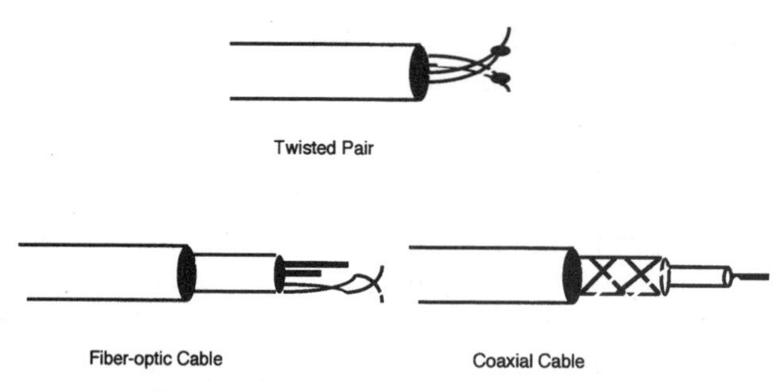

Figure 5.24 Types of data and voice transmission cable.

twisted pairs by encasing them in a metallic sheathing and connecting the sheathing to ground increases the cost and signal losses but reduces electrical noise.

Coaxial cable is in the middle as far as cost, but is immune to electrical noise when properly shielded. It has the ability to carry large amounts of data at high speeds over long distances. It has the disadvantage of being cumbersome and unwieldy to install.

Fiber-optic cable is the most expensive, but carries no noise and is very flexible and small. Fiber-optic cable consists of glass fibers that transmit data and voice by light signals. It transmits no electrical noise because it carries data and voice not by an electrical signal but by light signals. It is more expensive because it requires special terminal equipment to convert electrical signals to light signals and back again at the ends of the fiber-optic cable. However, it can carry more data at higher speeds than either twisted pair or coax cables.

Shielding

Ancient knights used shields to absorb the impact of swords, while today's police officer often wears a bulletproof vest to shield against the impact of a bullet. Modern electrical cables use electric shielding for a similar reason. They use shielding to reduce the effect of unwanted electrical noise. Electrical shielding absorbs or reflects electromagnetic interference (EMI) or radio-frequency interference (RFI) noise. Shielding of a communications cable consists of a metallic mesh that surrounds the data-carrying conductor. It performs two functions: First, it absorbs the noise emitted from the communication line, thus shielding equipment and power lines from receiving unwanted electrical noise. Second, it reflects noise emitting from nearby power lines or ground conductor and keeps the noise from affecting the communications signal.

There are other methods for shielding equipment from noise. They include metal conduits, surrounding a room with a metal shield, and spraying the inside of an equipment case with conductive paint. The metal conduits of power and communications cables can act as shields.

Surrounding a room with a metal shield is an effective way to keep unwanted signals from entering or leaving a room or enclosure containing sensitive electronic equipment. Even spraying the inside of a sensitive equipment case, like a hearing aid case, with conductive paint can be an effective shield against the RFI noise radiating from nearby fluorescent light ballasts. [This method of hearing aid shielding stopped a person's hearing aid from receiving and amplifying the signal from overhead fluorescent lights but caused the person's jaw to vibrate (see Power Electronics Applications Center's *Power Quality Testing Network Solution,* August 1995, Bulletin No. 4 for more information).]

Cable shielding is most effective when it is grounded both at the sending and receiving ends. This provides a low-impedance path for the unwanted signal. Next to proper wiring, correct grounding of power and communication cables and wires is the most cost-effective power quality solution.

Grounding Solutions

Many power quality problems are solved by the use of a solidly grounded system for sensitive electronic equipment. A solidly grounded system must provide a low-impedance path to ground. It is often difficult to provide a solidly grounded system for both power and communications circuits. What determines whether an electrical system is solidly grounded with a low-resistance path for power current and low-impedance path for noise signals to ground? What are the characteristics of a solidly grounded electrical system? What are the configurations of grounding systems? How should equipment be connected to these grounding systems?

A solidly grounded system for sensitive electronic equipment must provide a ground for power and a reference for signals, or a signal reference grid (SRG). There are basically three types of grounding electrode systems: (1) ground rods, (2) ground rings, and (3) ground grids. How does each one of these systems provide a solidly grounded electrical system? How do you use use these grounding systems to solve power quality problems without causing new problems?

Ground rods

Ground rods provide an effective low-resistance path to earth. Their effectiveness is dependent on the material, depth, and configuration of each rod and the number of rods, as well as the resistivity of the earth. They are usually made of metal, cylindrical in shape, and driven 8 to 10 ft into the ground. As can be seen from the dashed lines in Figure 5.25, the ground rod electrically connects to the earth by increasingly larger concentric cylinders emanating from the ground rod.

The resistivity of the earth depends on the temperature, moisture content, and chemical composition of the earth. Resistivity refers to the amount of resistance measured in ohms-linear centimeter (Ω-lin cm) of soil. Table 5.1 shows how the resistivity of topsoil and sandy loam vary according to the moisture content of the soil.

Table 5.2 provides an example of how sandy loam's resistivity varies with temperature.

Sometimes the resistivity of the soil is too great for the ground rod to be effective. Salting the earth with chemicals improves the effectiveness of ground rods. The three methods of applying metallic salts to the earth for the purpose of reducing the resistivity of the soil and improving the effectiveness of ground rods are: (1) the trench system, (2) the basin system, and (3) container system. Figure 5.26 illustrates these methods.

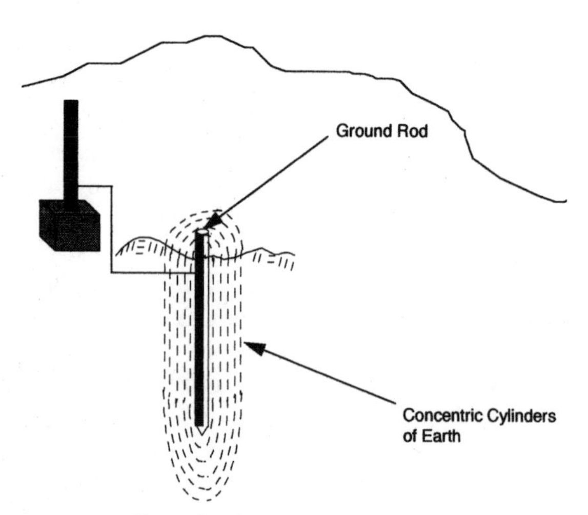

Figure 5.25 Ground rod.

TABLE 5.1 Soil Resistivity Dependency on Moisture Content*

	Resistivity, Ω-cm	
Moisture content, % by weight	Topsoil	Sandy loam
0	1,000,000,000	1,000,000,000
2.5	250,000	150,000
5	165,000	43,000
10	53,000	18,500
15	31,000	10,500
20	12,000	6,300
30	6,400	4,200

*Reprinted with permission from EC&M's "Practical Guide to Quality Power for Sensitive Electronic Equipment."

TABLE 5.2 Sandy Loam Soil (15.2% Moisture) Resistivity Dependency on Temperature*

Temperature		
°C	°F	Resistivity, Ω-cm
20	68	7,200
10	50	9,900
0	32 (water)	13,800
0	32 (ice)	30,000
−5	23	79,000
−15	14	330,000

*Reprinted with permission from EC&M's "Practical Guide to Quality Power for Sensitive Electronic Equipment."

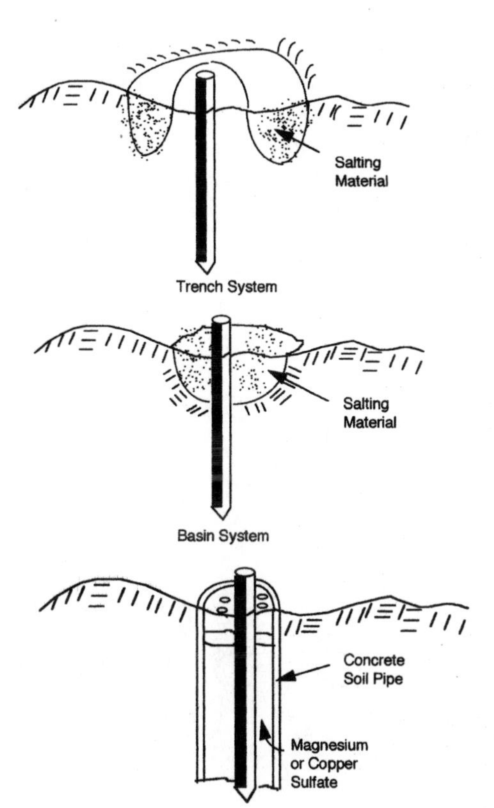

Figure 5.26 Salting methods for reducing earth's resistivity.

Ground ring

An exterior ground ring encircling a building provides a low-impedance path from the building's grounding system to the earth. It usually connects to ground rods at each corner of the building and possibly at the midpoint of the building. It usually consists of a no. 2 gauge conductor buried 2 ft below the ground or below the frost line and completely circling a building. IEEE Standard 142-1991, "IEEE Recommended Practice for Grounding of Industrial and Commercial Power Systems" (*The Green Book*) says "One of the most effective ground electrode systems is a ground ring tied to the building steel at suitable intervals."

Ground and signal reference grids

Ground grids are an effective way to provide grounding for 60-Hz power and high-frequency signals. At the 60-Hz power frequency, they reduce the ground current and magnetic density by spreading the ground current throughout the grid. At high frequencies, the ground grids keep the length of wire short and impedance low between the equipment bond and ground. This keeps the ground leads from resonating or becoming transmitting antennas for radio-frequency (RF) noise. They are used inside buildings and substations to solidly ground equipment.

In an office building, they are used effectively as signal references to ground and accessed through removable panels in the floor. They can be constructed two ways. One way, shown in Figure 5.27, is to place a grid of conductors below the raised floor in an office.

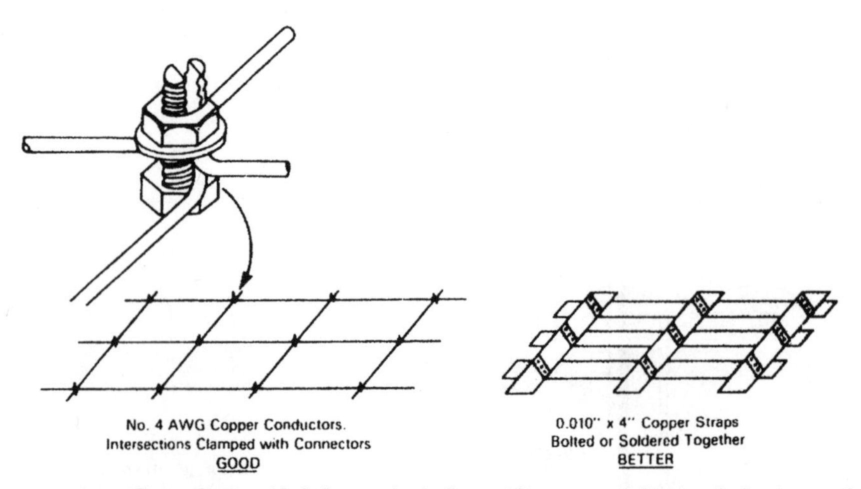

Figure 5.27 Grounding grid below raised floor. (*Courtesy of National Institute of Standards and Technology.*)

Another way is to use the risers for the raised floor as the support for a ground grid. This results in a 2- × 2-ft ground and signal reference grid. In an office environment, the grid is bolted to each riser, as shown in Figure 5.28.

Other grounding systems

In addition to the previously described grounding systems, most electrical installers use the cold-water pipes or metal columns as grounds and sometimes use a grounding plate. They ground all metal enclosures that surround electrical conductors for safety and electronic equipment performance. They sometimes encase grounding electrodes in concrete. They also use isolated grounds in an attempt to solve power quality problems.

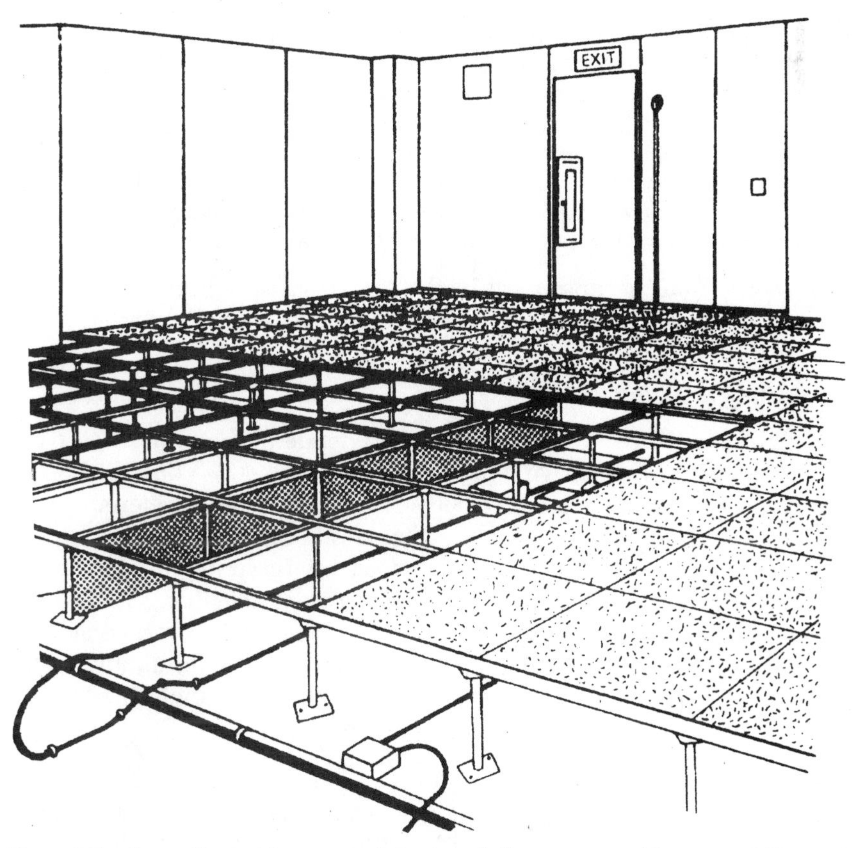

Figure 5.28 Grounding grid as part of the raised floor support. (*Courtesy of National Institute of Standards and Technology.*)

Isolated grounds

Electrical installers often misunderstand and misuse isolated grounds. They incorrectly connect computer equipment to separate grounds. They should instead provide an insulated conductor (green with a yellow stripe) that goes from an electrical outlet to the service entrance panelboard via a separate ground lead. They should not connect it to any conduit or the panelboard enclosure. The isolated ground is supposed to provide a means for reducing the electrical noise from entering sensitive electronic equipment via the grounding conductor. It does this by providing an effective ground path for noise from the connected equipment back to the power source at the service entrance. A receptacle painted orange or marked by an orange triangle identifies the isolated ground at the outlet. In regard to isolated ground receptacles, *NEC* 250-75, Exception No. 4* notes that:

> Where required for the reduction of electrical noise (electromagnetic interference) on the grounding circuit, a receptacle in which the grounding terminal is purposely insulated from the receptacle mounting means shall be permitted. The receptacle grounding terminal shall be grounded by an insulated equipment grounding conductor run with the circuit conductors. This grounding conductor shall be permitted to pass through one or more panel boards without connection to the panel board grounding terminal as permitted in Section 384-20, Exception so as to terminate within the same building or structure directly at an equipment grounding conductor terminal of the applicable derived system or service.

Most power quality experts prefer the term *insulated ground* to *isolated ground.* Figure 5.29 illustrates the wiring for an isolated ground.

Multipoint grounding

IEEE Standard 1100-1992, "IEEE Recommended Practice for Powering and Grounding of Sensitive Electronic Equipment" (*The Emerald Book*) recommends that all metallic objects crossing the single reference grid (SRG) be bonded to it, as shown in Figure 5.30. This is called *multiple-point grounding. The Emerald Book* does not recommend single-point grounding even though it is a common practice in telephone companies.

Multiple connections of neutral to ground should be avoided according to the *NEC*. They can provide multiple paths for faults and cause protective devices to misoperate. The solution is a single neutral-to-ground connection at the service entrance.

---

*Reproduced with permission from NFP-70-1999, *National Electric Code,®* copyright © 1998, National Fire Protection Association, Quincy, MA 02269. This reprinted material is not the completed and official position of the NFPA on the referenced subject, which is represented only by the standard in its entirety.

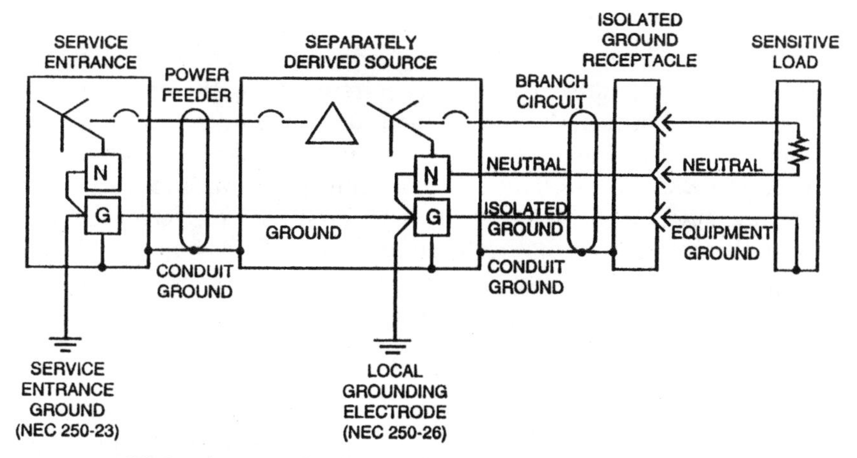

Figure 5.29 Wiring for an isolated ground. (*Courtesy of IEEE, Standard 1100-1992, Copyright © 1993 IEEE. All rights reserved.*)

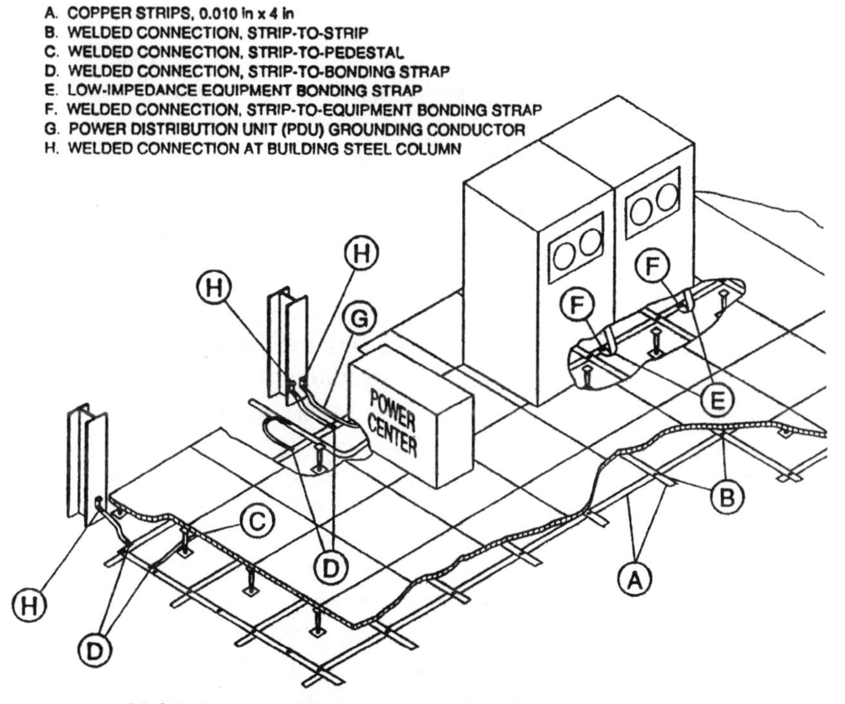

Figure 5.30 Multipoint grounding to a signal reference grid. (*Courtesy of IEEE, Standard 1100-1992, Copyright © 1993 IEEE. All rights reserved.*)

Separately derived source grounding

Separate grounding of sensitive electronic equipment can be accomplished without violating the *NEC* requirements by the use of a separately derived source. Chapter 4 mentions that an effective way to reduce unwanted noise is to provide a separately derived source from an isolation transformer. This isolation transformer provides an opportunity to install a new grounding system for sensitive electronic equipment without violating the *NEC* requirements. This can be applied to a single-phase as well as a three-phase system. Otherwise, separate grounding systems should be avoided.

Conclusions

Proper wiring and grounding to avoid or solve power quality problems is complicated by the fact that electricity behaves differently at the power frequency of 60 Hz than at the high frequencies required in communications systems. This fact must be taken into account when trying to solve power quality problems. Another complicating issue is the need to meet safety as well as power quality requirements. With sensitive electronic equipment it is important not only to ground equipment to meet the safety requirements of the *National Electrical Code* but also to meet the power quality requirements of FIPS 94 and the IEEE *Emerald Book.* Most power quality problems are caused by poor wiring and grounding practices (some surveys estimate 80 to 90 percent). It is a good idea to look at the wiring and grounding of a facility before deciding whether to buy and install power conditioning equipment. That is why most power quality experts recommend examining the wiring and grounding of a facility as the first step in performing a power quality survey or audit.

A power quality survey is critical to identifying power quality problems and their solutions. Even readers who are power quality experts will find it helpful to read about the fundamental requirements of a power quality survey. Experts can then better explain the steps of a power quality survey to their clients. Even nonexperts can decide to do their own minisurvey. Or they may decide to hire a power quality expert and need to oversee the performance of the power quality survey. What is a power quality survey? How do you perform a power quality survey? These and other questions will be answered in Chapter 7. But before performing a power quality survey, everyone needs to know how to pick and use the diagnostic tools for measuring power quality problems. These diagnostic tools are the various types of instruments for measuring and monitoring power quality disturbances. They are presented in Chapter 6.

References

1. Martin, Marty. "Two Modern Power Quality Issues—Harmonics and Grounding." URL address: *http://www.copper.org/pq/issuess.htm.* Available from Copper Development Association Inc., New York.

2*a*. Michaels, Kenneth. 1994. "Effective Grounding of Electrical Systems—Part 1." *EC&M Electrical Construction & Maintenance,* vol. 93, no. 1, January, pp. 47–51.

2*b*. ———. 1994. "Effective Grounding of Electrical Systems—Part 2." *EC&M Electrical Construction & Maintenance,* vol. 93, no. 2, February, pp. 59–63.

2*c*. ———. 1994. "Effective Grounding of Electrical Systems—Part 3." *EC&M Electrical Construction & Maintenance,* vol. 93, no. 6, April, pp. 57–62.

2*d*. ———. 1994. "Effective Grounding of Electrical Systems—Part 4." *EC&M Electrical Construction & Maintenance,* vol. 93, no. 4, June, pp. 67–70.

3. Lewis, Warren, and Frederic Hartwell. 1996. "Quality Grounding and Power Quality." *EC&M Electrical Construction & Maintenance,* vol. 95, no. 2, February, pp. 33–38.
4. Shaughnessy, Tom. 1998. "Types of Grounding Systems." *Power Quality Assurance,* vol. 9, no. 4, July/August, pp. 74–75.
5. Bush, William. 1991. "Telecom System Fundamentals." *Power Quality,* vol. 2, no. 6, November/December, pp. 30–37.
6. Shaughnessy, Tom. 1996. "Facility and Equipment Grounding." *PQ Today,* vol. 3, no. 1, summer, pp. 6, 7.
7. "Ground Loop Basics," 1999. URL: *http://www.hut.fi/Misc/Electronics/docs/groundloop/basics.html.*
8. Kowalczyk, Stan W. 1992. "Root Out the Silent Effects of Electrical Noise." *Chemical Engineering,* vol. 99, no. 6, June, pp. 145–148.
9. Melhorn, Chris. 1997. "Flickering Lights—A Case of Faulty Wiring." *PQ Today,* vol. 3, no. 1, Summer, p. 4.
10. Waggoner, Ray. 1993. "Lightning Disruption through Earth." *EC&M Electrical Construction & Maintenance,* vol. 92, no. 8, August, pp. 16–18.
11. Knisley, Joseph R. 1995. "Establishing an Electrostatic Discharge Control Program." *EC&M Electrical Construction & Maintenance,* vol. 94, no. 13, December, pp. 72–72.
12. Lafdahl, Craig S. 1996. "The Case of the Triple Threat." *EC&M Electrical Construction & Maintenance,* vol. 95, no. 2, February, pp. 90–91.
13. Lewis, Warren. 1996. "Understanding IG Receptacles (Insulated Grounding—Part 1)." *EC&M Electrical Construction & Maintenance,* vol. 95, no. 1, January, pp. 14–17.

13*b*. ———1996. "Understanding IG Receptacles (Insulated Grounding—Part 2)." *EC&M Electrical Construction & Maintenance,* vol. 95, no. 2, February, pp. 14–17.

Chapter

6

Power Quality Measurement Tools

Solving any diagnostic problem requires the right tools and the ability to use them. Like doctors trying to solve a health problem, power quality engineers and technicians need meters and other measurement tools to solve electrical facility health problems. The first step in solving a power quality problem is to determine the cause of the problem. Making either visual inspections or electrical measurements of the electrical distribution system can do this. Chapter 7, "Power Quality Surveys," shows how to make visual inspections of a facility's electrical distribution system. This chapter, "Power Quality Measurement Tools," explains how to perform and analyze power quality measurements using power quality measurement tools.

As shown in Figure 6.1, there are a myriad of power quality measurement tools available today. They include instruments that measure and display the basic electrical parameters of voltage, current, frequency, and impedance of an electrical distribution system. These tools include ammeters, voltmeters, multimeters, oscilloscopes, flicker meters, electrostatic voltmeters, infrared detectors, radio-frequency interference and electromagnetic interference meters, harmonic and spectrum analyzers, power quality monitors, and various types of wiring and grounding testers. These instruments measure, display, and store electrical parameters for the purpose of helping solve power quality problems. In addition to these electrical measurement tools, there are devices, such as video cameras and audiotape recorders, for recording the effects of power quality problems. With all these choices, power quality experts as well as novices must know how to choose and use the right instrument. How

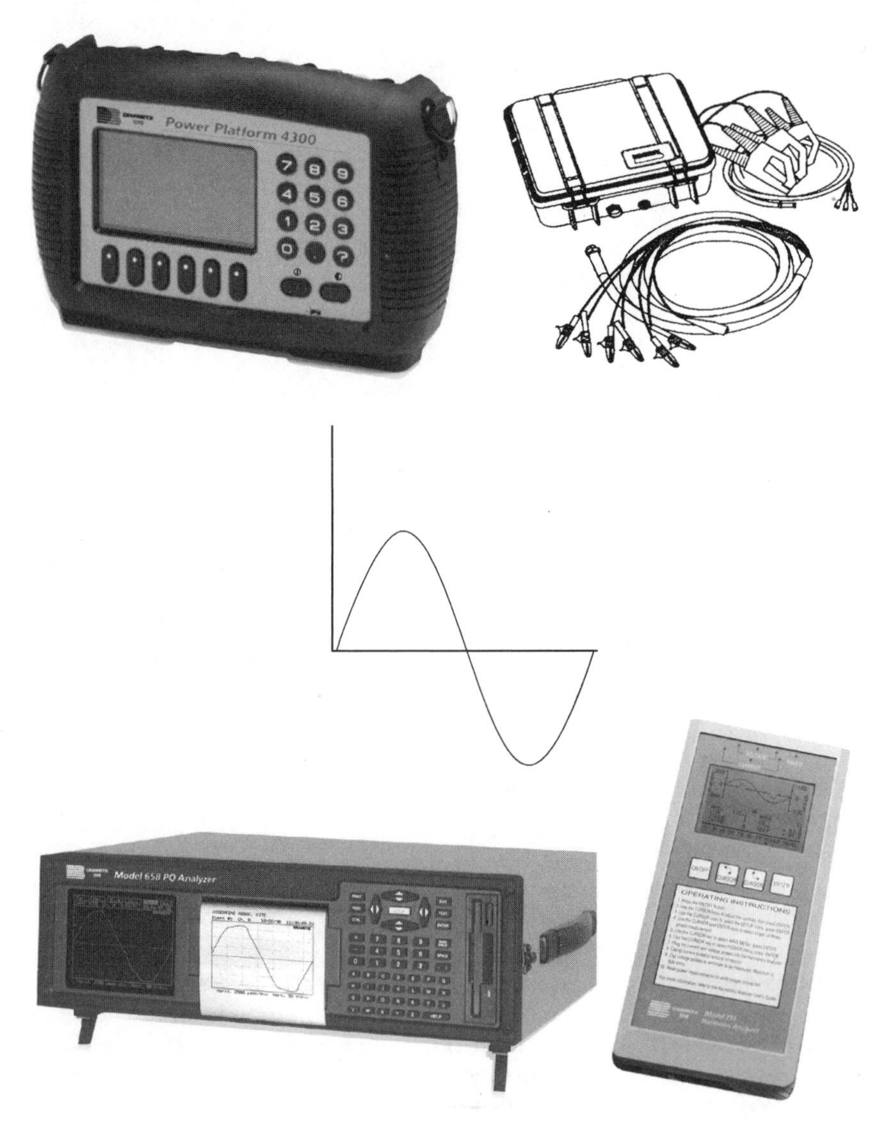

Figure 6.1 Measurement tools. (*Courtesy of Dranetz-BMI.*)

to choose the right tool to match a particular power quality problem seems like a difficult problem itself.

Knowing how to choose the right measurement tool is a three-step process. It first requires knowing the various types of power quality problems discussed in Chapter 2. They include voltage swells, voltage sags, various types of interruptions, overvoltage, undervoltage, harmonics, and transients. Secondly, it requires knowing the various types of instruments to measure those disturbances. The three primary types of instruments are multimeters, oscilloscopes, and analyzers especially designed to measure and record power quality disturbances. Thirdly, it requires knowing how to match the instrument to the power quality problem, as shown in Table 6.1.

How do these instruments work? How do they differ from standard, more familiar types of electrical meters like the kilowatt-hour meter?

Kilowatt-Hour Meter

The most familiar and fundamental meter is the watt-hour or kilowatt-hour meter located at the service entrance to a utility customer's house, office, or factory. This is called a revenue meter because it displays and records the amount of electrical energy the electrical company charges its customers. It measures the amount of electrical energy in kilowatt-hours consumed in each facility each month. It usually is an analog meter, and is called an analog meter because it uses the current and voltage to directly move the meter dials. The most common type, which has been in use for nearly 100 years, is the Ferraris kilowatt-hour meter. This type of kilowatt-hour meter uses an ac motor with two windings, one for voltage and one for current, to move the meter dials. The torque produced by the voltage and current causes the conducting disk mounted between the two windings to rotate. The number of revolutions of the rotating disk represents the amount of electrical energy consumed over a certain period of time. The rotating disk causes the meter dials to move and display the amount of kilowatt-hours being used. This is based on the principle that kilowatt-hours are equal to the current multiplied by the voltage multiplied by the time in hours divided by 1000. Figure 6.2 illustrates a standard kilowatt-hour meter.

Other types of standard nondigital (analog) electrical meters include ammeters that measure the current flowing in a wire in amperes, voltmeters that measure the voltage between two points in volts, and ohmmeters that measure the resistance in a wire in ohms. Often voltmeters combine with ohmmeters to form voltohmmeters or VOMs, as shown in Figure 6.3.

The two basic types of probes include current and voltage. Current probes, similar to the ones shown in Figure 6.4, convert voltmeters to ammeters. They surround a wire and through inductive action measure the current flowing in the wire. Voltage probes, similar to the one in Figure 6.5, connect to a meter or oscilloscope and attenuate the voltage to an acceptable level.

TABLE 6.1 Matching Measurement Tools to Type of Disturbance

Disturbance description	True rms multimeter	Oscilloscope	Harmonic analyzer	Disturbance analyzer	Infrared detector	Gauss meter	Static meter	Flicker meter
Impulsive transients		X		X				
Oscillatory transients			X					
Sags/swells		X		X				
Interruptions		X		X				
Undervoltages/overvoltages		X		X				
Harmonic distortion			X	X	X			
Voltage flicker								X
Static discharge							X	
Noise	X					X	X	
Wiring and grounding problems	X				X			

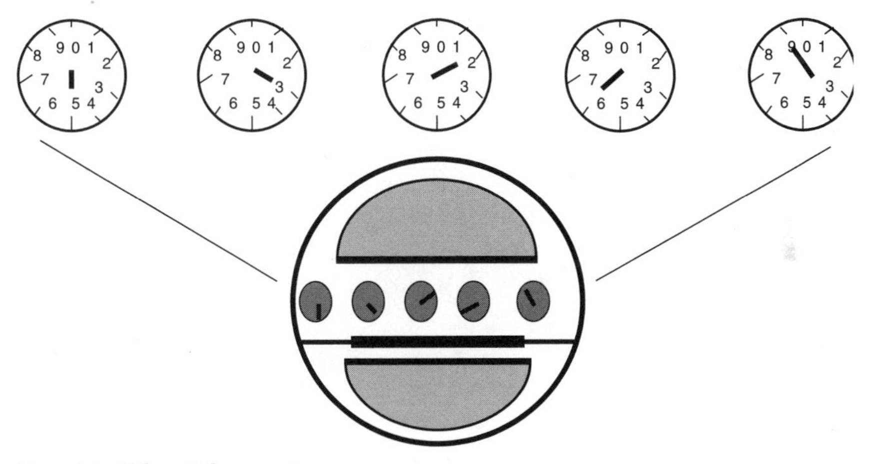

Figure 6.2 Kilowatt-hour meter.

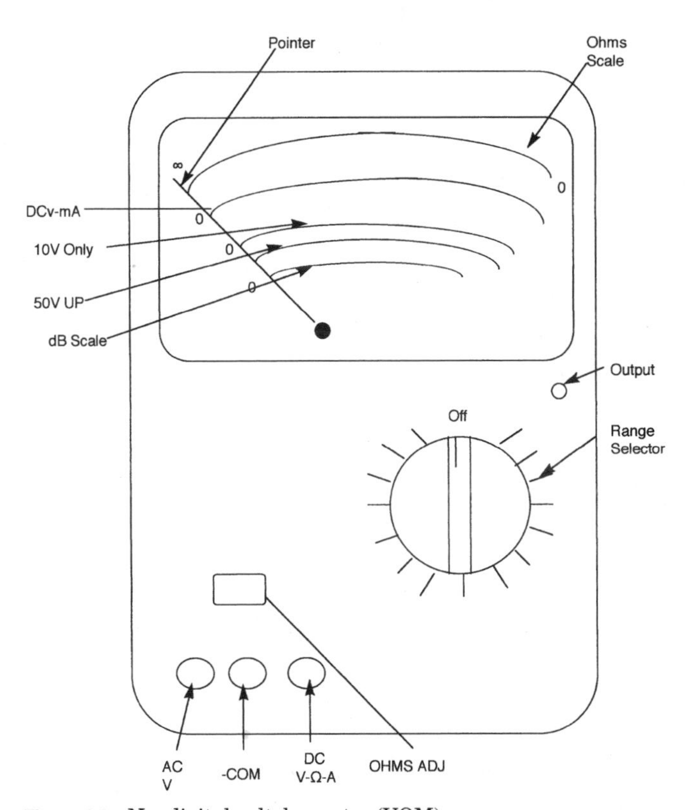

Figure 6.3 Nondigital voltohmmeter (VOM).

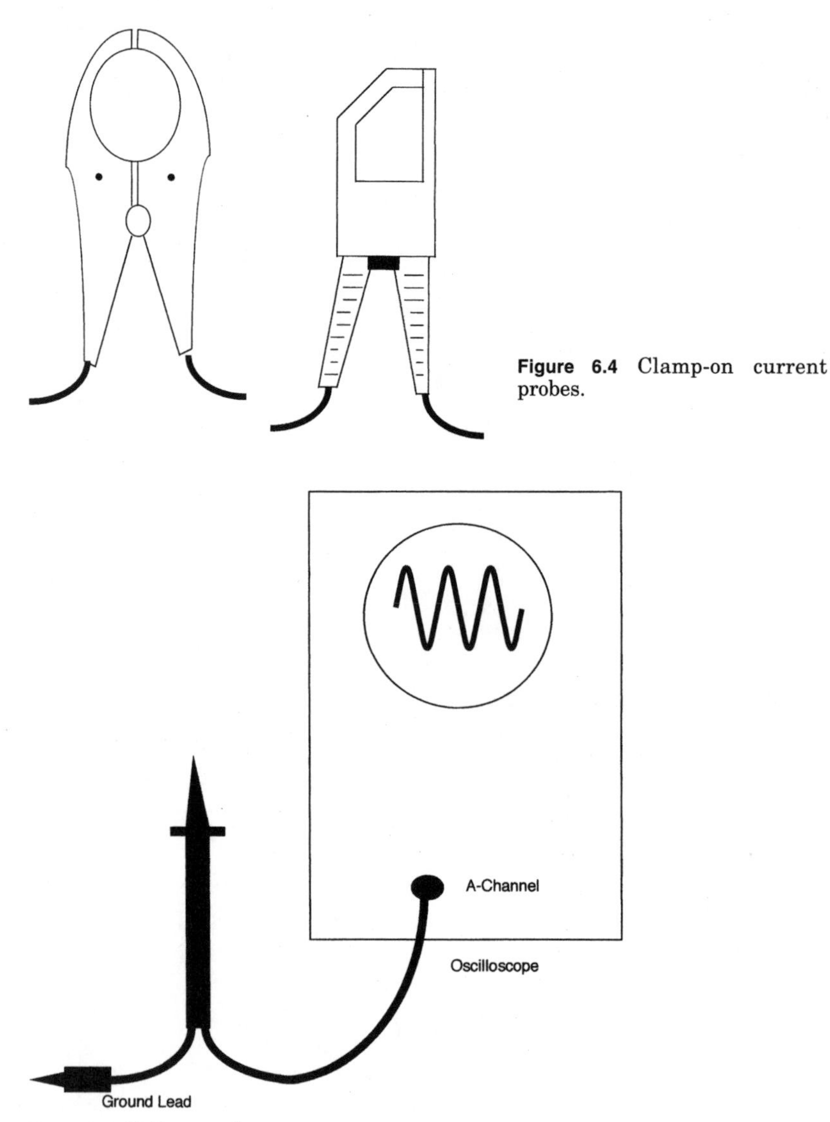

Figure 6.4 Clamp-on current probes.

Figure 6.5 Voltage probe.

The multimeter has replaced the VOM. It combines the ammeter, voltmeter, and ohmmeter into one meter.

Multimeters

Most multimeters today are digital and are often referred to as DMMs, or digital multimeters. Digital meters are more accurate and reliable than analog meters. Digital multimeters use an analog-to-digital converter to convert the electrical quantity being measured into a signal that can be displayed on a digital readout, as shown in Figure 6.6.

The constantly changing nature of alternating current poses a measurement problem to the design of multimeters. As shown in Figure 6.7, the alternating current magnitude begins at zero, increases in value, and reaches a maximum value before returning to zero and continuing downward to a negative peak value before increasing in value again. At 60 Hz, the alternating current repeats this cycle 60 times every second. Should multimeters measure the total area under the alternating current waveform? No, because they would always give a meaningless measurement of 0 A. Or should multimeters measure the peak value of the alternating current? It depends on the purpose of the measurement. Most users of multimeters want to measure the heating effects of current on electrical equipment and conductors. They want to know whether the equipment or conductors can take the heat. Therefore, multimeters need to measure the current that is proportional to its heating effect.

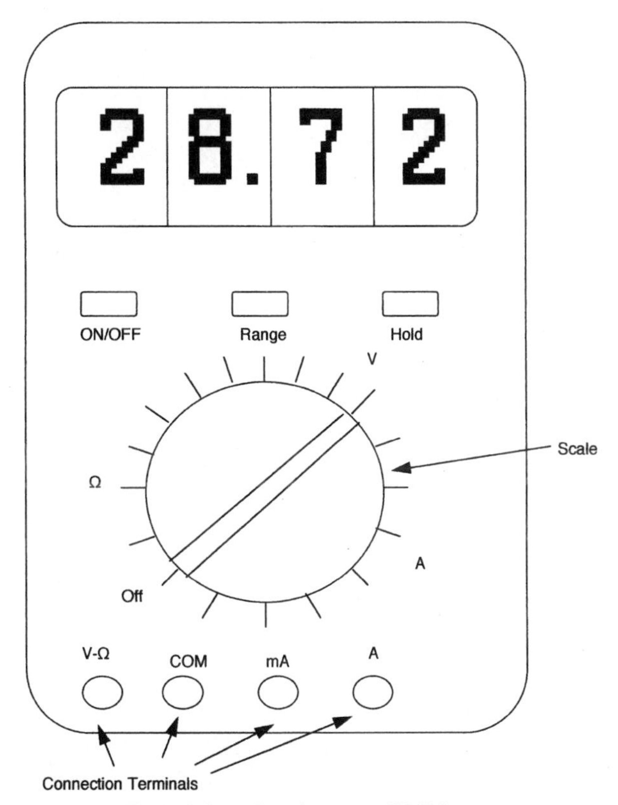

Figure 6.6 Typical digital multimeter (DMM).

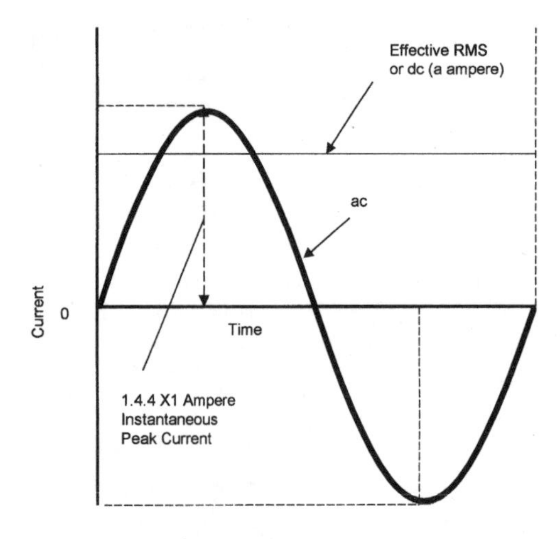

Figure 6.7 Current sine wave components.

Consequently, meter manufactures have designed multimeters to measure effective, or root-mean-square, amperes.

What is an effective ampere? An *effective ampere* is the alternating current (ac) equivalent to the direct current (dc) value of the sine wave. Both ac and dc have the common effect of generating heat when they flow through a resistor. The heat from dc is directly proportional to the magnitude of the dc losses. DC losses equal the dc amperes squared times the resistance in ohms, i.e., I^2R, with current represented by the letter I and resistance represented by the letter R. An effective ac ampere is equal to the amount of heat produced by a dc ampere flowing in the same resistor. In other words, one ac effective ampere or root-mean-square ampere flowing in a resistor will emit the same amount of heat as one dc ampere flowing in the same resistor. Effective amperes are called root-mean-square amperes because they equal the square root of the sum of the squares of the instantaneous values of the sine wave, as shown in the following formula:

$$I_{\text{effective}} = \sqrt{(I_1^{\,2}) + (I_2^{\,2}) + (I_3^{\,2}) + (I_4^{\,2}) + \ldots} \tag{6.1}$$

where I_1 represents the 60-Hz current and the subsequent currents represent current values at other frequencies. This formula shows that the effective current equals the average of the instantaneous current in the case of a pure sine wave. Average-responding multimeters assume a pure sine wave.

Standard multimeters or average-responding meters display the peak value of the electrical current and voltage or the average root-means-square (rms) value. They perform this measurement in two steps. First, they determine the rms amperes or volts by sampling the instantaneous voltage or current over one cycle and averaging it.

Second, they calculate the peak value by simply multiplying the average value by 1.414, or $\sqrt{2}$. The second step is based on the fact that in a perfect sine wave, as shown in Figure 6.7, the peak value equals 1.414 or $\sqrt{2}$ times the average value of the sine wave.

However, power quality problems do not fit into nice 60-Hz sine waves. By their very nature power quality problems involve some distortion of the sine wave. Power quality disturbances, like harmonics, sags, or swells in the voltage and current, distort the sine wave. The correct measurement tool for a power quality problem must accurately measure the characteristics of a distorted sine wave. Consequently, as shown in Figure 6.8, a multimeter user who wants to solve power quality problems must first avoid an average-responding meter and select, instead, a true rms meter. What is a true rms meter? How does a true rms meter differ from the average-responding, or peak rms, meter?

Average-responding versus true rms multimeters

As their name implies, true rms multimeters measure the "true" rms of a distorted sine wave. How do they accomplish this? They either use the heating effect of the voltage across a resistor or sample the signal's waveform with a microprocessor, calculate the rms value, and display the true rms value. Average-responding and peak-value multimeters, on the other hand, do not measure the true rms value of a distorted sine wave. They sample values of the alternating current over a cycle, determine the average value of the sine wave, and convert it to effective amperes or rms amperes. They convert alternating current to rms amperes by multiplying the average value of the waveform by 1.414 ($\sqrt{2}$) if they use the averaging method or 0.707 if they use the peak method. Average-responding rms meters measure distorted waveforms with readings that are 25 to 50 percent below the actual rms values. As shown in Figure 6.9, the average rms method

Figure 6.8 True rms digital multimeters. *(Reproduced with permission of Fluke Corp.©)*

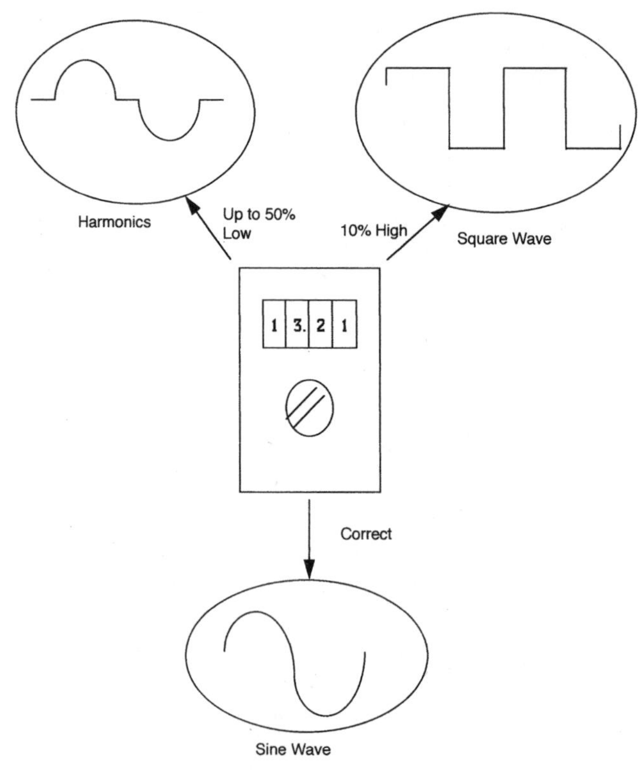

Figure 6.9 Average-responding rms inaccuracies.

results in inaccurate measurements of a distorted waveform because it measures the waveform over time and misses distorted waveform peaks. Even though true rms meters may cost twice as much as average-responding meters, only true rms meters provide accurate measurements of distorted sine waves, like those containing harmonics. Figure 6.10 shows the different rms measurements of harmonics in the same circuit from true and average-responding rms meters. True rms meters differ in their capabilities, such as measuring crest factor and bandwidth.

Crest factor and bandwidth

Selection of a multimeter with the wrong crest factor can cause inaccurate measurements of current and voltage. Crest factor equals the ratio of a waveform's peak or crest to its rms voltage or current. It provides an important description of a sine wave. It measures the maximum sine wave current or voltage applied to a particular piece of equipment. The crest factor for a sinusoidal wave always equals 1.414,

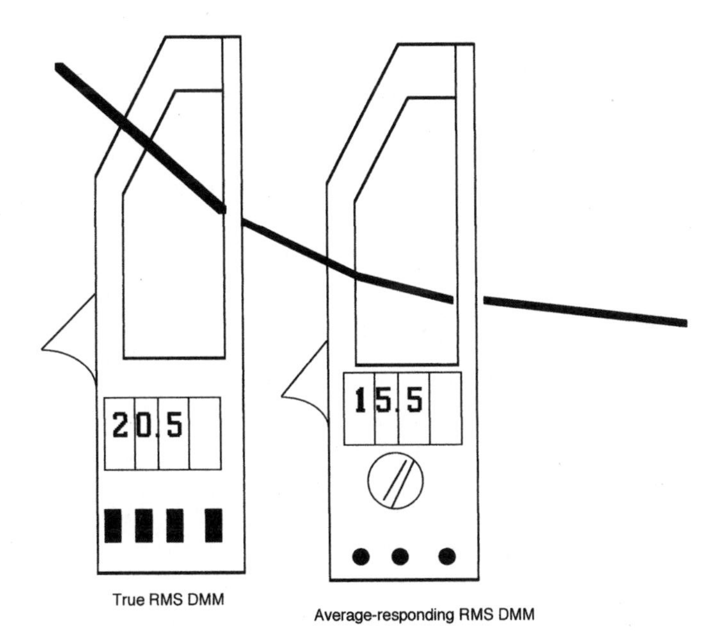

Figure 6.10 True rms versus average-responding DMM readings.

while the crest factor for a nonsinusodial wave will differ from one wave to another. Therefore, select a multimeter that measures a signal's crest factor accurately.

The crest factor of a multimeter can limit the true rms measurement. For example, harmonics typically have peaking signals with values higher than those of 60-Hz sine waves. Consequently, nonsinusoidal waves have higher crest factors than the crest factor of 1.414 for sine waves. Crest factors for true rms multimeters can vary from a low of 2 to a high of 7. Newer digital multimeters generate peak current, rms current, and crest factor readings at the press of a button.

Bandwidth or frequency response of the true rms is another important factor to take into consideration when selecting a true rms multimeter. Manufacturers of multimeters design multimeters to measure voltage and current within a certain frequency range or bandwidth. Selection of a multimeter to measure waveforms with frequencies outside the capability of the multimeter will result in incorrect measurements.

If a meter is a true rms meter, the front panel should read "true rms." Its performance specifications should provide the crest factor and bandwidth capability of the meter. Set the crest factor value in the midrange to get the best results.

Other selection considerations

Other considerations for selecting a multimeter include its ability to handle physical and electrical extremes. Manufacturers of multimeters have designed them to withstand various levels of harsh environments. They increase the cost of multimeters proportionate to their ruggedness. They have also designed them to withstand various levels of voltage and current spikes. They recommend selecting multimeters that match the expected voltage and current spikes in order to avoid destroying the multimeter and creating a safety hazard.

Multimeter users typically use clamp-on multimeters to measure current. They find them simple to use. Users like clamp-on multimeters because they do not have to disconnect any wires to perform current measurements. As shown in Figure 6.10, they simply open and close the clamp around the current-carrying conductor.

How does the clamp-on multimeter measure current? It measures the alternating current using either the current transformer or the Hall-effect method. In the current transformer method, the multimeter surrounds a conductor with a coil of wire that picks up the alternating current in the conductor. As shown in Figure 6.11, a transformer reduces the current to a magnitude that can be measured by an ammeter. In the Hall-effect method, the multimeter measures the current that passes through a semiconductor in the presence of a magnetic field.

Manufacturers of multimeters have developed graphical multimeters that combine the accuracy of a digital multimeter with the graphical display of an oscilloscope. They display the current and voltage waveform on a small screen, as shown in Figure 6.12. Power quality inspectors find that the graphic display helps them diagnose power quality problems. They also use another tool that provides a graphic display: the oscilloscope.

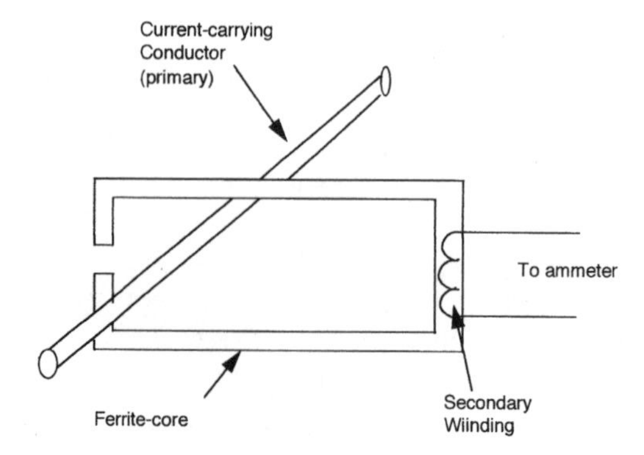

Figure 6.11 Current transformer method.

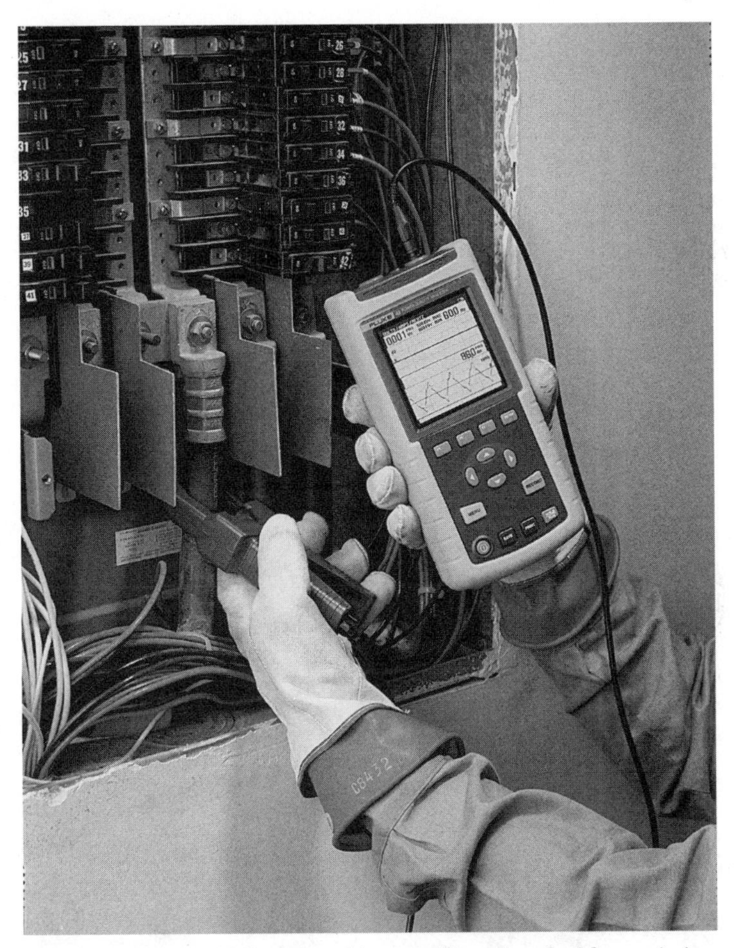

Figure 6.12 Graphical display multimeter. (*Reproduced with permission of Fluke Corp.©*)

Oscilloscopes

A German physicist by the name of Karl Braun invented the oscilloscope in 1897. He discovered that he could change the green fluorescence formed on a cathode-ray tube (TV tube) to follow the electromagnetic field of a varying current. He found that this image, called a *trace,* shows graphically the oscillations (thus the name oscilloscope) of alternating voltage. He invented one of the most valuable diagnostic tools for the power quality engineer or technician.

Oscilloscopes have progressed significantly since Braun's invention. Tektronix introduced the modern oscilloscope, Tektronix Model 511, in 1946. The first oscilloscopes used analog technology in which the electron beam traced on the oscilloscope's screen directly traced the input voltage's waveform. More recent oscilloscopes use digital technology that samples the waveform and uses an analog-to-digital converter to reconstruct the waveform on the oscilloscope's screen. In the past, analog technology had the advantage over digital technology of being able to sample the waveform at a higher rate with a lower-cost oscilloscope. Currently, lower-priced and smaller digital oscilloscopes can display sudden changes in voltages that occur with power quality disturbances. Today's digital oscilloscopes have the ability to store the voltage waveform for later display and analysis. They provide the user the ability to analyze the signal's frequency, i.e., spectrum analysis, and even make energy calculations. They are small enough to be held in a person's hand. They provide input to a personal computer for analysis of the display utilizing analytical software. Some manufacturers combine the features of analog and digital technology in one oscilloscope, as shown in Figure 6.13.

Oscilloscopes use both voltage and current probes. Voltage probes reduce the voltage to a level that oscilloscopes can measure. Current probes use current transformers to convert current to a voltage that standard oscilloscopes can measure. Oscilloscopes display voltage waveforms as a function of time and cannot measure current directly. They convert the voltage into a current display by using Ohm's law ($I = V/R$) and measure the voltage across a known resistance on the secondary side of the current transformer. Figure 6.14 shows how

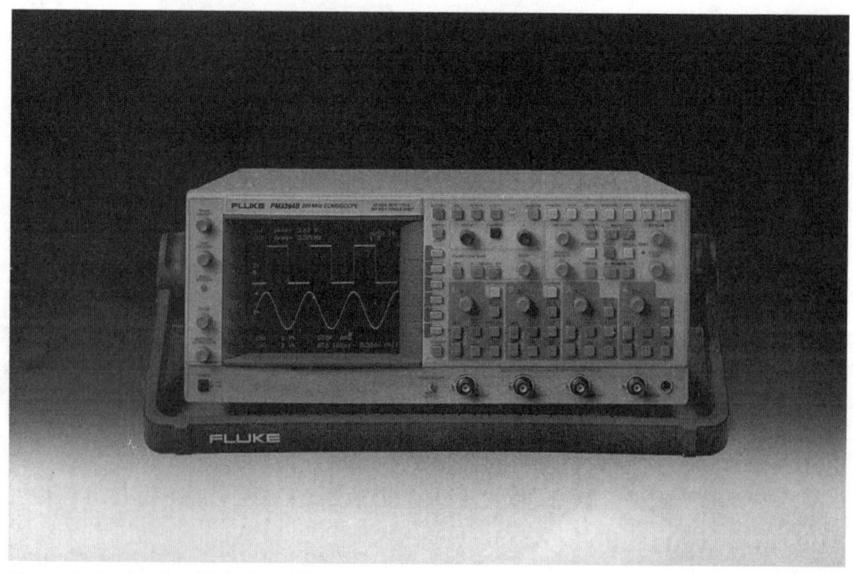

Figure 6.13 Analog-digital oscilloscope. (*Reproduced with permission of Fluke Corp.©*)

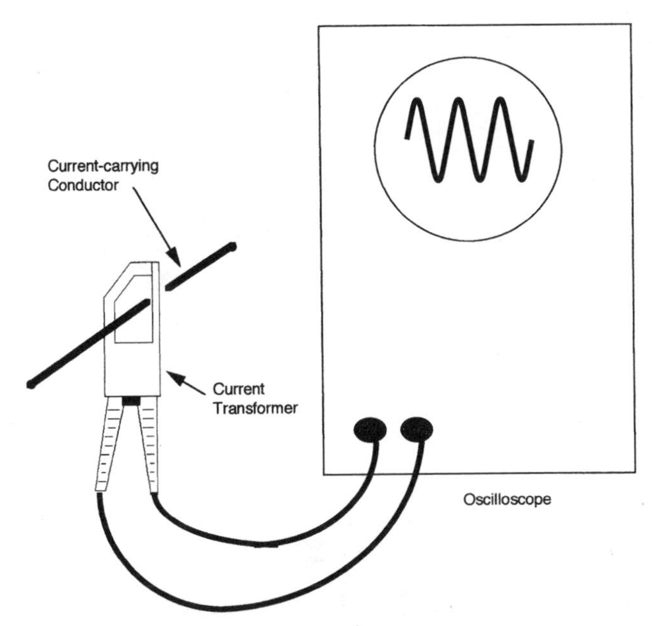

Figure 6.14 Oscilloscope measuring current.

an oscilloscope uses a current transformer to measure the current in a wire.

To select a digital oscilloscope properly, the user has to know the application. Oscilloscopes are categorized by bandwidth, rise time of the signal, and sample rate. Oscilloscopes should have a bandwidth 3 to 5 times the bandwidth of the signal being measured.

Disturbance analyzers

Disturbance analyzers provide measurements similar to oscilloscopes. However, they display information specifically needed to analyze power quality disturbances. They measure, store, and display a wide range of disturbances from voltage sags to voltage swells, as well as short-term transients. Whether installed permanently or temporarily, they measure and record disturbances. They capture the waveform and store it magnetically on a hard drive and display it graphically on paper, as shown in Figure 6.15.

Users have several choices of the method for retrieving information from disturbance analyzers. They can retrieve the waveform at the site of the meter via a floppy disk or remotely by a modem and a telephone line. As shown in Figure 6.16, they can use recent analyzers equipped with new software to notify them of a disturbance by numerical readings or a beeping signal on a pager. They can call a telephone

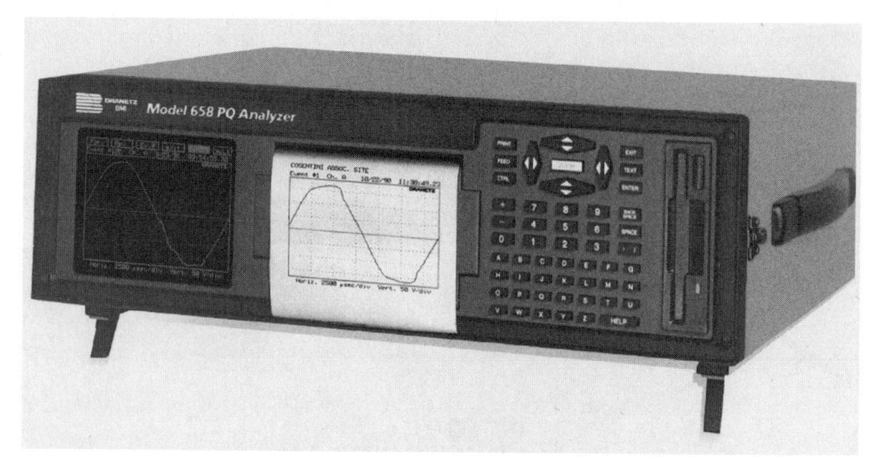

Figure 6.15 Disturbance analyzer. (*Courtesy of Dranetz-BMI.*)

Figure 6.16 Power quality pager. (*Courtesy of Dranetz-BMI.*)

number connected to the disturbance analyzer. They call the telephone number and receive a computer-generated verbal summary of the disturbances. Or they can obtain disturbance information from especially designed disturbance analyzers remotely via the Internet using a standard Web browser.

Disturbance analyzers typically cannot measure harmonics without special accessory equipment. Analyzers with special accessory equipment for measuring harmonics are called *harmonic analyzers*.

Harmonic Analyzers

Harmonic analyzers have several capabilities. They capture harmonic waveforms and display them on a screen. They calculate the *K* factor to derate transformers and the total harmonic distortion (THD) in percent of the fundamental (see Chapter 2 for an explanation of THD). They also measure the corresponding frequency spectrum, i.e., the harmonic frequency associated with the current and voltage up to the fiftieth harmonic. They display the harmonic frequency on a bar graph or as the signal's numerical values. Some measure single-phase current and voltage while others measure three-phase current and voltage. All of them measure the power factor (PF). The power factor provides a measurement of how much of the power is being used efficiently for useful work. Some can store data for a week or more for later transfer to a PC for analysis. This makes them powerful tools in the analysis of harmonic power quality problems.

Harmonic analyzers come as small hand-held units like the one shown in Figure 6.17 for on-the-spot power quality surveys or as larger power quality monitors for long-term or permanent installation. They offer the same retrieval capabilities that were described for the disturbance analyzers using floppy disks, pagers, or the Internet.

Power factor measurement

The power factor measurement capability is usually included with a harmonic analyzer. Power factor is determined by dividing the power reading by the product of the voltmeter and ammeter readings.

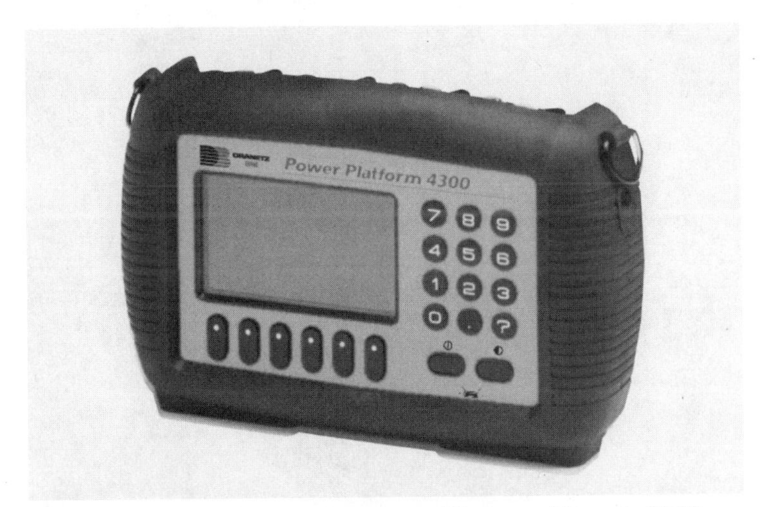

Figure 6.17 Hand-held harmonic analyzer. (*Courtesy of Dranetz-BMI.*)

Static Meters

There are two basic ways to measure an electrostatic discharge. One way requires a person to scrape his or her feet on the floor and touch a metal object to see if there is a discharge from a finger to the object. Not a good approach because it takes more than 4000 V for the human body to sense a shock. It takes much less than 4000 V to fry microprocessor-based equipment. Even an electrostatic voltage as small as 25 V can cause damage to microprocessors. A better way to measure an electrostatic discharge requires the use of an electrostatic discharge voltmeter. What is an electrostatic discharge voltmeter? How does it differ from a standard voltmeter?

An electrostatic discharge voltmeter, as shown in Figure 6.18, differs from a standard voltmeter in its ability to measure the static voltage without transferring the charge to the voltmeter. Electrons from a metal object attract the protons on a person's body and cause a static discharge. This often causes an electrostatic arc from the object to the person. Similarly, when a voltage probe touches an electrostatically charged body, it causes the static voltage to discharge to the probe and nullify any voltage measurements. Therefore, an electrostatic discharge voltage probe must measure static surface potential (voltage) without physical contact to the static charged object. How does it do that?

One must first place the electrostatic discharge voltage probe near the electrostatically charged surface. The capacitance between the voltage probe and the metal object transmits a small alternating cur-

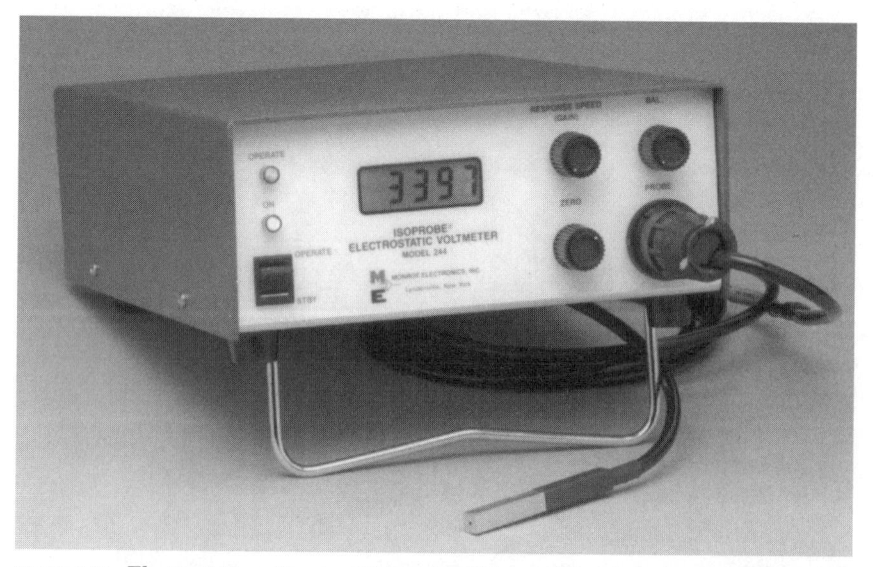

Figure 6.18 Electrostatic voltmeter. (*Courtesy of Monroe Electronics.*)

rent to the voltage probe. The voltage of the probe charges up to the same unknown voltage on the surface of the metal object. A light-emitting diode circuit senses the voltage and displays its value.

Electric Field Strength and Magnetic Gaussmeters

Electric field strength meters provide measurements of the electric field emanating from electrical equipment or conductors, while magnetic gaussmeters measure the strength of the magnetic field in gauss.

A digital multimeter acts like a field strength meter when its input/output jacks connect to a small-gauge wire looped 30 times. Figure 6.19 shows the loop in the area of the magnetic field. The lines of magnetic flux induce a voltage in the loop and register the induced voltage on the digital multimeter.

Infrared Detectors

Ironically, infrared detectors help identify conservation opportunities as well as locate power quality problems. The irony comes from the fact that many conservation technologies cause power quality problems. How do infrared detectors help find power quality problems?

Infrared detectors, as shown in Figure 6.20, detect overheated electrical components. Harmonics, a loose connection, unbalanced loading on conductors, and triplen harmonics in the neutral conductor all cause components to overheat. The detectors display the heat from these sources by first converting the radiated heat energy into electric current. Next, they amplify the current and convert it into an analog or digital display in degrees or British thermal units per hour.

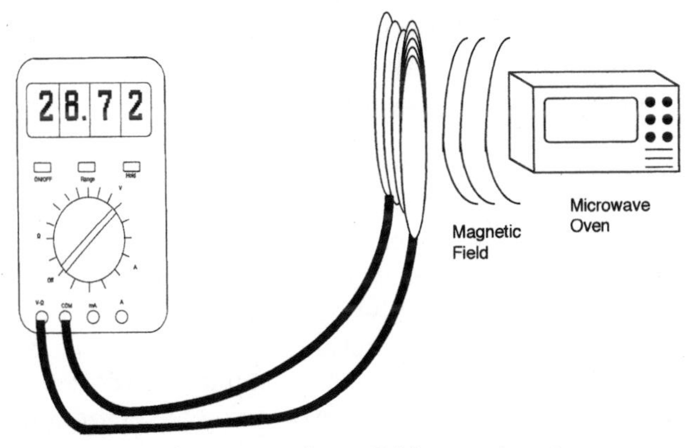

Figure 6.19 Multimeter used as a field strength meter.

Figure 6.20 Infrared detector (thermometer). (*Reproduced with permission of Fluke Corp.*©)

Infrared imagers or cameras provide a picture of the heat radiating from electrical components and the surrounding area. A black-and-white TV screen then shows temperature differences in various shades of gray. The tones become lighter as the temperature increases. A white image on a black-and-white TV screen identifies the overheated component. Alternatively, a color infrared camera and screen show the overheated component in red. A recording of the infrared images on videotape allows later analysis. Applications of these detectors include preventive maintenance as well as diagnostic analysis.

Flicker Meters

Flicker meters measure flicker in terms of the fluctuating voltage magnitude and its corresponding frequency of fluctuation. Electric arc furnaces and arc wielding usually cause lights to flicker. How to convert the voltage and the frequency of fluctuation into a standard parameter that defines the flicker limit becomes a problem. The difficulty comes from correlating the frequency of the flicker to what the human eye detects. Flicker tests illustrate this problem.

Results of flicker tests depend on two variables. The first variable is the subjective reaction of the person involved in the test. The second variable is whether the flickering light is incandescent or fluorescent. The frequency range of fluctuations identified by the human eye varies from 1 to 30 Hz. Consequently, the subjectivity of the flicker tests makes it difficult to develop flicker standards.

Presently, the power quality industry lacks an international standard on flicker. Many utilities have developed with their customers their own standards. Both the Institute of Electrical and Electronics Engineers (IEEE) and the International Electrotechnical Commission (IEC), have developed flicker curves for incandescent lamps, as shown in Figure 6.21.

The IEC has developed a flicker meter. Besides the IEC flicker meter, several instrument manufacturers sell flicker meters, like the one shown in Figure 6.22, commercially. They use software to convert the flicker voltage fluctuations into statistical quantities called P_{st} and

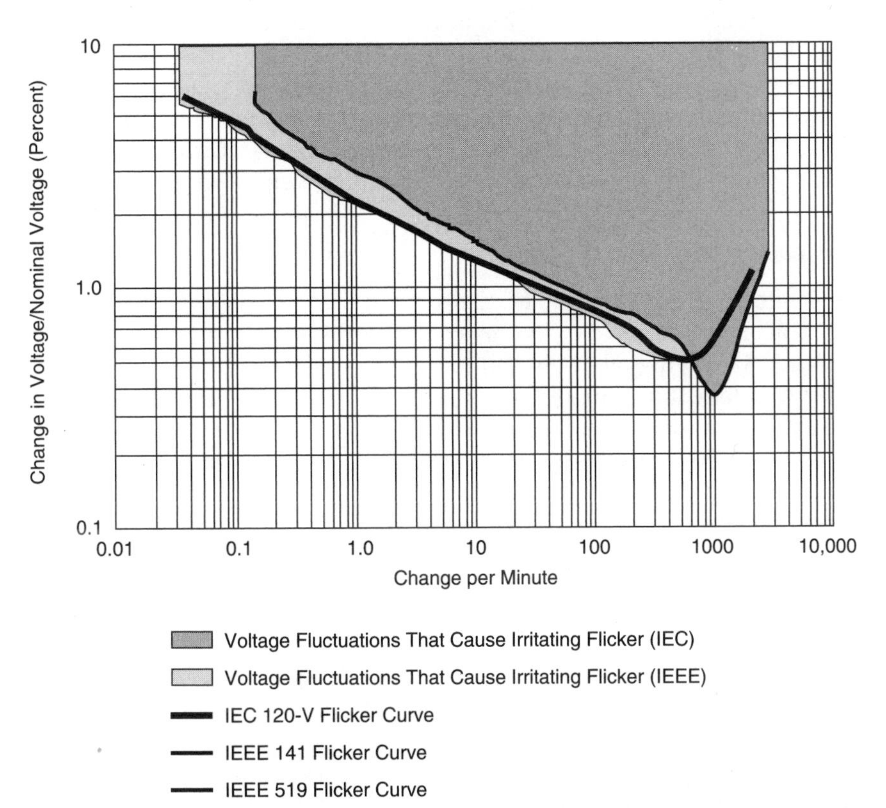

Figure 6.21 IEC and IEEE flicker curves for incandescent lights. (*Courtesy of EPRI, PQTN Case No. 1, "Light Flicker Caused by Resistive Welder."*)

P_{lt}. P_{st} is the short-term flicker severity index, while P_{lt} is the long-term flicker severity index. Flicker meters take measurements automatically at 10-min intervals. A single P_{st} is calculated every 10 min. A P_{st} greater than 1 indicates that the flicker will irritate 50 percent of the people exposed to it. The P_{lt} is a combination of 12 P_{st} values. IEC Standard 1000-3-7 has set standards for P_{st} and P_{lt} for medium voltage (MV) of less than 35 kV, high voltage (HV) of greater than 35 kV but less than 230 kV, and extra-high voltage (EHV) of greater than 230 kV, as shown in Table 6.2.

Wiring and Grounding Instruments

Wiring and grounding power quality problems require specialized measurement tools. Several types of measurement testers detect and identify the cause of a particular wiring and grounding problem. These instruments provide the power quality surveyor information on whether a circuit is open or incorrectly connected. The three main types of testers are the receptacle circuit (three-lamp circuit), ground impedance, and earth ground testers. They measure the impedance of the equipment grounding conductor. They detect isolated ground shorts, neutral-to-ground bonds, incorrect ground and neutral impedance, open grounds, open neutral, open hot wire, and reversed polarity. They also determine phase rotation and phase-to-phase voltages.

Receptacle circuit testers

Receptacle circuit testers, or three-lamp circuit testers, plug into a receptacle and measure wiring connections to the receptacle. They indicate the wiring errors in the receptacles by a combination of lights. They cost less than any other tester on the market. Some power quality experts doubt their accuracy. In fact, the "IEEE Recommended Practice for Powering and Grounding Sensitive Electronic Equipment," page 32 (*The Emerald Book*), says "These devices have some limitations. They may indicate incorrect wiring, but cannot be relied upon to indicate correct wiring."

TABLE 6.2 IEC Standard 1000-3-7 Flicker Levels*

	Planning levels	
Flicker symbol	MV	HV–EHV
P_{st}	0.8	0.9
P_{lt}	0.7	0.6

*Courtesy of IEC. See Ref. 2, p. 206.

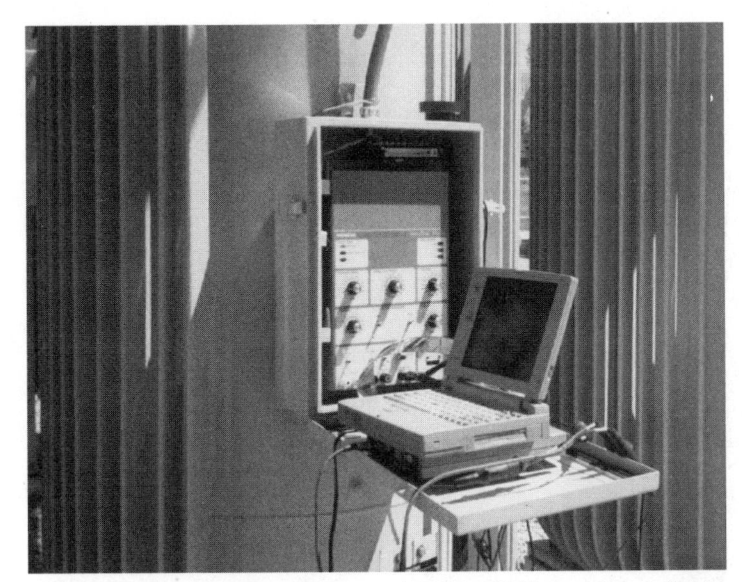

Figure 6.22 Flicker meter. (*Courtesy of EPRI, PQTN Case No. 1, "Light Flicker Caused by Resistive Welder."*)

Ken Michaels in the December 1998 issue of *EC&M* magazine doubts their usefulness in an article titled "Three-Lamp Circuit Tester: Valid Tester or Night-Light?" In this article, he concludes that circuit wiring capacitance and leakage current make these devices unreliable and no better than a night light. He recommends, instead, a ground impedance tester.

Ground circuit impedance testers

Ground impedance testers, shown in Figure 6.23, measure the impedance of a circuit from the point of test to the bond between the neutral and ground bond. Some can handle 120-V ac single-phase voltage while others can handle 600-V ac three-phase voltage. They also measure voltage and determine the presence of neutral-to-ground connections, isolated ground shorts, reversed polarity, and an open equipment grounding conductor.

Earth ground testers

Earth ground testers measure the ground electrode and earth resistance. They use the fall-of-potential method to measure the ground resistance. They determine the resistance using Ohm's law by passing a known current through an unknown resistance and measuring the voltage, i.e., fall of potential. They can also function as voltmeters. Figure 6.24 illustrates an earth ground tester.

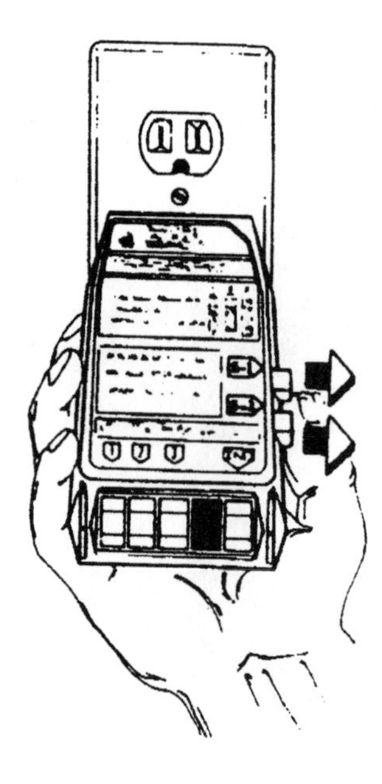

Figure 6.23 Ground circuit impedance tester. (*Courtesy of ECOS Instrument, Inc.*)

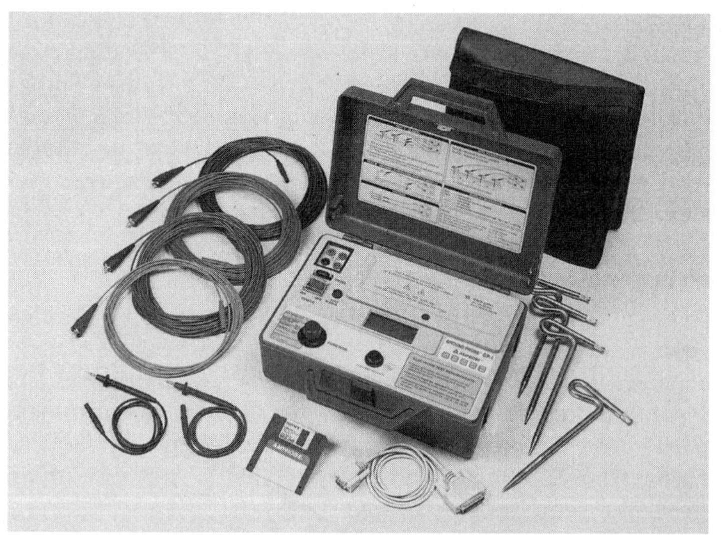

Figure 6.24 Earth resistance tester. (*Courtesy of Amprobe.*)

Permanent Power Quality Monitoring

The deregulation of the electric utility industry raises several questions about power quality measurement tools. What kind of new tools will be needed in the deregulated utility industry? Will power quality measurement tools used for power quality surveys be sufficient in the deregulated electric utility industry? Will the utilities need tools not only to troubleshoot power quality problems but to permanently monitor the quality of power continuously? What tools will end users need to determine the quality of power the utilities supply them? The characteristics and sensitivity of end-user equipment within customer facilities ultimately define power quality requirements. Improving the energy efficiency and productivity of industrial and commercial facilities can sometimes result in the use of technology that either causes power quality problems or becomes sensitive to power quality variations. Historically, utilities have monitored power quality problems only when their customers complain.

In the deregulated industry the roles of utilities and their customers become blurred. What are the power quality requirements at the interface between the transmission company and the distribution company? What base level of power quality should the distribution company provide its end-use customers? What kinds of enhanced power quality services can the energy service company offer to end-use customers? A permanent power quality monitoring program between the different entities resulting from the utility industry deregulation would answer many of these questions. This section describes the progression of power quality monitoring as the electric utility industry restructures and becomes more competitive.

Need for power quality monitoring

The same three factors that increase the need for solving and preventing power quality problems also increase the need for power quality monitors, as shown in Figure 6.25. They include the increasing use of power quality–sensitive equipment, increasing use of equipment that generates power quality problems, and the deregulation of the power industry. All these factors influence the utilities and their customer's competitiveness.

First, utilities need power quality monitoring when their customers use present-day highly sensitive computer and computer-controlled equipment that requires a power source of higher quality and more reliability than standard, less-sensitive electricity-consuming equipment. Traditionally, utilities did not get involved in power quality problems that occurred on a customer's system unless contacted. When contacted, the utility's approach was to determine the cause of the problem and who caused it. This approach no longer works in a competitive environment. Many utility customers expect high-quality power without problems. Many utilities and their customers find a permanent power quality monitoring system an effective tool that prevents and solves power quality problems. Although utilities can cause power quality problems, various surveys indicate that their customers cause 80 percent of their own power quality problems.

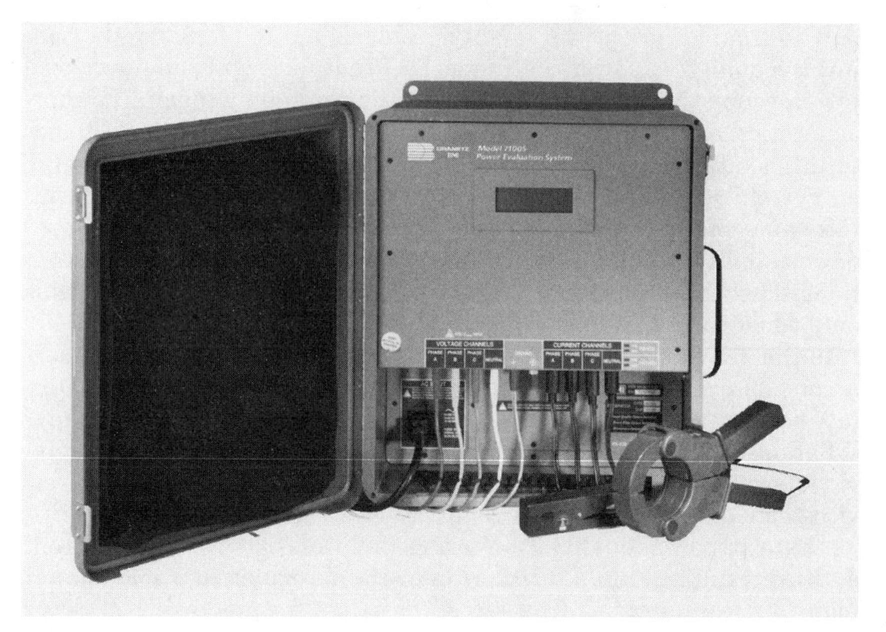

Figure 6.25 Permanent power quality monitor. (*Courtesy of Dranetz-BMI.*)

Second, programs to improve the efficiency of production have resulted in the use of nonlinear equipment, like adjustable-speed drives, or power factor–improving shunt capacitors. These devices often generate or amplify existing harmonics that distort the voltage wave. These distortions can get on the utility's system and affect interconnected customers. In addition, increased use of arc furnaces causes voltage flicker that in turn causes lights to flicker and irritate people. New types of loads, such as electron beam furnaces for melting titanium and induction furnaces for processing aluminum, generate harmonic voltage distortion. Large-horsepower motors cause voltage dips and large inrush currents during start-up. All these types of loads result in one customer causing power quality problems for another customer. Utilities cannot afford to have their customers causing power quality problems. This affects the utilities' and their customers' competitiveness. Utilities need to identify the customers causing the power quality problems and require them to fix it. Utilities often find permanent power quality monitoring an effective tool that helps prevent and solve power quality problems caused by their customers. The need for power quality monitoring will become even more intense when the utility industry becomes deregulated.

Third, the deregulation of the utility industry will cause many customers to choose utilities that can supply power that is high in quality as well as low in cost. Consequently, utilities will retain existing customers and attract prospective customers by showing them that they can deliver power with high quality. Utilities with power quality monitoring systems will need to convince existing and prospective customers that they see power quality not as a problem but an opportunity to provide customer service and become more competitive. Utility customers will become more competitive when they receive power high in quality and reliability. How will the deregulated competitive utility industry affect the evolution of power quality monitoring?

Evolution of power quality monitoring

A comparison of the evolution of power quality monitoring systems to the evolution of utility transmission and distribution systems reveals some interesting similarities. As the need for a more reliable and efficient electrical power system grew, the utility transmission and distribution system increased in complexity and voltage. As the need for higher power quality increases, utilities and their customers need prompt and immediate information about power quality at the transmission, distribution, and end-user levels. Power quality systems have become more complex and sophisticated. Utilities and their customers find it costly and inefficient to install power quality monitors after a power quality problem occurs. They need ongoing up-to-date power quality information. They will need this information even more when the electric utility industry becomes deregulated.

Deregulation's effect on power quality monitoring

The need to determine the source of power quality problems will become imperative when utilities break up into separate companies. Permanent power quality monitoring at the point of interconnection between GENCOs and TRANSCOs, TRANSCOs and DISTCOs, and DISTCOs and end users will be necessary for determining the source of power quality problems. Figure 6.26 shows suggested locations of these power quality monitors on the power system.

Historically, power quality engineers installed monitoring equipment at the point of common coupling. The point of common coupling is the point where the utility connects to the end-user customer. After the monitor collected the data for a week or two, someone would have to go to the site and download the data, take the data back to the office, and analyze it to determine the cause of the power quality problem. This approach was time consuming and inefficient. A more efficient approach involves the use of new types of meters. These meters allow power quality engineers to access the data remotely through a modem

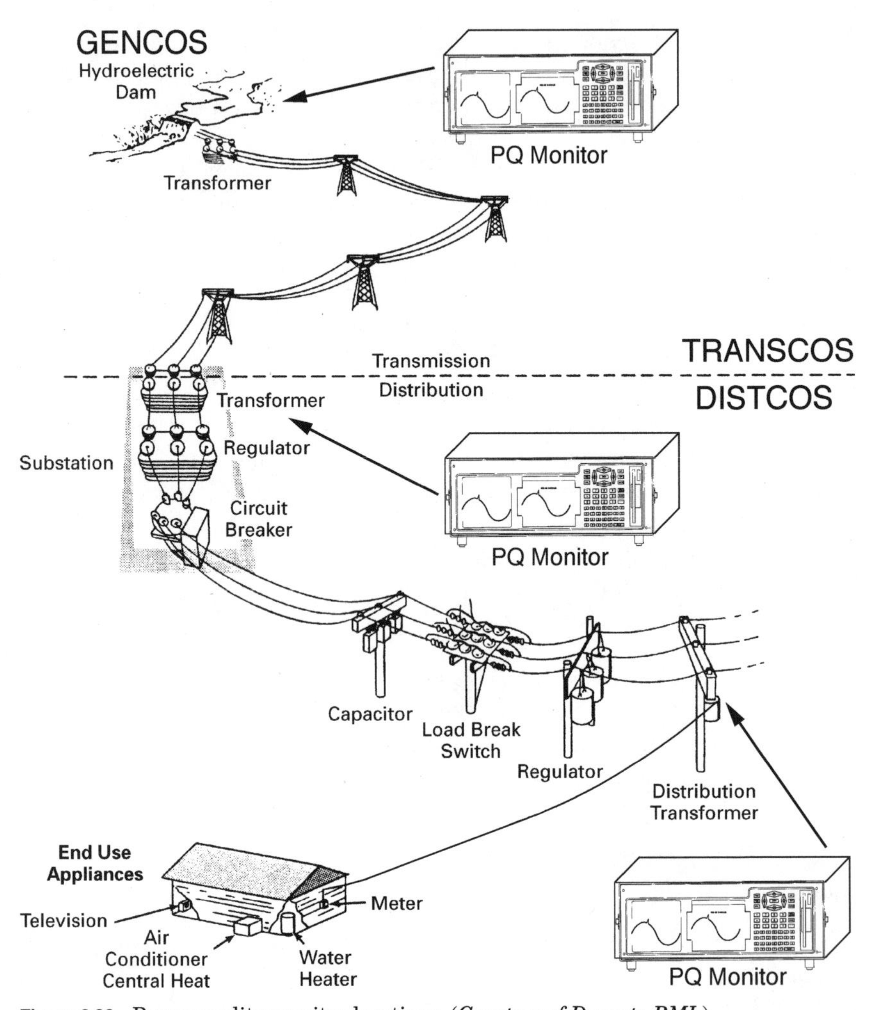

Figure 6.26 Power quality monitor locations. (*Courtesy of Dranetz-BMI.*)

and telephone line. Even this approach has its limitations. It worked when only a few temporary meters were required. Present and future power quality monitoring will require several meters at many sites installed permanently to monitor the power quality for statistical or diagnostic analysis. Engineers can use statistical analysis to determine the relationship between the power system configuration and the power quality to show deviations from power quality standards. They can use this diagnostic analysis to continually track the power quality state of a sensitive load. EPRI realized the need for continual and permanent power quality monitoring and developed an Internet-based power quality monitoring system.

Power quality monitoring system

After EPRI completed a 2-year study of power quality monitors at 250 end-user sites, it realized the need for a power quality monitoring system. EPRI decided to develop a system that would utilize the World Wide Web to transfer data from remote power quality meters installed at several sites. It wanted this system to be easily accessible by power quality engineers and their customers using standard Internet browsers, like Microsoft Explorer and Netscape Navigator. It developed the power quality monitoring system shown in Figure 6.27.

This system includes a server for storing power quality data, download stations for calling up power quality data from remote meters, and analytical software for viewing the data stored on the server. The power quality data comes from various measuring devices displayed in various forms. Figures 6.28, 6.29, and 6.30 show various types of data-reporting formats for statistical and diagnostic analysis.

In a competitive environment, utilities need to show their customers that they can supply power that meets the customers' power quality requirements. Many utilities throughout the world see permanent power quality monitoring as a necessary means to show their

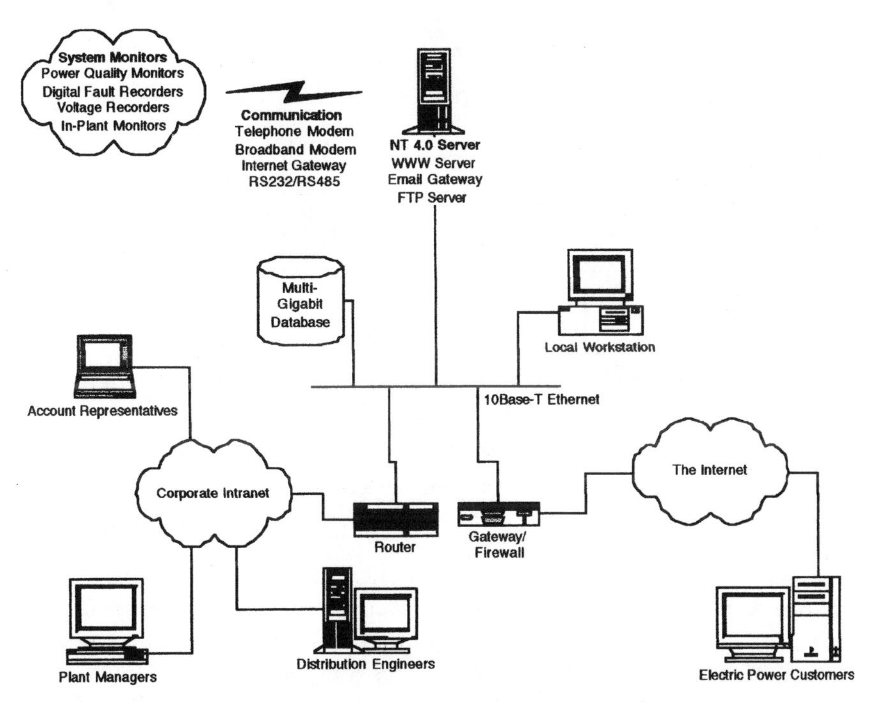

Figure 6.27 Power quality monitoring system using the Internet.

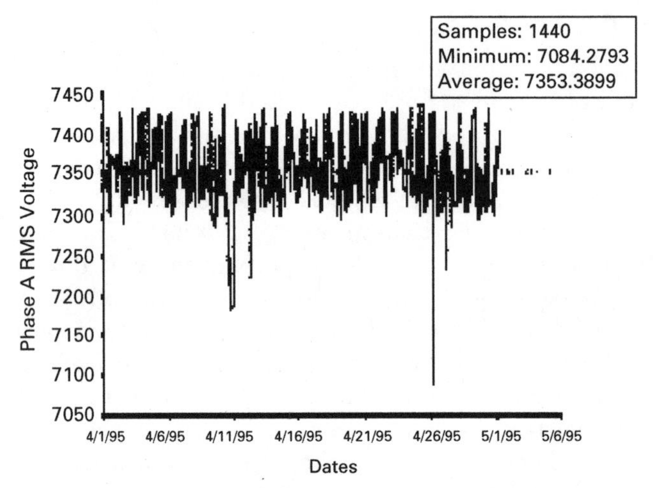

Figure 6.28 Trend of steady-state sampled data. (*Courtesy of Electrotek Concepts, Inc.*)

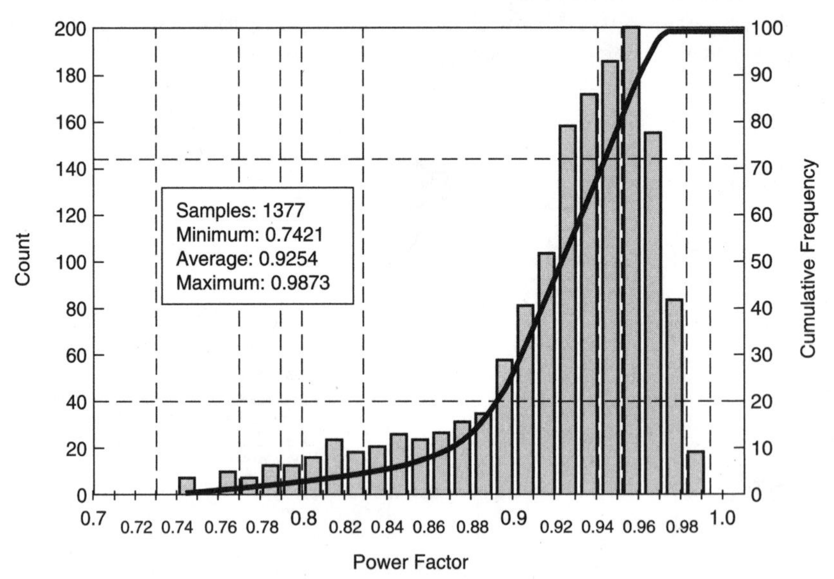

Figure 6.29 Histogram of steady-state sampled data. (*Courtesy of Electrotek Concepts, Inc.*)

customers the level of power quality they are providing. Utilities in France and the United Kingdom use power quality monitoring systems as a measure of power quality and their customers' satisfaction. For example, EDF-DER in France has adopted the Emeraude contract with its customers. In this contract, EDF will compensate its

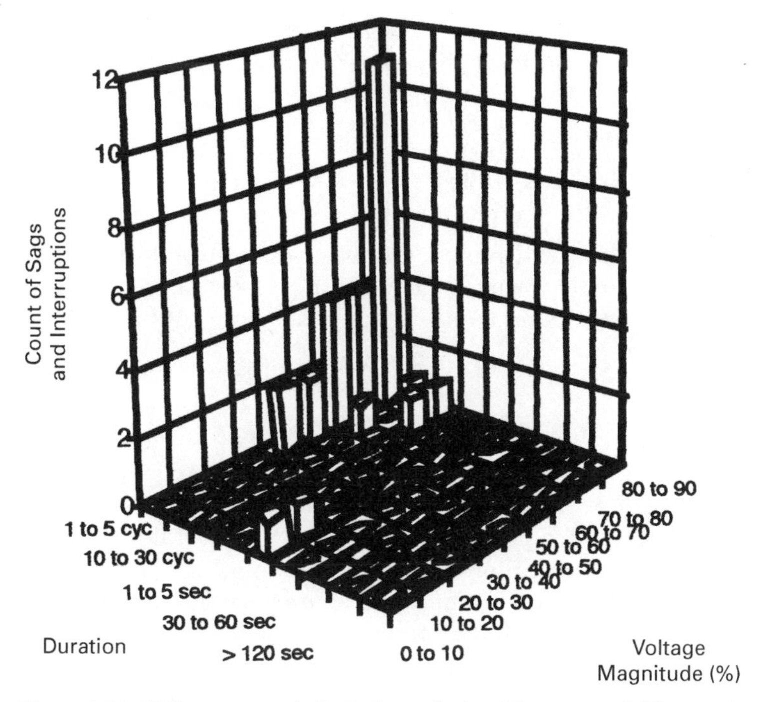

Figure 6.30 Voltage sag statistical analysis. (*Courtesy of Electrotek Concepts, Inc.*)

customers if it exceeds the thresholds agreed to by EDF. In the United Kingdom, the Regional Electricity Company has installed power quality monitoring to determine the quality of power it is providing its customers. Several U.S. utilities and service companies have adopted this approach.

Many utilities and service companies in the United States have purchased power quality monitoring systems. They include Tennessee Valley Authority (TVA), Consolidated Edison, and the Bonneville Power Administration (BPA). TVA has installed 65 meters at various sensitive loads to obtain statistical data on the level of power quality they are providing to their industrial customers. Consolidated Edison has installed power quality monitors throughout its system. BPA has utilized for its customers the power quality monitoring system provided by Electrotek Concepts Inc. and plans to offer access to it for its customers on a pay-as-you-use basis. The City of Richland, Washington, used BPA's power quality monitoring system to monitor one of its customer's adherence to IEEE 519-1992 standards. An energy service company, Western Resources, has purchased and installed meters for utilities and end users to monitor power quality throughout the United States. Utilities and their customers will need permanent power quality systems to assure compliance with power quality contracts in the deregulated and competitive electric utility environment.

Monitoring and Analysis to Evaluate Compliance

Power quality contracts will require some means of evaluating compliance. This will usually involve a combination of system monitoring and analytical tools. System monitoring provides data to characterize system performance.

Monitoring to characterize system performance

Power quality contract participants first require system performance data to determine baseline power quality levels. They obtain this data from the power quality monitors. This is a proactive mode of power quality monitoring. They use the baseline characteristics and ongoing performance measurements to identify problem areas and assure adequate performance.

Monitoring to characterize specific problems

Many power quality service departments or plant managers solve problems by performing short-term monitoring at specific customers or at difficult loads. This is a reactive mode of power quality monitoring, but it frequently identifies the cause of equipment incompatibility. This often provides information that leads to a solution.

Monitoring as part of an enhanced power quality service

Several retail marketers (RETAILCOs) and energy service companies (ESCOs) provide enhanced power quality services as part of their service offerings. Distribution companies (DISTCOs) or transmission companies (TRANSCOs) may enter into contracts with performance-based rates. They may offer differentiated levels of power quality to match the needs of specific customers. A provider and customer can together achieve this goal by modifying the power system or by installing equipment within the customer's premises. In either case, monitoring becomes essential to establish the benchmarks for the differentiated service and to verify that the supplier achieves contracted levels of power quality.

Summary

With the proliferation of electronic loads and the increased competition expected in the deregulated electric utility industry, power quality will become increasingly important. Past utility practice of getting concerned about power quality problems only when customers complain will not work in the competitive deregulated utility environment. With the breakup of vertical integrated utilities into generating, transmission, distribution, and service companies, utilities and their customers will need permanent power quality monitoring to record the source of power quality problems and to prevent problems before they become complaints. Utilities have had to build more complex transmission and distribution systems to deliver power to their customers reliably and efficiently. Utilities will also have to build more complex power quality information systems to deliver power quality information to their customers accurately and effectively. Power quality monitoring systems that utilize the Internet will provide readily available and accurate power quality information to utilities and their customers. Power quality monitoring systems that minimize the amount of worker-hours to maintain and utilize them will provide deregulated and regulated utility companies and their customers a competitive advantage in the deregulated competitive utility industry environment.

The evolution of power quality monitoring has come full circle. The old watthour meter has changed from an analog to a digital meter. It has combined with the features of a power quality monitor to provide not only energy consumption information but power quality data as well. Using microprocessors to record and store power and power quality measurements, some manufacturers have combined the features of a watthour meter and power quality meter into one meter. This allows the utility and the utility's customer to monitor the power use requirements of the customer's facilities not only for revenue purposes but for power quality purposes as well. These meters record both real and reactive power use. They capture power quality measurements, like harmonics, sags and swells, power factor, waveforms, crest factor, and calculate the corresponding K factor, as shown in Figure 6.31.

References

1. Piehl, Dick. 1995. "Using the Right Meter for Power Quality Troubleshooting." *EC&M Electrical Construction and Maintenance,* vol. 94, no. 10, October, pp. 70–73.
2. Lewis, Warren. 1998. "Handheld Instruments and Test Equipment; What to Use and How to Use It (Power Quality Advisor)." *EC&M Electrical Construction and Maintenance,* vol. 97, no. 10, September, pp. PQ 18–23.
3. Williamson, Ron. 1992. "True-RMS Meters and Harmonics (Root-Mean-Square; Discussion of Digital Multimeter Performance in Measuring Nonsinusoidal Signals)." *EC&M Electrical Construction and Maintenance,* vol. 91, no. 4, April, pp. 31–32.

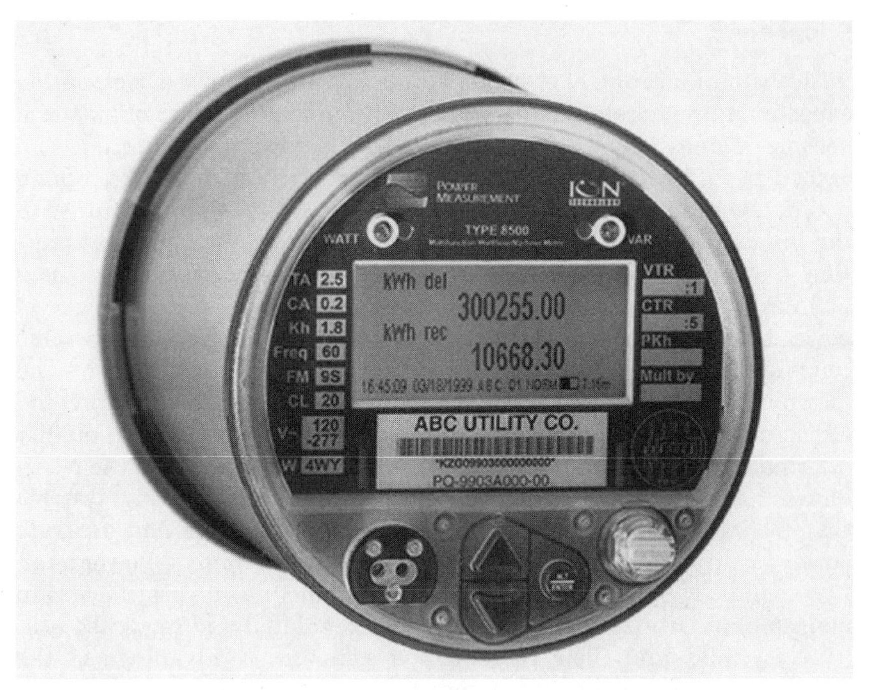

Figure 6.31 Power quality monitor and revenue meter. (*Courtesy of Power Measurement.*)

4. Andrews, Gene. 1988. "Understanding Digital Storage Oscilloscopes." *Electronic Design,* vol. 36, no. 21, September 22, pp. 135–139.
5. Boyer, Jim. 1990. "Measurement Applications for Digital Radio Using Spectrum Analyzers." *Telecommunications,* vol. 24, no. 8, August, pp. 39–41.
6. Newcombe, Charles. 1997. "Instrumentation for Accurate Harmonics Measurement." *Plant Engineering,* vol. 51, no. 13, December, pp. 122–124.
7. Cowling, H. 1998. "Check for Live ac Wires with this Electrostatic Voltage Probe." *Electronics Now,* vol. 69, no. 123, December, pp. 53–56.
8. Lewis, Warren. 1995. "Troubleshooting for Electrical Noise—Part 1." *EC&M Electrical Construction and Maintenance,* vol. 94, no. 5, May, pp. 20–22.

8*a*.———. 1995. "Troubleshooting for Electrical Noise—Part 2." *EC&M Electrical Construction and Maintenance,* vol. 94, no. 6, June, pp. 18–21.

9. Erfer, Lynn. 1996. "Measuring Temperature with Infrared Sensors." *Machine Design,* vol. 68, no. 124, August, pp. 90–92.
10. Michaels, Ken. 1999. "Ten Easy Steps for Testing Branch Circuits," *EC&M Electrical Construction and Maintenance,* vol. 98, no. 1, January, pp. 16–17.

10*a*.———. 1998. "Three-Lamp Circuit Tester: Valid Tester or Night-Light?" *EC&M Electrical Construction and Maintenance,* vol. 97, no. 13, December, pp. 16–17.

11. Kennedy, Barry W. 1998. "Power Quality Monitoring as a Competitive Advantage in the Restructured Competitive Utility Industry." *Proceedings of PQA '98 North America,* June 8–11, Phoenix, AZ.
12. IEC 1000-3-7, Flicker Limits, 1995. The author thanks the International Electrotechnical Commission (IEC) for permission to use the following material. All extracts are copyright © IEC, Geneva, Switzerland. All rights reserved. All IEC publications are available from *www.iec.ch*. IEC takes no responsibility for damages resulting from misinterpretation of the reference material due to its placement and context of its publication. The material is reproduced with their permission.

Chapter

7

Power Quality Surveys

A power quality survey is the first step in the process of finding a solution to the problem. What is a power quality survey? What is the purpose of the survey? Who performs the survey? Does the utility, end user, or a consultant perform the survey? How do you conduct a power quality survey? How do you choose the right measurement tool for the survey? How do you analyze the results of the survey and determine the most cost-effective solution to the power quality problem? This chapter will answer these and similar questions.

When end users experience power quality problems, they need answers to several questions. Whom do they call when they have a power quality problem? Do they call their local utility, an electrical contractor, or an engineering consulting firm? Or do they try to solve the power quality problem themselves? It depends on the type of problem and type of end user. Problems caused by events external to the end user need to involve the utility. End users with small staffs probably lack the expertise to solve power quality problems and need the help of their local utility. It also depends on the status of the restructured electrical utility industry. In a restructured utility industry, the local utilities may not have responsibility for the quality of power. They may care only about the reliability of their distribution system. End users may have to seek help from an energy service company or an engineering consulting company specializing in power quality. Certainly, end users who know the process for conducting a survey and solving a power quality problem will get better service. They will know how to locate and assist power quality experts. End users as well as power quality experts need to know how to plan, perform, and analyze a power quality survey.

Purpose of a Power Quality Survey (Checkup or Examination)

A power quality survey serves the same purpose as a doctor's checkup. It determines what is wrong and how to fix it. It provides a step-by-step procedure for isolating the problem, its cause, and its solution.

End users usually call a power quality expert from the local utility or engineering consulting company. Power quality experts make "house calls." End users need to schedule an appointment to have the power quality expert visit their facility. At the facility, the power quality expert performs a physical and electrical checkup of the electrical power system. This checkup is called a *power quality survey* and has four purposes or objectives, as shown in Figure 7.1. They are:

1. To assess the "health" or condition of the power system (especially the wiring and grounding system)
2. To identify the "symptom of the sickness" or power quality problem (usually an ac voltage quality issue)
3. To determine the "disease" or cause of the power quality problem (source of the power disturbance)
4. To analyze the results of the power quality survey in order to determine the "cure" or cost-effective solution to the power quality problem

How does the power quality expert assess the "health" or condition of the end user's electrical power system? How does the end user interact with the power quality expert to make sure the power quality expert assesses the end user's electrical power system adequately and accurately?

Assess the power quality (health)

When patients visit their doctors, they don't want the doctor to look at them and say "Go home, take an aspirin, and get plenty of sleep." They want a thorough assessment of their health. They expect the doctor to ask questions to determine the extent and type of their sickness. In addition to asking them questions, they expect the doctor to give them a thorough but appropriate physical exam. However, the doctor can help them better if they communicate clearly the condition of their health. The same thing applies to assessing the condition of an end-user power systems.

Power quality experts need to ask questions to determine the scope of the power quality problems. They follow their questions with a thorough but appropriate on-site power quality survey of the facilities.

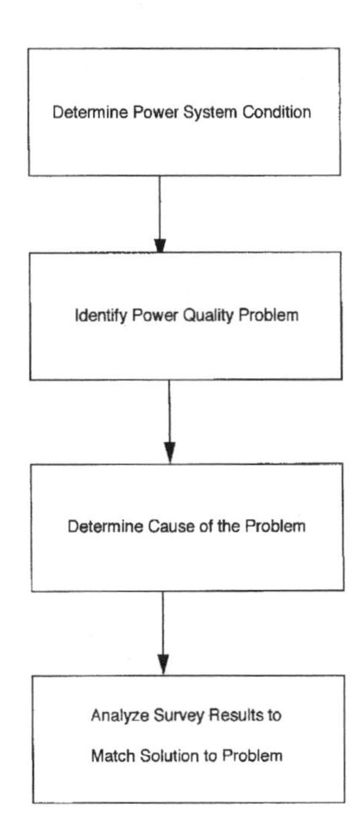

Figure 7.1 Power quality survey process.

Power quality experts or local electric utility representatives can better help their customers solve their power quality problems when their customers have a basic understanding of the condition of their power systems.

End users don't want power quality experts to jump quickly to solutions that may result in costly purchases that don't really solve their power quality problems. Power quality experts don't want to give their customers costly and potentially wrong solutions. However, this often happens when power quality experts recommend buying power conditioning equipment before assessing the causes of the power quality problems. The power conditioning equipment may not solve the power quality problems. It may actually make them worse. Figure 7.2 shows how a UPS can distort power and make it worse. It's similar to when doctors prescribe drugs whose side effects cause more problems than the original disease. Avoid these mistakes by first determining the scope of the power quality problem. Determine the problem's scope by quantifying and measuring its effect on sensitive equipment.

Before conducting an on-site power quality survey, whether an expert or not, a person needs to know and document the scope of the problem and the steps in performing a power quality survey, as shown in Figure 7.3. This involves determining what, when, and how the problem occurred. The surveyor needs to:

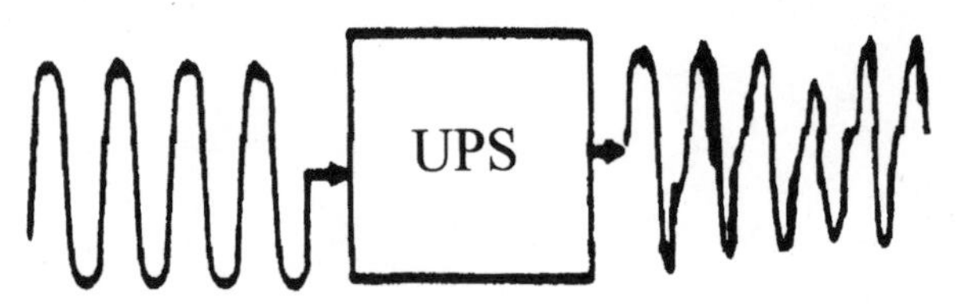

Figure 7.2 UPS distorted power.

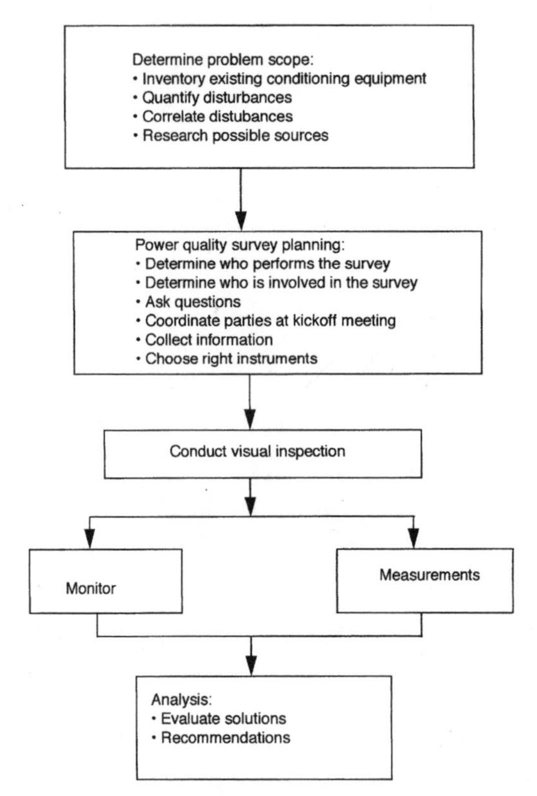

Figure 7.3 Power quality survey procedure.

- Determine the existing sensitive and power conditioning equipment and its location
- Quantify the number, time, location, and types of disturbances using the terms described in Chapter 2 and defined in the Glossary
- Determine whether any disturbances occur at the same time
- Research possible sources that could cause the problem

This information will help in assessing the extent and cost of a power quality survey. Once the scope of the power quality survey has been determined, the surveyor needs to identify the type of power quality problem.

Identify the power quality problem (symptom)

When people go to the doctor, they know that their time with the doctor has limits. In order to use their time effectively, they prepare an accurate and succinct description of their sickness. Similarly, end users need to prepare an accurate and succinct description of their power quality problems. They should avoid imprecise and ambiguous terms. They shouldn't use nontechnical and nondescriptive terms, like glitch, spike, wink, blackout, blink, or dirty power. These meaningless terms offer no clues as to the cause of a problem. They need to instead describe their problems accurately. They need to describe problems with their sensitive electronic equipment. Are their computers or computer-related equipment losing data, freezing, or experiencing component failure? Do their lights dim when they turn on certain equipment? Do their lights flicker? If so, when do they flicker? Are their relays tripping out at certain times? If so, what relays? Where are they located? Do any of their motors and adjustable-speed drives become overheated and trip off line? Do they hear any unusual noises associated with the problem? Do they smell any burning, smoke, or ozone? Is there any coincident effect on other equipment, like telephones or small appliances like microwave ovens or coffee makers? They need to keep a log that describes the power quality problem and when it occurs. They need to use the reporter's list of questions when identifying power quality problems: what, when, how, and who? They save money and time when they help the power quality expert identify the power quality problem. They need to describe their maintenance procedures and provide maintenance logs. A precise description of the power quality problem will help immeasurably in matching the power quality problem to the cause of the problem.

Determine the cause (disease)

If patients describe accurately their sickness, their doctor may diagnose immediately the cause of their sickness. Otherwise, the doctor will have to ask more questions, examine the patients, and perform laboratory tests. Likewise, power quality experts try to diagnose the cause of their client's power quality problems through questions and answers. If they cannot determine the cause through the preliminary review, they will examine the power system, make measurements, and run a few tests to determine the cause of the power quality problem.

Analyze the results of the survey (diagnose) to determine a solution (cure)

When doctors finish examining their patients and evaluating the results of the laboratory tests, they report to their patients the illness and the cure. Similarly, qualified power quality experts provide an analysis of the results of the power quality survey, including a recommendation of how to solve the power quality problem cost effectively. They should have experience in troubleshooting the relationship between failure of electronic equipment and power disturbances. They need to troubleshoot events on both sides of the end user's and its neighbor's power meter. They need to look at the total power system, including the electric utility power system, malfunctioning equipment power system, and adjacent end-user power system. They should approach the analysis of the power quality survey thoroughly and systematically, as shown in Figure 7.4. They should look for the following: (1) power disturbances coincident with equipment malfunction, (2) power disturbances that exceed equipment specifications, and (3) visual inspection observations as they relate to equipment problems. They should not have a bias toward certain solutions, like trying to sell a certain approach or product.

Many local electric utilities provide power quality surveys as part of their customer service program. Most of these utilities provide free preliminary surveys. Their customers can take advantage of this service by calling their local utility's customer service department. In the past, utilities and their customers tried to point an accusatory finger at each other. Often utility customers assumed utilities caused power quality problems, while utilities assumed their customers caused their own power quality problems.

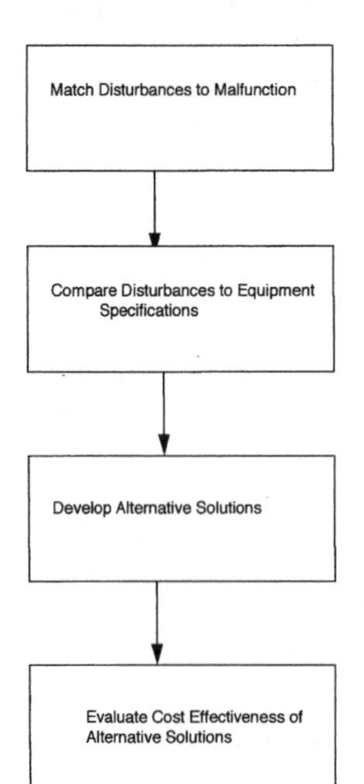

Figure 7.4 Power quality survey analysis.

Georgia Power performed a survey to determine how the utility and its customers perceived who is causing power quality problems. Besides natural causes, the survey showed that the utility and its customer had diverse viewpoints as to the source of power quality problems. Figure 7.5 shows that the utility perceives its customer causes 25 percent of the power quality problems, while Figure 7.6 shows that the customer perceives that the utility causes 17 percent of the problems.

Customer service–oriented utilities give a higher priority to finding the cause of the problem rather than who caused the problem. Not only

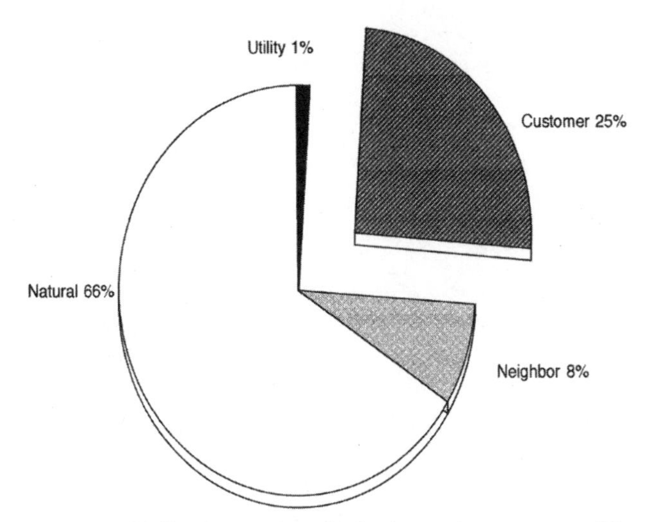

Figure 7.5 Utility perception of who is causing power quality problems. (*Courtesy of Georgia Power.*)

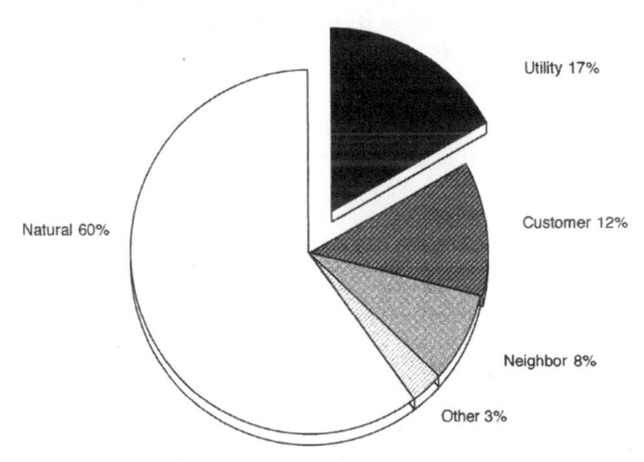

Figure 7.6 Customer perception of who is causing power quality problems. (*Courtesy of Georgia Power.*)

the utility but everyone involved in the survey should have the same attitude. They should realize that the best power quality survey analysis requires a team effort that includes the power quality expert, customer, and the local electric utility. After the survey's completion, they should expect a technical report that summarizes the findings of the survey and recommends solutions.

The report should include a summary of the data collected in the survey, description of the power quality problems, answers to questions, measurements taken, recommended power conditioning equipment, and any additional power quality monitoring and engineering analysis. A more complicated analysis may use a computer simulation program to evaluate alternative problem scenarios and solutions. The analysis should include an evaluation of the cost of the power quality problem and solution. Before performing a power quality survey or analysis, everyone needs to plan the power quality survey carefully.

Planning a Power Quality Survey

A well-planned power quality survey of a facility having power quality problems minimizes incorrect electrical changes or unnecessary purchases of power conditioning equipment. It often pays for itself in terms of reduced lost production. It first requires a decision as to the scope of the survey. Does the situation require a basic or comprehensive survey?

The "IEEE Recommended Practice for Grounding of Industrial and Commercial Power Systems" (*The Emerald Book*) refers to three survey levels, as shown in Figure 7.7. Level 1 requires tests and analyses of the ac distribution and grounding system that supplies the surveyed equipment. Level 2 includes level 1 and monitoring the ac voltage that supplies the surveyed equipment. Level 3 includes levels 1 and 2 plus monitoring the site environmental conditions. The EPRI *Power Quality for Electrical Contractors Applications Guide* refers to a basic survey as one that involves "testing and analysis of the power distribution and grounding system." Good surveys include the following tasks at all levels of the power quality survey:

1. Determine who performs the survey.
2. Prepare a list of questions to be answered by the end user and the servicing electric utility.
3. Coordinate parities involved.
4. Collect documents about the facility's electrical distribution system.

Each one of these tasks will help the utility, power quality expert, and end user come to a quick and effective solution to any power quality problem.

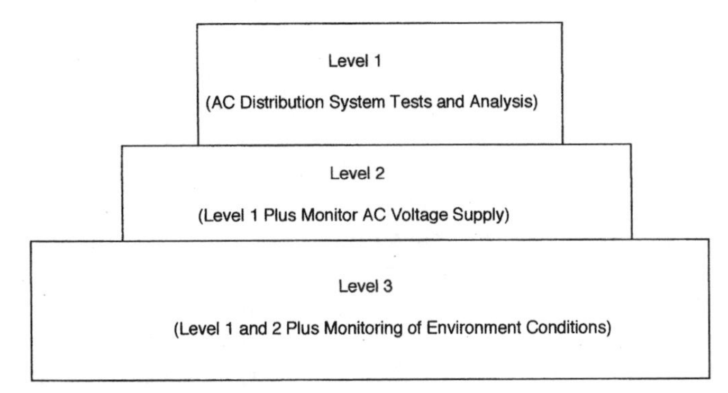

Figure 7.7 Power quality survey levels. *(Courtesy of IEEE, Standard 1100-1992, Copyright © 1993.)*

A well-planned power quality survey requires identifying the participants and performer of the survey. Who participates in the survey? Who performs the survey?

Identify the participants and performer of the survey

Chapter 1 discussed several organizations that have a stake in power quality. Many of the same organizations should participate in the power quality survey. They include the end user or owner of the affected equipment, the equipment manufacturer, the independent power quality consultant, power quality meter and monitor suppliers, the electrical contractor or facility electrician, power conditioning equipment suppliers, and the electric utility company. The deregulation of the electric utility industry in the United States has added the energy service company to these organizations. Many of these organizations perform power quality surveys.

Some electric utilities have comprehensive power quality programs. Many of these utilities employ experts who perform power quality surveys. Several electrical engineering consulting firms perform power quality surveys. Member utilities of EPRI use EPRI's Power Quality Electronics Applications Center or PEAC experts to perform power quality surveys. Many small utilities look for power quality expertise from their larger parent energy supplier company. For example, the Bonneville Power Administration (BPA) provides electrical energy to 150 electrical utilities. Many of these electrical utilities include small cooperatives, public utility districts, or municipalities that service rural areas and lack the staff to perform a power quality survey. BPA employs a full-time power quality engineer and technician who help their electric utility customers help their end-user customers with a power quality survey. Newly formed energy service companies have experts that perform power quality surveys. Facility electricians and electrical contractors can assist the power quality expert, but usually lack the experience and qualifications to perform the power quality survey themselves. Figure 7.8 shows that end users have several alternative choices for power quality surveyor.

Clearly, end users that experience a power quality problem should first contact the customer service department of their local utility. A customer service representative will help them determine who should perform the power quality survey. End users benefit when they clearly understand their servicing utility's responsibilities. Many utilities will help their end-user customers solve power quality problems at no charge. However, with the deregulation of the utility industry and the effort by many utilities to cut cost and to unbundle services, i.e., to charge for services that in the past were included in the price of the electricity, many utilities charge for power quality services. End users need to determine initially how much help the local utility will provide. Some utilities will concern themselves only with who caused a power quality problem. They want to reduce their liability and determine the responsible party to any power quality problem. End users and utilities can protect themselves if they clearly understand the types of power quality problems utilities and end users cause. Figure 7.9 identifies the types of problems caused by utilities and their customers on both sides of the meter.

The scope and type of power quality service electric utilities provide their customers varies from one utility to another. Some utilities provide full power quality service that includes power quality surveys, a report with recommendations for solutions, and an offer to lease and/or sell power conditioning equipment to their customers. These

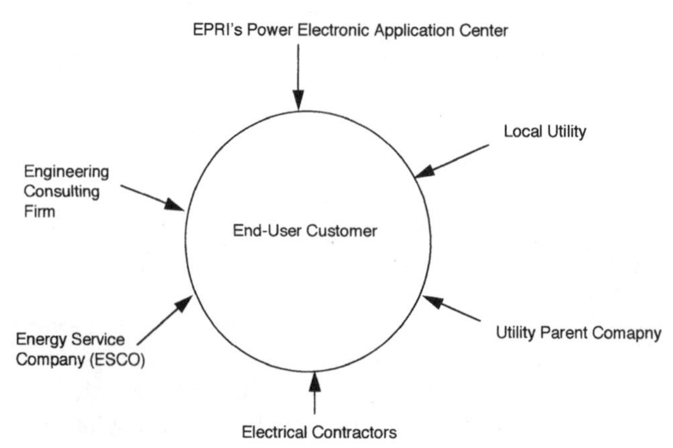

Figure 7.8 Power quality surveyors.

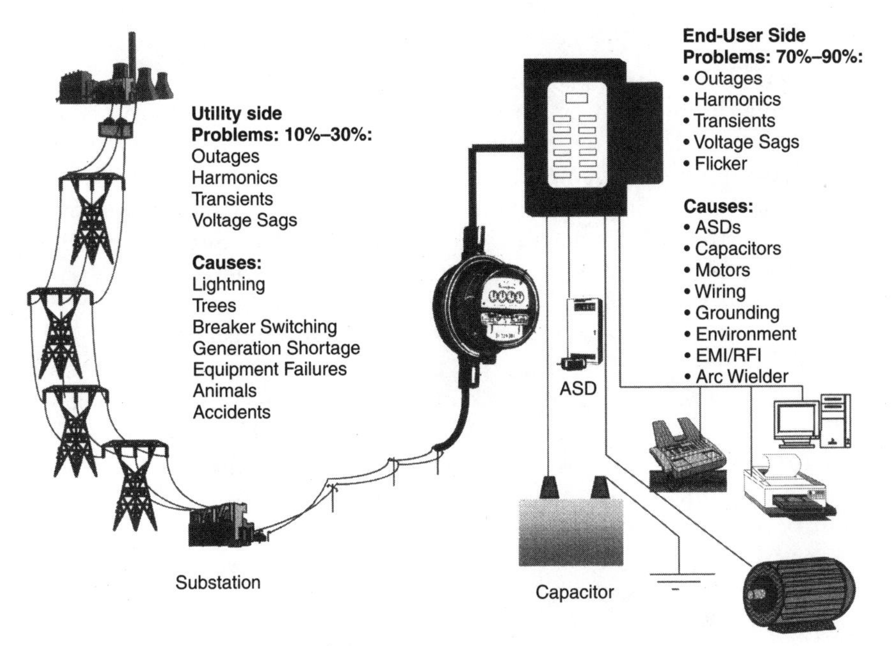

Figure 7.9 Power quality problems on both sides of the meter.

utilities not only provide an essential service to their customers but increase their revenues by selling this service. (Some regulators prevent utilities from charging for these types of services. They restrict utilities from charging for consulting and selling power conditioning equipment because of a concern that utilities will use their regulated business to subsidize these services.) Many other utilities, however, make a business decision to provide minimal power quality service restricted to only determining who causes a power quality problem.

Many different companies besides utilities provide power quality products and services. How do you determine specifically who provides these products and services? Each year *Power Quality Assurance* magazine in its November/December issue provides a list of utility power quality programs, power quality consultants, manufacturers of power conditioning equipment, and suppliers of power quality measuring instruments. Many of these companies have Web sites on the Internet. The Bibliography at the end of the book identifies power quality documents located at various Internet Web sites.

The Northwest Power Quality Service Organization (NWPQSC) provides a series of brochures on how to perform power quality surveys in the home, office, factory, or farm. It also provides training modules based on its brochures. These brochures, like any power quality survey, provide a list of questions.

Ask questions

Surveyors know that good questions and answers are essential to ferreting out the cause of power quality problems. They prepare a list of questions and forms for participants to answer and complete. They know two main questions that need answers before an on-site power quality survey can begin: How did the equipment work before the power quality problem? How does it work after the power quality problem? All other questions stem from these two basic questions.

They will try to obtain answers to the basic reporter questions of what, when, how, where, and who. What is the characteristics of the power quality problem and the sensitive equipment experiencing the problem? When did the problem and coincident problems occur? How did the problem occur? Where did the problem occur? Who observed the problem? Who are looking for solutions to the problem and do they have any ideas as to the possible sources of the power quality problem? What kind of power conditioning equipment, if any, is presently being used?

Coordinate parties

The leader of the power quality survey coordinates the parties affected by the survey. The leader assembles a team that includes representatives of the local electrical utility, sensitive electronic equipment manufacturers, and electrical contractors that installed and maintained the equipment, as well as facility engineers and electricians. The leader clearly defines the roles of these various team members, as shown in Figure 7.10.

Everyone involved in the survey should understand each other's roles. The end user provides information on the power quality problems and answers to the power quality surveyor's questions. The equipment supplier provides specifications for the installation, operation, and maintenance of the affected equipment. Electrical contractors provide useful information on the installation of equipment but should not perform the power quality survey. The power quality consultant who coordinates everyone may work for an engineering firm, the local utility, or an energy service company. The consultant needs to perform the survey in an objective and unbiased manner and not try to sell a particular product or method for solving the power quality problem. In an 1998 article titled "Power Site Survey: A Case of What Went Wrong" in *EC&M Electrical Construction & Maintenance* magazine, Ken Michaels says "Beware of consultants and companies that may have a financial stake in the resolution of your power quality problem. Also, make sure their personnel follow recommended investigative techniques/measurement practices."

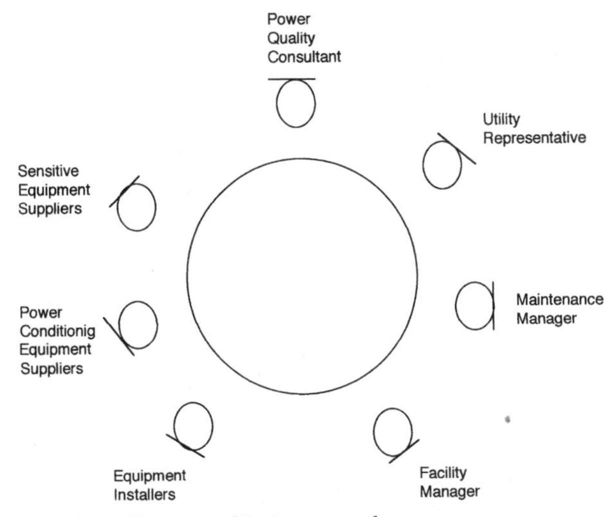

Figure 7.10 Power quality team members.

The success of the survey depends on everyone's participation, including the local utility's. The utility engineer can determine whether the source of the power quality problem comes from the utility or the end user. As mentioned before, many utilities offer power quality surveys as a customer service. These utilities usually understand the value of good customer service and know how to keep their customers happy. But not all utilities provide this service. For their part, end users need to know and provide information on the layout and condition of their facilities.

Know facilities

Power quality surveyors collect all the appropriate information and documentation about the facility's electrical and telecommunication system. This includes schematics, maintenance records, electrical changes, location of sensitive electronic equipment, and existing power conditioning equipment. They conduct the survey more efficiently, effectively, and safely when they have ready access to this information.

Survey forms

Power quality experts use various types of power quality survey forms. Electric utilities as well as power quality consultants develop these forms. Some forms require detailed information about the end user and the power quality problem. Most forms require the name, address, and phone number of the end-user customer, name of a contact person, and a short description of the power quality problem. Some forms even ask the end user's interest in power quality training. Figure 7.11 provides an example of the basic power quality survey form.

In addition to the initial customer and power quality introductory form, the power quality industry has developed test data forms. The "IEEE Recommended Practice for Grounding of Industrial and Commercial Power Systems" (*The Emerald Book*) contains several of these test data forms. They include the following sets of forms: power distribution verification test data and power distribution and grounding summary data. Figure 7.12 provides an example of a test data form.

Choosing the Right Power Quality Instruments

Whoever performs the power quality survey needs to decide what power quality instruments to use in the survey. The surveyor should first use the best and most important power quality instruments available: four of their five senses. They are free and readily available. Surveyors can learn a great deal about a power quality problem by what they see with their eyes, hear with their ears, feel with their fingers, and smell with their nose.

First they need to look at the electrical schematic to see if any sensitive electronic equipment connects to equipment that causes power quality problems, like adjustable-speed drives or fluorescent lights. Next they look carefully at the service panel, as shown in Figure 7.13. They always follow safety rules. They look for loose connections, incorrect wiring, and reversed conductors. They look for burnt connections

Date: 8-Jul-98 Time: 8:30am Project #: 1234-56

Customer		Account Address	
Name:	Acme Industrial	Name:	
Address:	Knoxville, TN	Address:	
Tel No:	(123) 456-7890	Tel No:	
Contact:	Bill Acme	Contact:	

PROBLEMS Low voltage power factor capacitor failures corresponding to the installation of two 100 HP dc drives.

	YES/NO	WHO
Are you a H.V. Customer?	No	
Have you contacted your local Utility?	Yes	Light and Gas
Have you contacted your Equipment Manufacturer?	No	

Figure 7.11 Typical power quality survey form. (*Courtesy of Electrotek Concepts, Inc.*)

Collect Data

Transformer ID			Panel ID	
kVA			Rated Amps	
Connection			Main CB Amps	
Primary Volts			Feeder Cable	
Secondary Volts				

			PANELBOARD		HARMONIC SPECTRUM DATA		
	Transformer Primary		or Transformer Secondary		Circle one:	% fund. or	% RMS
	RMS (V or A)	THD (%)	RMS (V or A)	THD (%)		Voltage dist. (%)	Current dist (%)
$V_{A\text{-}B}$					THD (%)		
$V_{B\text{-}C}$					Phase		
$V_{C\text{-}A}$							
$V_{A\text{-}N}$					Harmonic		
$V_{B\text{-}N}$					3		
$V_{C\text{-}N}$					5		
$V_{N\text{-}G}$					7		
I_A					9		
I_B					11		
I_C					13		
I_N					other		
I_G							
Note: *Fill in as much info as you can*					**Note**: *Enter phase voltage and current from highest voltage THD*		

Figure 7.12 Typical harmonic test data form. (*Courtesy of PowerSmiths International.*)

and other hot spots. They might use the infrared instruments discussed in Chapter 6 to overcome their eyes' inability to see hot spots. They follow the connections from the service panel to any sensitive equipment. They look for wiring of sensitive equipment to other equipment on the same circuit. They follow data lines and see if they locate any near lighting ballasts or other devices that could interfere with data communication. They look at the failure log and maintenance record and see if there is any correlation. They look for damaged circuit breakers. They look for any NEC violations and whether the neutral connects to ground. They examine the wires to see if they can handle the load. They especially focus on the neutral wires to see if they can carry the three-phase electronic load. They keep good records and document what they see.

They listen and smell for any arcing. Arcing from loose connections causes many power quality problems. Arcs make a distinctive noise and ozone odor.

Surveyors should touch the outside of electrical equipment, like transformers, to check for overheating caused by harmonics. If they feel any excess heat, they then verify and measure the amount of overheating with infrared sensing equipment.

Figure 7.13 Typical service entrance panel. (*Reproduced with permission of Fluke Corp. ©.*)

If they have not isolated the cause of the power quality problem using their four senses, they then need to choose the right tool from the many power quality instruments available today. Chapter 6, "Power Quality Measurement Tools," presents many types of instruments for troubleshooting power quality problems. They include ammeters, voltmeters, multimeters, oscilloscopes, flicker meters, static meters, infrared detectors, radio-frequency interference and electromagnetic interference meters, harmonic and spectrum analyzers, power quality monitors, and various types of wiring and grounding testers. Do surveyors own all these instruments? They usually buy one, try it, and buy another if it doesn't answer their questions. How do they decide what meter to use when performing a power quality survey without wasting their money and time?

The type and frequency of the problem determines the type of instrument best suited for isolating the cause of the problem. Prudent surveyors keep the cost of the instrument in line with the cost of the problem. They do not buy instruments with features that they will not need. They can spend less than $100 or more than $10,000. It depends on the instrument's features and complexity. Some surveyors start with simple wiring and grounding testers and work up to more complicated power quality monitors that cost as much as $12,000. It is best to start with simple, low-cost instruments that measure frequently occurring power quality problems. Someone getting started with power quality surveys probably should begin with low-cost wiring and grounding instruments. Remember wiring and grounding problems make up over 80 percent of power quality problems.

What kind of instruments do surveyors need to detect grounding problems? Grounding includes proper bonding between the neutral and the ground, adequate sizing of the neutral conductor, and correct grounding and neutral impedance. They can use a $15 analog ohmmeter to make these measurements. If they need to measure the grounding electrode, they require a more expensive device.

What kind of instruments do they need to detect wiring problems? They can measure the voltage at the outlet using a two-pronged voltage indicator that costs about $35. They insert the two prongs in the outlet socket to measure the voltage between the 120-V line and neutral and between the 120-V line and ground. The measurements should not exceed a 2-V difference between neutral and ground.

If they suspect the presence of harmonics when they feel the heat from some equipment, like transformers, they have an inexpensive way to determine the presence and level of harmonic distortion. They should avoid using average-responding rms digital multimeters (DMMs), which measure incorrect voltage and current in the presence of harmonics. They should, instead, use a true-rms DMM to measure the correct voltage and current even when harmonics distort them. One way to verify the presence of harmonics in the neutral conductor is to use both an average-responding rms and a true rms DMM to make current measurements. If the meters give different results, as shown in Figure 7.14, they indicate the presence of harmonics. The ratio of average-responding value divided by true rms value provides a benchmark of the size of the harmonics. A ratio of less than 0.75 indicates the neutral probably contains enough harmonics to cause problems. A more expensive harmonic analyzer measures the THD and TDD.

Surveyors may observe that some equipment acts as either an electromagnetic or a radio-frequency interference source to sensitive equipment. For example, the electronic ballasts from fluorescent lights

Figure 7.14 Average-responding rms to true rms ratio of less than 0.75. The meter on the left registers 59.2 and the meter on the right registers 40.5. (*Reproduced with permission of Fluke Corp.* ©.)

radiate electromagnetic fields that cause nearby computer monitor screens to wiggle. The same fields can cause the programmable logic controller to malfunction. A gaussmeter will verify and measure the interference.

They shouldn't rely on their senses or bodies if they suspect the presence of electrostatic charges. It only takes 2.5 V to damage a sensitive chip in a computer, while it takes over 4000 V for a spark to discharge from a person's body to a grounded object. They should, instead, use a static discharge meter to measure the amount of static discharge.

Power quality surveyors use various types of analyzers to detect voltage disturbances and their source. Infrequent power quality problems require low-cost instruments, while frequent power quality problems require more complex and expensive instruments that can cost as much as $12,000. They try to match the instrument to the problem, as shown in Figure 7.15.

Conducting a Power Quality Survey

Of course, surveyors' four senses and appropriate meters are not the only important tools. A logical and systematic process for conducting the survey provides another important tool to a successful survey. This requires that they conduct the survey systematically. With the right tools for the job, they need to approach the survey in a logical step-by-step manner. They will minimize wasted activities, avoid erroneous conclusions, and prevent dangerous, unsafe practices. What steps characterize a successful and safe power quality survey?

As shown in Figure 7.16, there are four basic steps to conducting a power quality survey:

1. Collect information
2. Visually inspect the site
3. Set up test instruments
4. Collect test data

Each step provides a systematic approach to conducting the survey.

Step 1: Collect information at coordination meeting

Surveyors collect information by first having a coordination kickoff meeting with the parties involved in the power quality problems. As shown in Figure 7.17, this meeting should include the people from the departments experiencing the problems, representatives of the

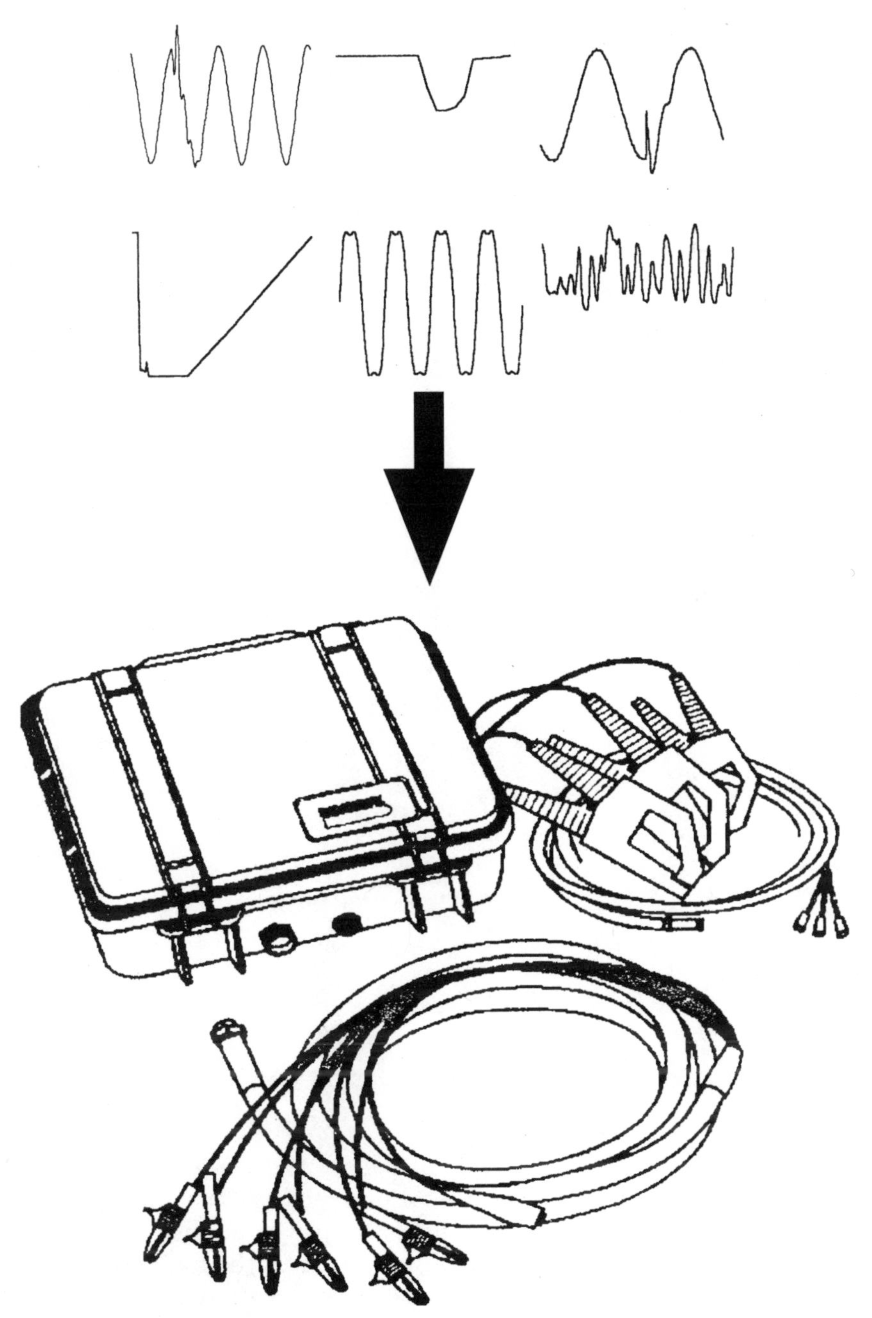

Figure 7.15 Matching type of power quality problem to meter. (*Courtesy of Dranetz-BMI.*)

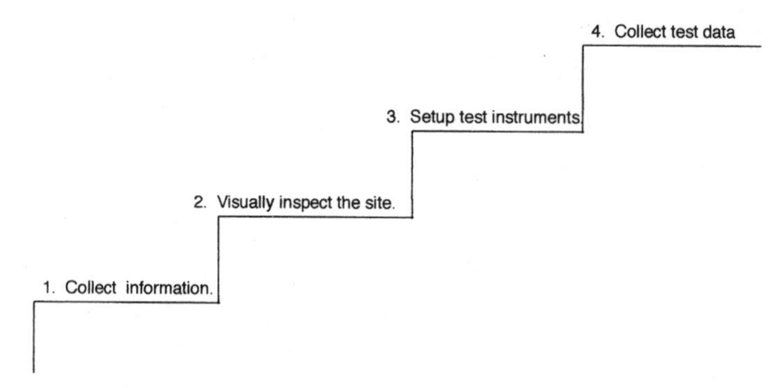

Figure 7.16 Steps in a power quality survey.

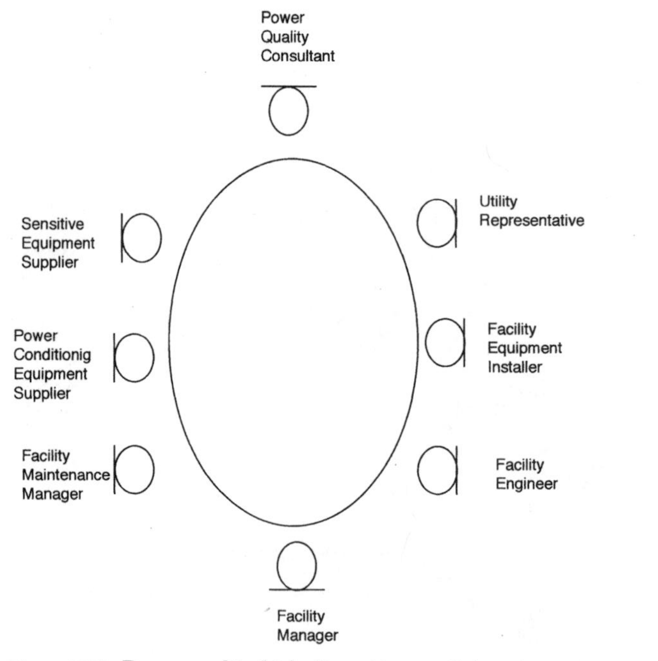

Figure 7.17 Power quality kickoff meeting participants.

manufacturers of the sensitive equipment and the existing power conditioning equipment, power quality expert or customer service representative from the local utility, the power quality consultant performing the survey, and anybody involved in the maintenance of the affected equipment. Either the end user experiencing the power quality problems or the power quality expert performing the survey runs this meeting The purpose of this meeting is to discuss the power quality problems and collect information that will help in isolating the cause of the problems. Whoever runs the meeting should ask the what, when, how, and who questions about the power quality problems and hand out power quality questionnaire forms.

The surveyors need to obtain several documents in the kickoff meeting. These documents include the one-line diagram of the facility electrical distribution system as well as that of the electric utility supply system. They require the specifications of any malfunctioning equipment. Also, they need a copy of the maintenance logs as well as any logs that describe the problems.

Finally, at the coordination meeting everyone needs to agree on when the surveyors will perform the on-site inspection and set up the necessary test equipment. From this meeting the power quality consultant determines the scope of the problems and prepares a proposal on the cost, steps, and process for conducting the power quality survey.

Step 2: Conduct on-site visual inspection

Surveyors need to visually inspect the site before they set up their instruments and do any testing. Why? A visual inspection is cheap and simple. They might even find the cause or causes of the power quality problems before they set up their test equipment. It also helps them plan their testing setup.

They need to examine visually three main areas of the facility. First, they begin at the location where the power quality problem started. Then they follow the electrical distribution system back to the electrical service entrance. They stop and look at the service entrance panel. They then go outside the facility and inspect the electric utility service.

They need to inspect the facility thoroughly. The inspection includes examining the equipment and electrical connections near the power quality problem and noting the physical location of the equipment. They look for devices that generate transients, voltage sags, inrush currents, and harmonics affecting sensitive electrical equipment. Table 7.1 provides a list of the various types of devices and the types of power quality disturbances they cause.

They observe whether the air or wires transmit the power quality disturbance. Air could transmit the disturbance if the source is near the sensitive equipment. Wires could transmit the disturbance if a common circuit supplies the source and the sensitive equipment.

Good inspectors keep good records. They take still pictures or, better yet, videos. They even look inside the equipment. They look for answers to certain questions. Are there any adjustable-speed drives? Adjustable-speed drives could be a source of harmonics. Are there any power factor improvement capacitors in the facility? They are a potential cause for amplifying harmonics. It depends on their size and when

TABLE 7.1 Matching Device to Type of Disturbance

Disturbance description	Motors	Capacitor	Adjustable-speed drive	Arc furnace and welders	Electronic ballasts	Switched-mode power supply	Dimmer switch	Photocopiers and laser printers
Impulsive transients		X						
Oscillatory transients		X	X					
Sags/swells	X							
Interruptions								
Undervoltages/overvoltages	X	X						X
Harmonic distortion		X	X		X	X	X	
Voltage flicker				X				X
Static discharge								
Noise				X		X	X	
Wiring and grounding problems								

they are switched on. They observe the environment around the sensitive electronic equipment. Is it hot, dirty, or humid? Is the equipment vibrating? They look at data connections as well as electrical connections. They check the wiring with the one-line diagram and note any changes. They use the one-line diagram as a reference for setting up their test equipment and later reporting on the results of their survey.

Now they take a look at the breaker panel. They open it carefully and check for damaged breakers. They determine if any nonlinear loads connect to this panel. They examine the wires connected to the panel and check whether wire size matches the one-line diagram. They remember to look and smell for any signs of arcing or loose connections.

They next look at the main transformer servicing the facility. They determine if the transformer is rated properly by comparing the nameplate rating to its loading. Figure 7.18 shows a typical transformer nameplate. They calculate the transformer loading by adding the load of all the equipment connected to the transformer. They answer questions about the transformer and its service. Is the wiring to the transformer the correct size? Are the transformer taps set correctly? Does the transformer feel hot or sound noisy? An overloaded transformer gets hot and noisy. Harmonics that add to the normal load currents cause transformers to overload.

Finally, they look at the main service and the electric utility service at the point of common coupling. They look for power factor correction capacitors in the utility substation and adjustable-speed drives in the facility. They know that adjustable-speed drives generate harmonics and that capacitors can amplify the magnitude of the harmonics. This

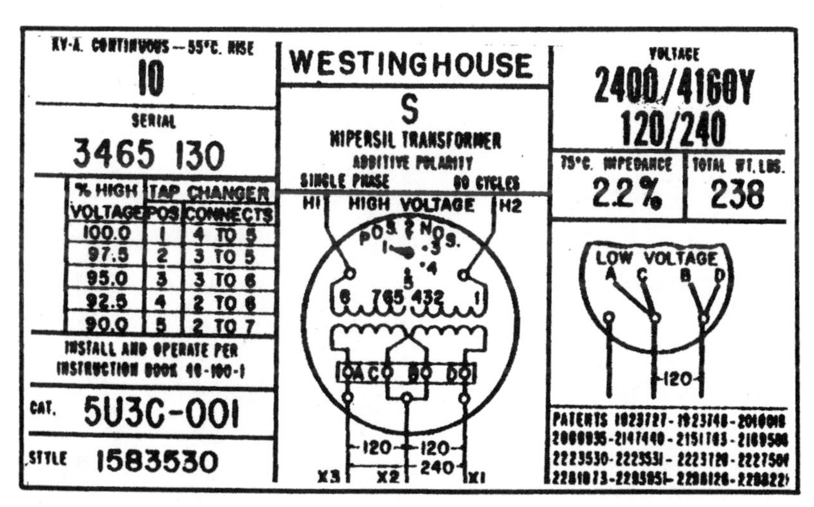

Figure 7.18 Transformer nameplate. (Reprinted from *Electrical Transformers and Power Equipment,* 3d ed., by special permission of The Fairmont Press Inc., 700 Indian Trail, Lilburn, GA 30047.)

problem gets worse when the utility and the end user both have capacitors. Depending on the size of the utility capacitors and when they switch on, they could interact with the facility capacitors. The utility capacitors could cause the end-user capacitors to resonate and amplify any harmonics in the facility. This problem could affect nearby end users.

The inspection should include any nearby substations served by the utility. It should determine the types of loads on the neighboring buses. As shown in Figure 7.19, neighboring loads could back-feed power quality disturbances into the site being surveyed.

Before concluding their visual inspection, surveyors need to determine how the critical loads get electrical service. Do dedicated feeders service only the critical electronic loads? Or do feeders that service electronic loads also service other loads? Other loads may interact with the critical electronic loads and cause problems. If they cannot determine the cause of the power quality problems from the visual inspection, they need to set up their test equipment.

Step 3: Set up test instruments

Before the surveyors setups the test instruments, they should review the safety guidelines. These guidelines tell all them to wear safety goggles and gloves and work with a certified electrician if they don't have a electrician's license. They need to get some safety training even if they have a license. (One man didn't wear safety gloves and glasses while working in an electrical panel. He caused an arc that flashed in his face that severely burned his hands and damaged his eyes.)

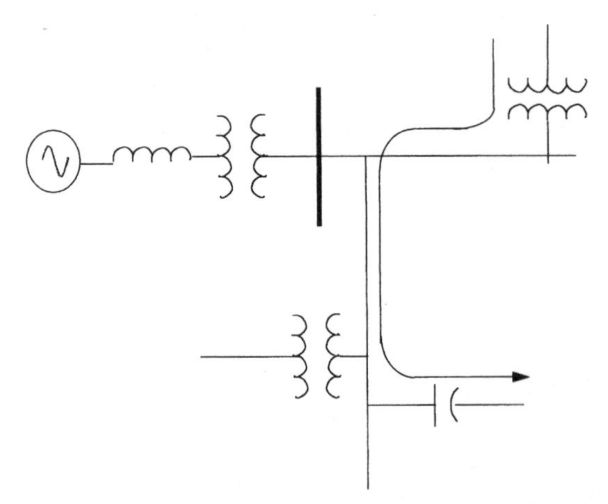

Figure 7.19 Neighboring loads back-feed power quality disturbances.

Finally, they need to secure test equipment and open panels. When not using the test equipment, they should cover it with an insulated rubber blanket.

In this step, surveyors need to decide what instruments to use, where to connect them, how long to leave them there, and what measurements to record. The instrument tests should supplement the visual survey. They should connect their test equipment at sites that were suspicious in their visual inspection. They usually perform test measurements at the sensitive equipment, as shown in Figure 7.20, and work their way toward the electric utility supply service. They start with simple instruments, like hand-held true rms DMMs, that measure voltage and currents. They check for harmonics, noise, EMI, and RFI. They progress to more complicated monitors, like harmonic analyzers and oscilloscopes, if necessary. In addition to checking transformers, receptacles, and electrical panels, they measure the resistance of the grounding electrode system with a grounding tester.

They use the test equipment to isolate the type of power quality problem and its source. The voltage measurement usually identifies the type of power quality problem, while current and voltage measurements help identify the cause of the power quality problem. They set the threshold of their measuring equipment to record only disturbances that will affect the sensitive equipment. They set the time interval to record background events. In the case of a random power quality problem, they may want to leave the monitoring equipment connected for at least 2 weeks.

They need to measure environmental conditions as well as electrical parameters. This includes temperature and humidity. High temperatures can cause overheating and failure of sensitive electronic

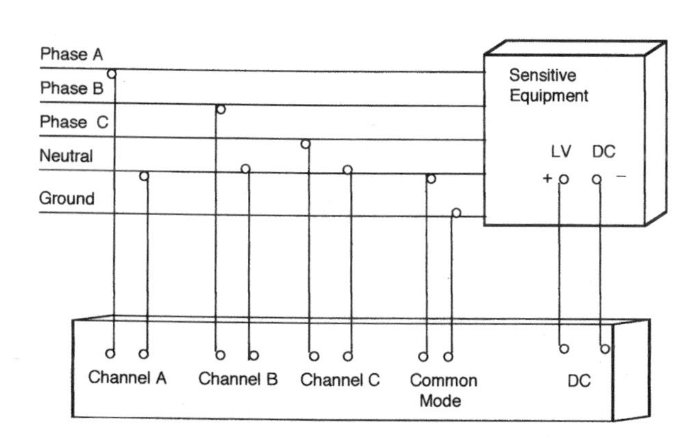

Figure 7.20 Power quality monitor setup at sensitive equipment.

components. High humidity can cause condensation and erroneous connections on electronic circuit boards, while low humidity contributes to electrostatic discharges.

Step 4: Collect test measurements

When the surveyors get to the service entrance panel, they need to perform a series of wiring and grounding measurements. These measurements include the following: (1) rms entrance voltage, (2) green wire ground current at the source, and (3) neutral conductor current. They record the voltage measurements of extraneous neutral-ground bonds at the main service panel, using a wiring and grounding tester. They record load phase and neutral currents at the service panel. They measure the impedance of the sensitive equipment's grounding conductor, using a ground impedance tester. They record the impedance of the neutral conductor from the sensitive electronic equipment to the source neutral bonding point. They measure the resistance of the grounding electrode, using an earth ground tester. They measure phase currents at the service panel with sensitive equipment turned off to determine if the sensitive equipment shares a circuit with other loads. They determine the presence of separately derived systems by recording impedance measurements with a ground impedance tester. They measure with a ground impedance tester the equipment grounding conductors, any isolated grounding conductors, neutral conductors, and connections to metal enclosures.

Analyzing Power Quality Survey Results

A systematic approach to analyzing the results of the power quality survey provides the best results. The purpose of the analysis is to determine the cause of the various power quality problems. The first challenge is to determine who is responsible for causing the power quality problems. There may be a combination of factors causing the power quality problems. Is it coming from inside the facility? Or is it coming from the utility power supply? Is it coming from a neighboring end user and being transmitted to the affected site via the utility's transmission system? Or is there a combination of the utility and the affected site causing the problems? It helps to analyze the data from the three parts of the survey at the sensitive equipment location, the low-voltage service entrance, and the utility side of the meter. Figure 7.21 provides a breakdown in pie chart form of typical sources of power quality problems.

The three steps to any good analysis are: (1) categorize the types of power quality problems, (2) categorize the causes of the problems, and (3)

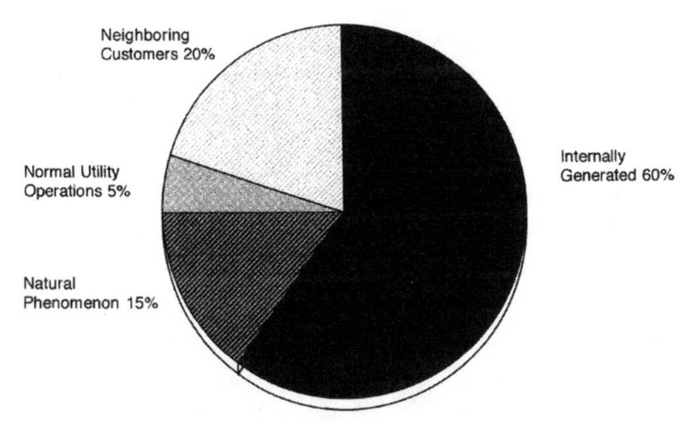

Figure 7.21 Power quality problem sources. (*Courtesy of Florida Power Corp. and the Edison Electric Institute.*)

match the cause to the problem. These steps require reviewing the records from the visual and test surveys along with any failure and maintenance logs. First group the power quality disturbances into the categories, like transients, harmonics, flicker, voltage sag, described in Chapter 2. Look for disturbances that might have caused sensitive equipment to malfunction by comparing the equipment specification to survey measurements. Analyze the results using the following three steps: (1) correlate power disturbances to equipment malfunction, (2) identify power disturbances that exceed equipment specifications, and (3) correlate problems found in the visual inspection with problems found in the measurement of equipment symptoms. Listing the causes and the effects on a spread sheet helps perform this analysis. Table 7.2 helps match a power quality disturbance to possible causes of the disturbance.

The analysis of the wiring and grounding measurements at the low-voltage side of the service entrance involves first looking for extraneous neutral-ground bonds if a current flows in the green wire. This includes examining the neutral conductor currents. Do they exceed the phase current? If so, they may indicate the presence of triplen harmonics. Another grounding concern is the magnitude of the grounding electrode resistance. This resistance should not exceed 25 Ω.

The surveyors must work closely with the local utility when they analyze the survey of the high-voltage utility supply. They look for power factor capacitors that the utility may switch on at the same time harmonics are observed inside the facility. They identify transient disturbances from the utility system that may cause power quality problems inside the facility. Their disturbance monitor at the point of common coupling provides data on utility-caused power quality problems. They may want to input the data from the survey into a diagnostic computer model.

TABLE 7.2 Matching Cause to Type of Disturbance

Disturbance description	Motor starting	Capacitor switching	Loose connections	Lightning and load switching	Undersize wiring	Tree limbs and auto accidents	Power surge	Utility system faults
Impulsive transients		X		X		X	X	X
Oscillatory transients		X	X					
Sags/swells	X							X
Interruptions			X	X		X		
Undervoltages/overvoltages	X			X				X
Harmonic distortion		X						
Voltage flicker	X			X	X			X
Static discharge								
Noise	X			X		X	X	
Wiring and grounding problems			X		X			

Input data into diagnostic model

In addition to performing an analysis by looking at the data collected during the survey and relying on the experience of the power quality expert, there are many diagnostic tools available today to aid in the analysis. EPRI has developed tools for diagnosing wiring and grounding problems as well as general power quality problems. EPRI calls its software package the Power Quality Toolbox.

The Power Quality Toolbox contains voltage sag analysis and wiring and grounding modules. These modules take the user through step-by-step procedures for analyzing power quality problems caused by voltage sags and improper wiring and grounding. They run on personal computers that use the Microsoft Windows operating system and are stored on CD-ROM disks. They are available free to EPRI members and for a fee to non-EPRI members. For more information about the EPRI Power Quality Toolbox, contact Electric Power Research Institute, Attn: Marek Samotyj, 3112 Hillview Ave., Palo Alto, CA 94304, (605) 855-2980, or on the Internet at *www.epriweb.com.*

Most everyone does not want to spend the money and time simulating transient or harmonic conditions on the computer. However, it is often necessary to simulate transient and harmonic conditions on the computer in order to determine what conditions cause power quality problems. It is too costly to allow actual situations to develop in order to determine when transients or harmonics cause power quality problems. The damage to equipment and production shutdowns often exceed the cost of computer simulations.

Several programs are available for simulating the flow of harmonics and transients under various conditions. These simulation programs provide an analysis of power quality problems under different situations and assumptions. They provide an effective means of determining how events happening on a utility's system are interacting with events on an end user's system. For example, a simulation will provide answers to questions about whether capacitors on the utility's system are interacting with capacitor's on an end user's distribution system and causing a resonant condition. The most popular transient simulation program is called the Electromagnetic Transients Program (EMTP), originally developed by the Bonneville Power Administration. While there are several harmonic simulation programs on the market, one of the more popular harmonic programs is called Super HarmFlow, developed by Electrotek Concepts, Inc.

Identify alternative solutions

The many solutions presented in Chapter 5 should be examined and evaluated as to their applications to the specific causes identified in the analysis. Alternative solutions include:

- Do nothing if the solution is too costly
- Correct improper wiring and grounding
- Relocate equipment away from damaging environments
- Buy more robust equipment
- Modify the size of the capacitors that cause resonance
- Add filters to filter out harmonics
- Add power conditioning equipment

Power conditioning equipment should come after making wiring and grounding corrections. It is best to avoid general solutions. The best approach requires a specific solution to a specific problem. There are often lower-cost and more lasting solutions than adding power conditioning equipment. Each solution needs to be evaluated as to its cost effectiveness. Lower-cost solutions should be selected over higher-cost solutions. Compare the cost of the problem to the cost of the solution. The cost of the power quality problem should include the cost of lost production, scrap, restart, labor, repair, replacement, process inefficiency, and energy inefficiency. The cost of the solutions should be less than the cost of allowing the problem to continue. For example, Figure 7.22 shows a comparison of ride-through solutions to a voltage sag problem in a plastic extruder factory.

Preventing power quality problems

Ben Franklin said, "An ounce of prevention is worth a pound of cure." He demonstrated this saying when he installed his own invention, the lightning rod, on his house. His house was struck by lightning, but the lightning rod protected it from damage. The same is true in preventing power quality problems. It is usually less costly to prevent a power problem than trying to cure it after it happens. That is why it is worth the cost to perform a power quality survey before a problem occurs. This is true in a home, office, factory, or farm. This is especially true in installing electronic equipment, like computers or computer-controlled equipment, that is sensitive to power quality problems. It is just as true when installing any equipment—like adjustable-speed drives, electronic ballast, fluorescent lights, or power factor improvement capacitors in a factory or laser printers and copiers in an office or home—that tends to cause power quality problems. It is important that the one-line diagram of the end user's electrical distribution system is up-to-date. Preventive maintenance reduces power quality problems. Good records of any failures or equipment malfunction helps expedite the analysis. Minimize power quality problems when

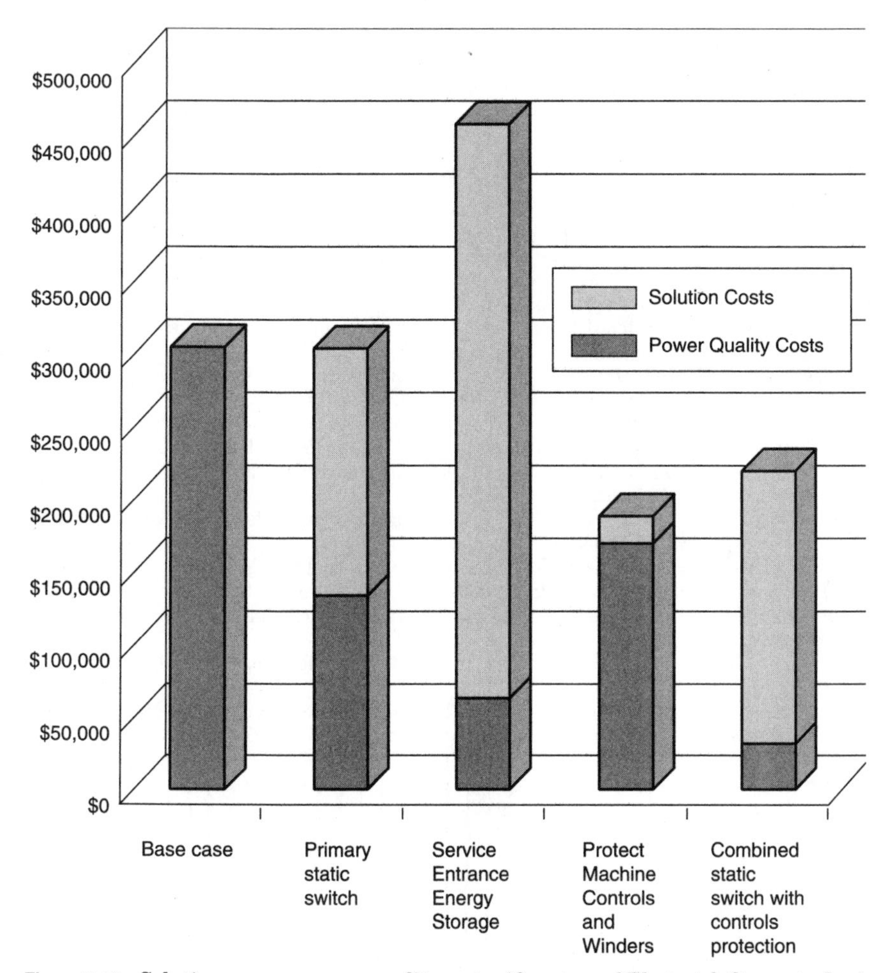

Figure 7.22 Solutions versus power quality costs. (*Courtesy of Electrotek Concepts, Inc.*)

installing equipment by using certified electricians who are experienced in power quality problems. Prudent end users buy power conditioning equipment that is UL1449-certified.

The end user will need to become knowledgeable on how to solve power quality problems as the electric utility industry becomes more deregulated. The next chapter presents the effect of deregulation on power quality.

References

1. Bingham, Richard P. 1998. "Planning and Performing a Power Quality Survey." *Power Quality Assurance*, vol. 9, no. 3, May/June, pp. 14–22.

1*a*.———. 1999. "The Power Quality Survey: Do It Right the First Time." *NETA World,* Summer, pp. 1–5.
2. Lamendola, Mark, and Jerry Borland. "Coming to Terms with Power Quality: Understanding What Power Quality Experts Say Can Be a Boon to Your Success in the Electrical Field (Power Quality Advisor)." *EC&M Electrical Construction and Maintenance,* vol. 98, no. 2, February, p. PQ-3.
3. Watkins-Miller, Elaine. 1997. "Don't Get Zapped (Office Technology and Power Quality Problems)." *Building,* vol. 91, no. 10, October, pp. 68–69.
4. "PQAudit™." 1999. URL address: *http://www.electrotek.com//PS-STUDY/indust/pqaudit.htm.* Available from Electrotek Concepts.
5. Beaty, Wayne. 1994. "Clean Power Requires Cooperative Effort." *Electric Light & Power,* vol. 72, no. 8, August, pp. 20–24.
6. McEachern, Alexander. 1995. "Power Quality Survey by Walking Around." *Power Quality Assurance,* vol. 6, no. 4, July/August, pp. 20–25.
7. DeDad, John. 1997. "Power Quality Site Analysis Step-by-Step." *EC&M Electrical Construction and Maintenance,* vol. 96, no. 3, March, p. 26.
8. Waggoneer, Ray. 1995. "Conducting a Power Quality Site Analysis—Part 1." *EC&M Electrical Construction and Maintenance,* vol. 94, no. 9, September, pp. 18–20.
8*a*.———. 1995. "Conducting a Power Quality Site Analysis—Part 2." *EC&M Electrical Construction and Maintenance,* vol. 94, no. 10, October, pp. 14–15.
8*b*.———. 1995. "Conducting a Power Quality Site Analysis—Part 3." *EC&M Electrical Construction and Maintenance,* vol. 94, no. 11, November,
9. Lonie, Bruce. 1994. "There's More to Power Quality Than Meets the Eye." *EC&M Electrical Construction and Maintenance,* vol. 93, no. 10, October, pp. 69–71.
10. Michaels, Ken. 1998. "Troubleshooting Industrial Power Quality Problems." *EC&M Electrical Construction and Maintenance,* vol. 97, no. 8, July, pp. 16–17.
11. Bowden, Royce. 1998. "The Spectrum of Simulation Software." *IIE Solutions,* vol. 30, no. 5, May, pp. 44–46.
12. Lawrie, Robert J. 1996. "Top Testing Tips for Better Maintenance Management." *EC&M Electrical Construction and Maintenance,* vol. 95, no. 91, September, pp. 38–42.

Chapter

8

Power Quality Economics

Many utilities and end users have discovered that they need to assess the economics of power quality improvements before they decide what power quality improvement to implement. They need to determine the costs of the disturbances and the improvement. They need to determine the level of power quality they wish to achieve. They need to decide where to make the power quality improvements—on the utility or the end-user side of the meter, or a combination of both. On the utility side of the meter, power quality improvements can include the addition of a static switch, custom power equipment such as a dynamic voltage restorer (DVR), and changes to power system operation like capacitor switching. On the end-user side of the meter, power quality improvements can include the addition of power conditioning equipment, a change in the equipment specifications and design, or an improvement in the wiring and grounding inside the end-user facility. Or both the utility and the end user can together make power quality improvements.

Everyone wants to determine the optimum solution or solutions. To do this, you need to evaluate solutions at all levels of the power system from the utility's transmission and distribution system to the end user's secondary system. You need to compare the cost of these improvements to the benefits to determine the cost-effectiveness of the improvements.

The benefits include reduced cost of the power quality problem. On the end-user side of the meter, the cost of a power quality problem can include lost production and revenue, cost of scrap, labor and overtime cost, and the cost to repair or replace damaged equipment. On the utility side of the meter, the cost of a power quality problem can include lost power revenues and disgruntled customers. In the case of reducing or eliminating harmonics, the benefits include the resulting reduction in losses. The goal is to minimize the total power quality improvement cost.

Total Power Quality Improvement Cost

In evaluating alternative power quality improvements the total power quality improvement cost is calculated by the following formula:

$$\text{TPQIC} = \text{PQIB} - \text{PQIC} \tag{8.1}$$

where TPQIC = total power quality improvement cost
PQIB = power quality improvement benefit in $/year
PQIC = power quality improvement cost in $/year

The power quality improvement benefit equals the reduced cost of the problem resulting from the power quality improvement. The following formula shows how to calculate PQIB:

$$\text{PQIB} = \text{PQC}_i - \text{PQC}_r \tag{8.2}$$

where PQIB = the power quality improvement benefit in $/year
PQC_i = the initial or base cost of the power quality problem in $/year
PQC_r = the reduced cost of the power quality problem in $/year

The power quality improvement cost equals the annual cost of purchasing, installing, and maintaining power conditioning equipment, custom power alternatives, or changes in the utility's or end user's power system. The resulting numbers calculated in the TPQIC formula can be used to choose the most cost-effective power quality improvement. Some power quality improvement purchasers, because of the uncertainty of the assumptions and values used in the TPQIC formula, choose to use sensitivity analysis. A sensitivity analysis involves determining how sensitive the TPQIC results are to assumptions and values in the TPQIC formula. How to perform a sensitivity analysis will be discussed later in this chapter.

The power quality improvement that provides the minimum TPQIC is the most cost-effective improvement. As shown in Figure 8.1, the power quality improvement costs increase as more expensive improvements result in larger reductions in the cost of the power quality problems. In most cases, it is not cost-effective to reduce all power quality problems to zero. It is better to find the cost-effective power quality improvement

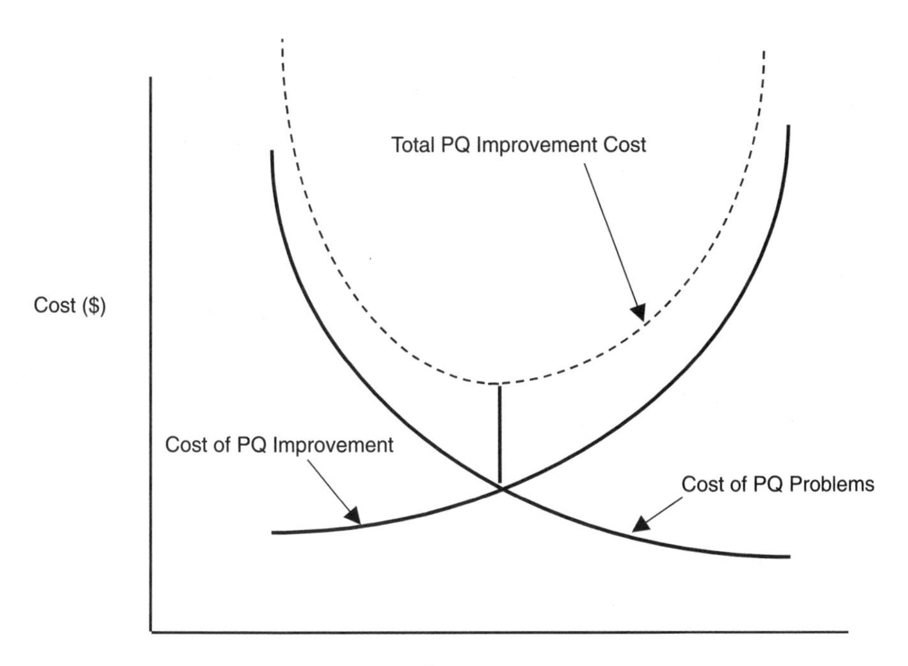

Figure 8.1 Minimizing power quality improvement costs.

that minimizes the TPQIC or total power quality improvement cost. In order to determine the cost-effective alternative it is important to perform the analysis in a step-by-step systematic approach.

Steps in Performing an Economic Analysis

Of course, the decision maker's determination of the TPQIC requires collecting data before completing a power quality economic analysis. A logical and systematic process for performing the economic analysis provides another important means of selecting the most cost-effective solution to a power quality problem. This requires that the data and the analysis of the data be done systematically. With the right data and procedures, the decision maker can approach the analysis in a logical step-by-step manner. This will reduce the chances of making erroneous and costly conclusions, and prevent the utility and end user from wasting time and money. What steps are required to perform a comprehensive power quality analysis? As shown in Figure 8.2, there are five basic steps to performing a power quality economic analysis:

1. Determine power quality problem base cost.
2. Determine the cost of power quality improvement alternatives.

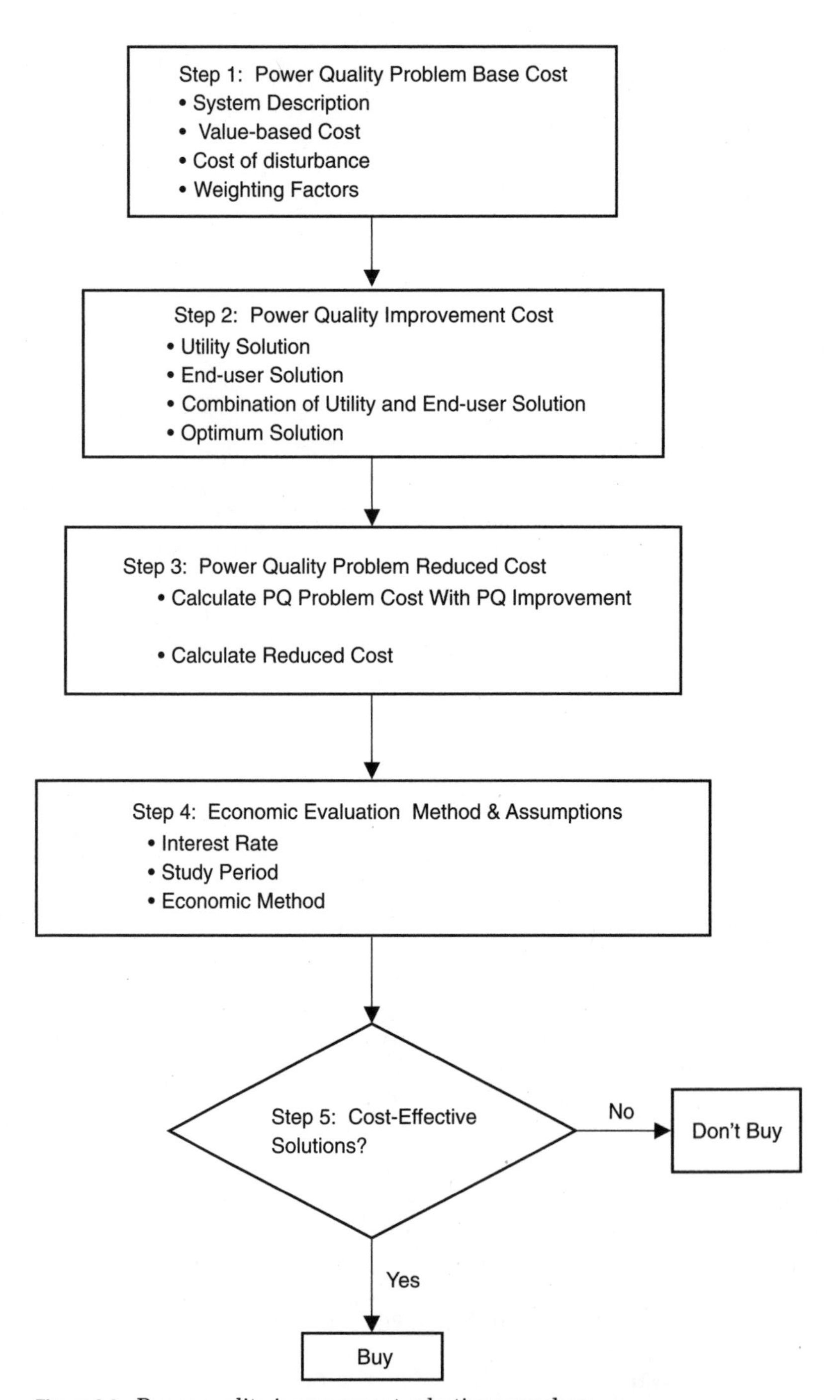

Figure 8.2 Power quality improvement selection procedure.

3. Determine the reduced cost of the power quality problem for each power quality improvement alternative.
4. Determine the economic evaluation method and assumptions.
5. Determine the cost-effective solution.

Each step provides a systematic approach to conducting the analysis.

Step 1: Determine Base Power Quality Problem Cost

Power quality experts determine the cost of the power quality problem by first describing the power system affected by the problem. This information can be obtained from the power quality survey described in the previous chapter. As shown in Figure 8.3, this information should include a diagram of the system experiencing the problems, including the utility substation and feeders and end-user facility distribution system and equipment. Next, data describing the event or events causing the power quality problems need to be examined. Different types of disturbances will have different cost impacts on the same facility. The four basic types of disturbances are interruptions, voltage sags, harmonics, and flicker. The purpose of this data is to determine the cost impact to

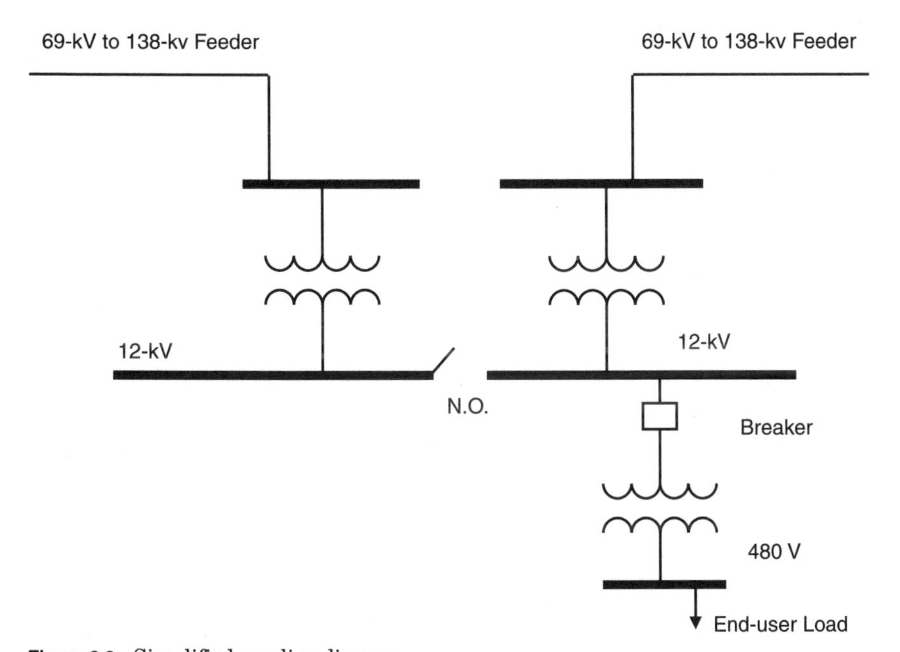

Figure 8.3 Simplified one-line diagram.

the utilities and their customers of these four basic types of disturbances. A value-based economic analysis provides the optimum solution regardless of the source of the power quality problem.

Value-based economic analysis

Traditionally, analysts separate utility and end-user power quality problems and solutions when performing an economic analysis. However, power quality problems and solutions do not recognize the location of the revenue meter. Consequently, many analysts use a value-based approach to their economic analysis.

A value-based economic analysis recognizes that the utility and the end user place different values on power quality problems. It also recognizes that different end users value power quality differently. For example, industrial and commercial end users normally value power quality higher than residential users of power. This type of analysis takes into account that the end user and the utility are connected financially as well as electrically.

As shown in Figure 8.4, both the utility and its end-use customer experience the cost impact of a power quality problem. They also mutually benefit from power quality improvements that take into consideration the value of the improvements to the utility and its customer. In a competitive deregulated situation, the cost to the utility because of poor power quality could possibly result in losing a customer to another utility or legal claims against the utility. A value-based approach requires the analyst to estimate the cost of the power quality problem and the benefits of the power quality improvement to the utility as well as to the end user.

End-user perspective. Historically, the impact of power quality problems has been the concern of industrial and commercial users of electricity. This is changing with the increase in home-based businesses that use computers, faxes, and laser printers. However, the current focus of utilities is on their large industrial customers who experience cost impacts from $3000 to $10,000 per event per customer. There have been reports of as much as $250,000 per event in a semiconductor plant. Even disturbances of less than a second can have large cost impacts when they happen to critical loads.

Critical loads include facilities like computer centers, paper mills, semiconductor factories, arc furnace foundries, and plastic plants that depend on electricity that is free of power quality problems to produce their products. These types of loads experience large cost impacts often because of two factors: the high cost of production, usually measured in dollars per hour, and the extended time it takes to bring equipment

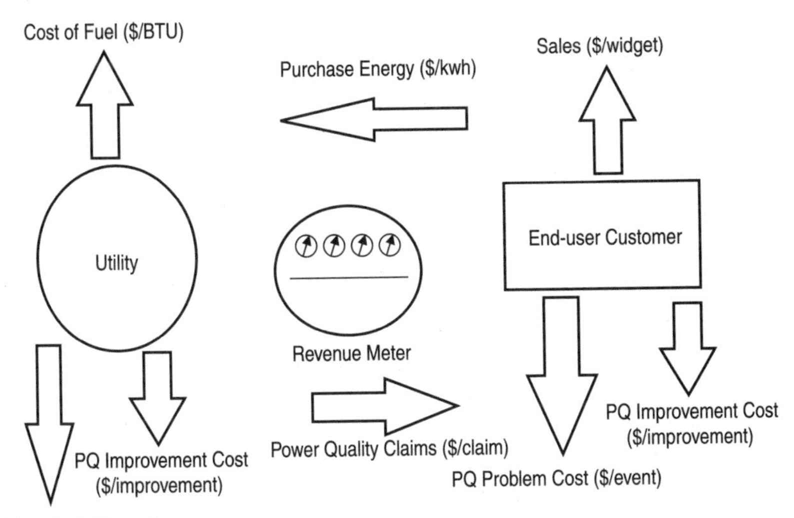

Figure 8.4 Utility and end-user power quality cash flow.

on line after a disturbance. For instance, Table 8.1 summarizes an EPRI report that compares the downtime cost impacts of various types of industries caused by the two most common types of disturbances: interruptions and voltage sags. In addition to the cost impact to end users, disturbances cost utilities money as well.

Utility perspective. Traditionally, most utilities have a different perception of power quality than their residential, commercial, and industrial customers. They have evaluated its cost impact on the basis of loss of revenue during a disturbance, liability claims, and cost to maintain and repair damaged equipment.

While industrial and commercial end users usually quantify the power quality cost impact in dollars per event, many utilities prefer to evaluate the cost impact in dollars per kilowatt-hour. They calculate this by dividing the cost of lost production by the electrical energy not consumed during the disturbance. In a paper presented at the IEEE 1994 Industry Applications Society Annual Meeting entitled "Impact of Fast Tripping of Utility Breakers on Industrial Load Interruptions," the authors estimate typical values to be $1 to $4/kWh of lost revenue. Besides the loss of revenue during a disturbance, utilities experience other cost impacts from poor power quality.

Utilities, like any manufacturer of goods and services, can experience the cost of liability claims due to poor power quality. Liability cost hinges on whether the courts treat electrical power as a product or a service. If it is deemed a product, then the utility can be held responsible for the

TABLE 8.1 Cost Impact of Various Types of Interruptions and Voltage Sags on Critical Loads.

Type of load	Disturbance	Cost impact
Computer center	2-second interruption	$600,000
Large machining plant	0.1-second voltage sag	$200,000
Paper mill	0.005-second voltage sag	$50,000
Semiconductor fabricator	Voltage sag	$1,000,000

SOURCE: Courtesy of EPRI.

reliability of that product. Most court cases specify that electricity does not become a product until it passes through the revenue meter. However, this definition of a product is ambiguous when the end user can have its own source of electricity via a UPS. Besides, electromagnetic interference can affect sensitive equipment that is not even connected to a utility's distribution system, let alone a revenue meter. Utilities can experience large liability claims because of poor power quality. For instance, in September 1992, a California judge awarded a mushroom grower $5.5 million, to be paid by the power company, for damages caused by a power outage.

With the advent of restructuring of the utility industry and increased competition, another major cost to utilities is the potential loss of a customer because of frequent power quality problems. There is also the potential loss of future customers because a utility has a reputation for poor power quality. Certainly, the utility often bears the cost of resolving a power quality problem. The potential of losing customers because of poor power quality increases as the utility industry is restructured and becomes more competitive. These costs vary from one utility and situation to another. These costs are sometimes intangible and difficult to quantify.

Because of the threat of losing major customers through poor power quality, some utilities have chosen to pay their customers for the cost of poor power quality. For example, in 1995, Detroit Edison signed a power quality agreement with three of its major automobile manufacturing customers, Chrysler Corporation, Ford Motor Company, and General Motors. In these contracts, Detroit Edison agrees to pay these customers if the power they deliver does not meet certain voltage interruption and voltage sag target values. Since 1995, Detroit Edison has paid these customers millions of dollars for interruptions. As shown in Figure 8.5, from 1995 to 1998, Detroit Edison's increased maintenance on specific lines, replacement of problem equipment, and installation of animal deterrents has resulted in 36 percent less interruption in 1998 than 1995. At the same time, the utility has been able to keep voltage sags to a minimum and paid only $230 for voltage sags in 1998.

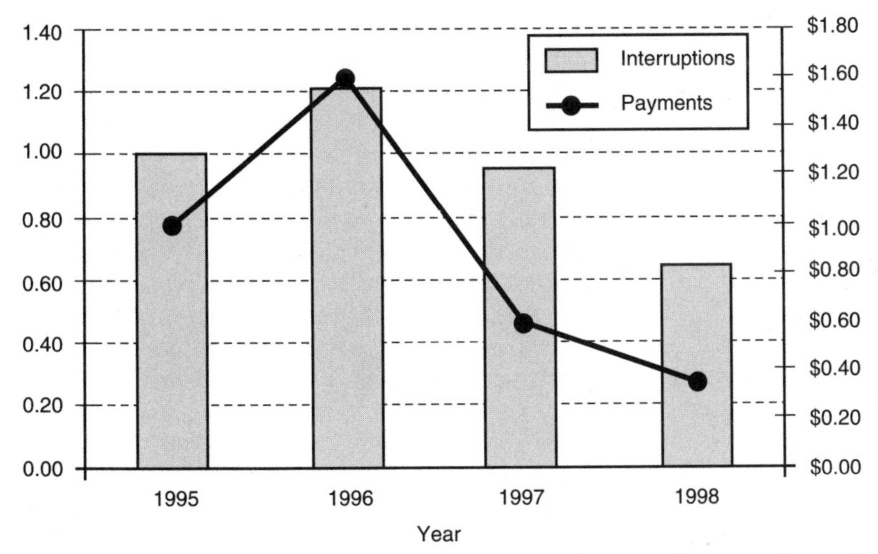

Figure 8.5 Detroit Edison interruption and payment history. (*Courtesy of Detroit Edison.*)

There is always the threat of regulatory penalties to utilities that provide poor power quality. Some regulators have introduced penalties for poor power quality. For example, Argentina has developed a formula for poor power quality caused by flicker and harmonics.

Another threat to utilities is customers that are not satisfied with the power quality they are receiving and resort to cogeneration or self-generation. Many high-tech companies have the resources to install generators and use the utility as backup. Many paper and lumber mills can use steam or wood as fuel to cogenerate electricity. Other businesses that are prone to use self-generators include electric equipment manufacturers, heavy machinery fabricators, computer software and hardware companies, and insurance companies.

If poor power quality is defined as "any power problem manifested in voltage, current, or frequency deviations that results in failure or misoperation of utility or end-user equipment," then the primary cost of the disturbance is the effect it has on both the utility and end-user equipment. The major contributing factor to the cost of poor power quality to both the utility and its customers depends on the type of disturbance and load.

Cost of the disturbance

There are two basic ways to calculate the cost of a disturbance. One way is to add up the financial losses associated with each load that is impacted by the disturbance. The other way is to perform a survey and statistical analysis of several end users and develop the cost-of-disturbance values associated with certain types of disturbances and loads. Both of these methods require the collection of cost data associated with the disturbance.

The cost of a disturbance involves the following three major losses: (1) product production, (2) labor, and (3) damaged equipment. "IEEE Standard 1346-1998, IEEE Recommended Practice for Evaluating Electric Power System Compatibility with Electronic Process Equipment," Annex A, provides a detailed description of how to calculate the cost of a disturbance. It suggests, as in a power quality survey, a need to involve all participants to determine the cost of the disturbance. This includes management as well as financial, operational, maintenance, and sales staff. Each one of these participants needs to be involved in completing a form similar to Table 8.2.

There are four basic types of disturbances that have a financial impact on utilities and their customers: interruptions, voltage sags, harmonic distortions, and flicker. Interruptions include various types of outages. Interruptions or outages can be initiated by transients that cause utility breakers and switches to operate. Next to voltage sags, they are the most common type of disturbance and usually have the largest cost impact.

Remember the importance of distinguishing an interruption from a voltage sag. Interruptions involve a complete loss of voltage in usually less than 1 second, while voltage sags are a reduction in voltage for less than 1 minute. Sometimes these two types of disturbances are confused, because they both may have a similar effect on sensitive equipment. However, the costs for mitigating voltage sags can be less than for mitigating interruptions. The cost impacts of voltage sags can affect one or two phases of a three-phase system, while interruptions usually affect all three phases of operation.

Interruptions

IEEE Standard 446-1995, "IEEE Recommended Practice for Emergency and Standby Power Systems for Industrial and Commercial Applications" (*The Orange Book*), pp. 41–42, provides a formula for calculating a rough estimate of the cost of an interruption:

$$\text{Total cost of a power failure} = E + H + I \tag{8.3}$$

where E = cost of labor for employees affected (in dollars)
H = scrap loss due to power failure (in dollars)
I = cost of start-up (in dollars)

The values of E, H, and I may be calculated as follows:

TABLE 8.2 Sample Cost-of-Disruption Evaluation Form

Item	Cost
Downtime related	
Increased buffer inventories (value of incremental inventories—WACC*)	______
Lost work	
Idled labor	
Disrupted process (worker-hours unloaded labor rate)	______
Starved process (worker-hour unloaded labor rate)	______
Lost production	
Lost profits (unbuilt product profit margin)	______
Makeup production	______
Overtime labor + premium	______
Overtime operating cost	______
Expedited shipping premiums	______
Late delivery fees	______
Cost to repair damaged equipment	
Repair labor	______
Repair supplies	______
Cost of replacement part availability	______
Or carrying cost of parts	______
Cost of recovery	
Secondary equipment failures (treat as repairs)	______
Recovery labor inefficiency	______
Product quality	
Replacement value of scrap (BOM† value + labor value)	______
Blemished product lost profit margin	______
Rework cost	
Labor	______
Manufacturing supplies	______
Replacement parts	______
Miscellaneous	
Customer's dissatisfaction	
Lost business	______
Avoided customers due to longer lead time	______
Fines and penalties	______
Other	______
Total	______

*Weighted average cost of capital.
†Bill of materials.
SOURCE: Courtesy of IEEE, Standard 1346-1998

$$E = ADC$$

$$H = FG$$

$$I = JKC + LG$$

where A = number of productive employees affected
B = base hourly rate of employees affected (in dollars)
C = fringe and overhead hourly cost per employee affected (in dollars) = $1.5B$

D = duration of power interruption (in hours)
F = units of scrap material due to power failure
G = cost per unit of scrap material due to power failure (in dollars)
J = start-up time (in hours)
K = number of employees involved in start-up
L = units of scrap material due to start-up

Interruption costs associated with outages and voltage sags should include the savings that occur during the disturbance. Otherwise there will be a tendency to overestimate the total cost of the disturbance. The savings during the disturbance include the cost of unpaid wages, unused raw materials, unused fuel, and damaged scrap material. The total cost of the disturbance can be summarized and simplified in the following formula:

$$\mathrm{DC} = \mathrm{LR} + \mathrm{OC} - \mathrm{OS} \tag{8.4}$$

where DC = disturbance cost (in dollars)
LR = lost revenue or lost sales of products resulting from outage (in dollars)
OC = outage costs like cost of restart, damage, and makeup (in dollars)
OS = savings during shutdown caused by the outage (in dollars)

Another way to calculate the cost of a disturbance is by the use of a survey of several utilities and their industrial and commercial end users. One survey in 1992 involved 210 large commercial and industrial end users and used the components of Eq. (8.3) to calculate the cost of disturbances for various types of outages and voltage sags. In this survey, it was found that the average cost of an outage was $9400. By regression analysis, the results of this survey and others were used to project the cost of a particular disturbance to a specific type of load. As shown in Table 8.3, the average total cost of outages varied from $7694 to $74,835.

Surveys of the cost of interruptions have been conducted in the last 10 years by Pacific Gas and Electric, Duke Power Company, Southern Companies, Southern California Edison, Niagara Mohawk, Bonneville Power Administration, and Saskatchewan Power. These surveys have been used to assess the value of increased reliability of the power system to the utility's customers.

Voltage sags

Calculating the cost impact of voltage sags is not as simple as calculating the cost impact of interruptions. A voltage sag's effect on sensitive

TABLE 8.3 Means of Components of Outage Costs for Large Commercial and Industrial Customers by Scenario

	Scenario				
Coat element	4-h outage, no notice	1-h outage, no notice	1-h outage, with notice	Voltage sag	Momentary outage
Production impacts:					
Production time lost (hours)	6.67	2.96	2.26	0.36	0.70
Percent of work stopped	91	91	91	37	57
Production losses:					
Value of lost production	$81,932	$32,816	$28,746	$3914	$7407
Percent of production recovered	36	34	34	16	19
Revenue change	$52,436	$21,658	$18,972	$3287	$5999
Loss due to damage:					
Damage to raw materials	$13,070	$ 8,518	$ 3,287	$1163	$2051
Hazardous materials cost	$ 323	$ 269	$ 145	$ 90	$ 136
Equipment damage	$ 8,421	$ 4,977	$ 408	$3143	$3239
Cost to run backup and restart:					
Cost to run backup generation	$ 178	$ 65	$ 65	$ 22	$ 22
Cost to restart elect. equip.	$ 1,241	$ 1,241	$ 171	$ 29	$ 29
Other restart costs	$ 401	$ 368	$ 280	$ 74	$ 149

TABLE 8.3 Means of Components of Outage Costs for Large Commercial and Industrial Customers by Scenario (*Continued*)

	Scenario				
Coat element	4-h outage, no notice	1-h outage, no notice	1-h outage, with notice	Voltage sag	Momentary outage
Savings:					
Savings on raw materials	$ 1927	$ 645	$ 461	$ 114	$ 166
Savings on fuel and electricity	$ 317	$ 103	$ 85	$ 9	$ 12
Value of scrap	$ 2337	$ 874	$ 450	$ 140	$ 228
Labor management recovery:					
Percent using overtime	33	26	25	6	7
Percent using extra shifts	1	1	0	1	1
Percent working labor more	3	4	4	4	7
Percent rescheduling operations	4	5	5	0	1
Percent other	1	2	2	0	1
Percent not recovering	59	62	64	89	84
Labor costs and savings:					
Cost to makeup production	$ 4854	$ 1709	$ 1373	$ 60	$ 254
Cost to restart	$ 665	$ 570	$ 426	$ 114	$ 192
Labor savings	$ 2159	$ 644	$ 555	$ 0	$ 0
Total	$74,835	$39,459	$22,973	$7694	$11,027

SOURCE: M. J. Sullivan, T. Vardell, and M. Johnson, "Power Interruption Costs to Industrial and Commercial Consumers of Electricity," *1996 Industrial and Commercial Power Systems Technical Conference,* May 6–9, 1996,

equipment depends on the magnitude and duration of the voltage sag as well as the type of sensitive equipment. Less severe voltage sags will affect less equipment and have a lower cost impact than more severe voltage sags. Consequently, factors for weighting the effect of various levels of voltage sags for specific loads have been developed.

In calculating the cost of a voltage sag or interruption, it is important to realize that the process for determining the cost of the disturbance not only depends on the type of load but the type of disturbance as well. Weighting factors need to be determined for various types of interruptions and voltage sags and their effect on sensitive electronic equipment.

Weighting factors for interruptions and voltage sags

The power quality analyst needs to determine the impact of certain types of power quality disturbances on various types of sensitive equipment. This means giving a weighting factor to various types of events. For example, various rms values of voltage sags will have different impacts on different types of sensitive electronic equipment. They can determine the weighting factors by taking measurements of disturbances within a specific time period. Then the magnitude as well as the duration of the event can be categorized and weighed as to its effect on equipment sensitivity. Table 8.4 illustrates the use of weighting factors for interruptions and voltage sags for a plastic extruder plant where each event costs $20,000.

Finally, creation of a sensitivity chart and table will illustrate the magnitude and duration of various power quality disturbances and provide a means to evaluate alternative power quality improvements and their locations on the utility and end-user systems. From these charts and tables, the analyst can determine the cost of various types of power quality improvements. Another type of disturbance that has cost impacts on industrial and commercial end users is harmonic distortion.

TABLE 8.4 Voltage Sag Weighting Factors for Economic Analysis

Category of event	Weighting for economic analysis, %	Expected number per year	Equivalent interruptions per year
Interruption	100	6	6.0
Sag below 50%	100	0	0.0
Sag between 50 and 70%	50	6.5	3.3
Sag between 70 and 80%	20	8.5	1.7
Sag between 80 and 90%	10	42.5	4.3
Total		63.5	15.2

SOURCE: *Courtesy of Electrotek Concepts, Inc.*

Harmonic distortion

As shown in Figure 8.6, various nonlinear loads, such as computers with switched-mode power supplies, motors with adjustable-speed drives, and fluorescent lights with electronic ballast, produce harmonics that combine and flow through distribution transformers to the utility's distribution system. They can even flow onto adjacent end user's distribution systems. Therefore, these harmonic currents can have a cost impact to the utility as well as its customers.

End users with nonlinear loads usually generate harmonics. They find that cost impacts from harmonics are not as easy to determine as cost impacts from interruptions. The cost to end users comes when the harmonic currents add to the normal load and increase losses and loading on their distribution systems. The increased losses reduce the capacity of the system, including conductors, transformers, and motors. The increased loading generates heat and accelerates the aging of power equipment, like transformers and motors. Other cost impacts of harmonics include noise and vibration, reduction in motor torque, decreased power factor, decreased performance of television sets and relays, and inaccurate readings from induction watt-hour meters. For instance, as shown in Table 8.5, a case study of a building with 240 distributed computers and other electronic equipment operating 12 hours per day, 365 days per year with a load of 60 kW harmonics produced increased losses of 4802 W at a cost of $2101 per year (based on a cost of energy of $0.10/kWh).

Electrical utilities incur costs from harmonic currents similar to end-user costs. They experience voltage distortions that affect the operation of their equipment and cause increased power loss on overhead conductors, underground cables, and transformers. The increased loading from harmonic currents also accelerates the aging of utility transformers and generators. In fact, utilities typically derate their transformers and generators up to 25% because of the additional heating from harmonics. Some utilities are setting harmonic limits for their customers based on IEEE Standard 519-1992. Others are installing special revenue meters to charge their customers for harmonics. Utilities and end users can spend $4000 to $5000 or more to perform the engineering study to analyze harmonic problems and determine cost-effective solutions. Another power quality problem whose cost impact is difficult to determine is flicker.

Flicker

Flicker is a subjective phenomenon. Consequently, it is difficult to determine the direct cost of its effect. It affects the fundamental quality of utility service—that is, the ability to provide lighting that is steady and consistent. Certainly it can affect production in an office or factory that needs steady lighting for its employees to be productive. The cost

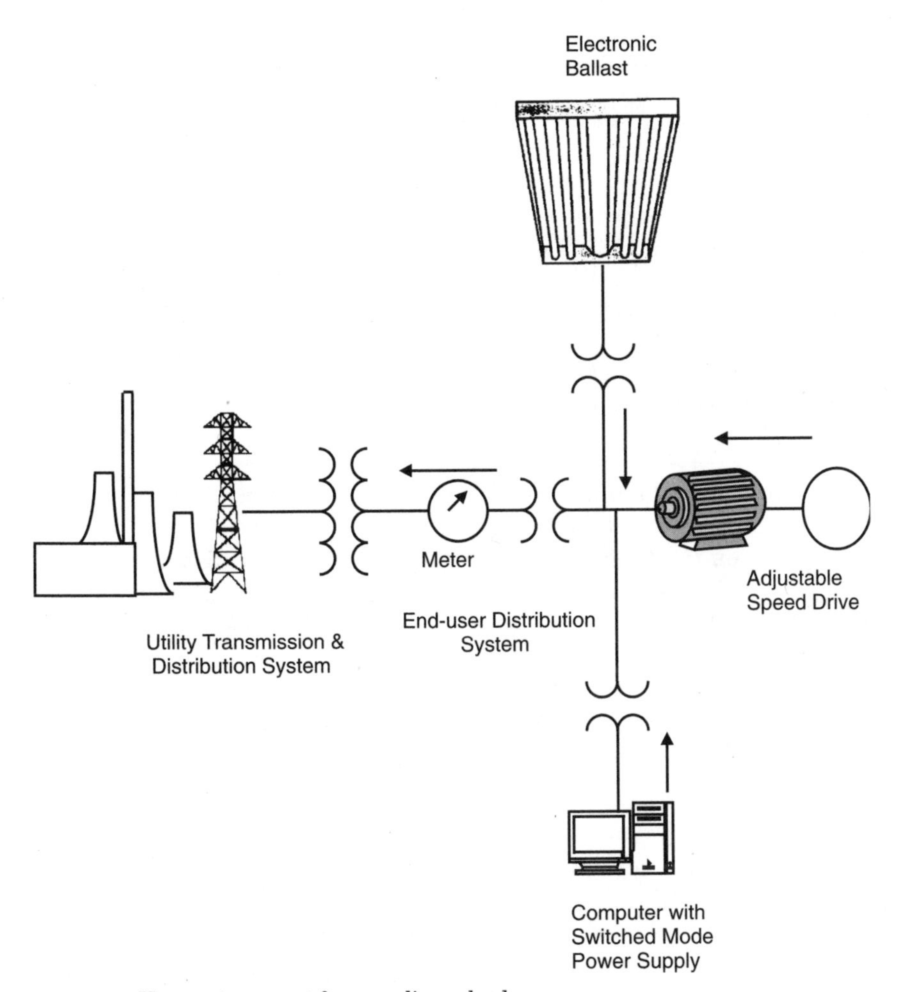

Figure 8.6 Harmonic current from nonlinear loads.

of flicker is usually based on the cost of mitigating it when the complaints become significant. This may involve curtailing or shutting down the source of the flicker, like an arc furnace, welder, or large motor starting up. For example, Southern Indiana Gas and Electric Company had 107 different residential customers complain about flicker from a new resistive spot welder and sought to remedy the situation.

Step 2: Determine Power Quality Improvement Cost

Analysts need to next evaluate the performance of various power quality improvement alternatives. Why? They need to identify effective

TABLE 8.5 Summary of Harmonic-Related Losses and Costs per Year

	Current THD, %	Cable length, ft	Harmonic loss, W	Harmonic cost/year, $
Cable 1	100	200	1320	578
Cable 2	100	50	712	310
Transformer	100—primary 30—secondary	N/A	2747	1203
Cable 3	30	150	23	10
Total			4802	2101

SOURCE: Tom Key and Jih-Sheng Lai, "Costs and Benefits of Harmonic Current Reduction for Switch-Mode Power Supplies in a Commercial Office Building," *1995 Industrial Application Society Annual Meeting,* October 1995, Orlando, Florida,

improvements and eliminate ineffective improvements. A specific power quality improvement will have different levels of effectiveness, depending on the type and level of the power quality disturbance.

They need to examine power quality improvement alternatives that include power conditioning equipment. First, they begin by determining the cost of wiring and grounding improvements, power conditioning equipment, and energy storage devices. Depending on the type of power quality problem, they need to examine power quality improvements in both the end user's and the utility's systems.

End-user power quality improvements

Analysts need to identify the type of power quality improvements needed inside the end user's facilities or at the service entrance. Usually, the lowest-cost improvements can be implemented at the end-use equipment. This may include isolating the critical loads and installing protection controls that protect them from power quality disturbances, like outages and voltage sags. It may involve the installation of dc capacitors to provide power to critical equipment during a disturbance. This involves an engineering analysis in three parts. First the analysis identifies the critical loads. Second, it specifies protection and sizes it correctly. Finally, it coordinates the protection scheme with the entire process. After evaluating the end-use equipment, analysts examine the cost and effectiveness of power quality improvement technologies at the service entrance.

Interruption and voltage sag improvement at the service entrance. The type of power quality improvements at the service entrance will again depend on the type of disturbance. Certainly, once the critical loads are identified and isolated, an appropriate backup source can be provided for them. This can be a separate feeder with an automatic switch or a UPS. For example, possible alternative improvements at the service entrance for a voltage sag or power interruption would include a UPS or backup feeder. Table 8.6 provides a list of the various types of service entrance UPSs and their costs.

Harmonic mitigation cost. The two methods for mitigating harmonics are installation of filters in the circuit and redesign of nonlinear loads that produce harmonics. Both of these methods have cost impacts. In the case of harmonics caused by switched-mode power supplies in a commercial building, the modification of the switched-mode power supply design so that it does not inject harmonic currents into the building's power system is the least costly of these options. It is not usually implemented unless mandated by regulators, because it increases the price of computer equipment. For example, in Europe, by the year 2001, manufacturers of information technology equipment will be required to meet the harmonic limits set by the International Electrotechnical Commission (IEC). This requirement is expected to increase the cost of information technology equipment 2 to 5 percent.

The cost to filter out the harmonics depends on the types of filters and their location. One case study found that the location of filters in the neutral wire of the service panel provided the lowest-cost filter alternative. Table 8.7 illustrates the cost of various passive methods for eliminating harmonics. Table 8.8 illustrates the cost of various active methods for eliminating harmonics.

Flicker elimination cost. The cost of eliminating flicker caused by arc furnaces is usually quite high. It depends on the size of the arc furnace load and the strength of the utility power system serving it. The cost

TABLE 8.6 Cost of Service Entrance Technologies for Power Quality Improvement

Power conditioning technology	Typical cost, $/kVA	Comments
UPS	700	Full protection
Synchronous motor-generator with flywheel	500	2-second ride-through without diesel option
Energy storage technologies	800	Shorter ride-through than UPS
Secondary static switch	100	Requires independent supply

SOURCE: Mark McGranaghan, et al. 1997. "Economic Evaluation Procedure for Assessing Power Quality Improvement Alternatives." *Proceedings of PQA '97 North America,* March 3–6, Columbus, Ohio.

TABLE 8.7 Cost of Passive Harmonic Elimination Equipment Options

Mechanism	Equipment	Features	Approximate price/kVA
Dilute or absorb harmonics	Existing power system capacity to dilute and absorb	■ Uses power system natural tolerance and diversity ■ Relies on system restricting and canceling effects ■ Higher watt losses and reduced capacity	
Restricts harmonics	Series inductor at load generator, low pass, 1 or 3 phase	■ Simple and relatively low cost ■ Reduces voltage at load	$30
Cancels specific harmonics	Phase-shifting transformer at load	■ 3 phase, multibridge ■ Complex structure, bulky	$100
Traps a specific harmonic	Series or parallel single-tuned filter at or near load	■ Compensates single harmonic ■ Possible under- or overcompensation, bulky	1 phase $200–$400 3 phase $30
Traps several harmonics	Series or parallel multituned filter at or near load	■ Normally tuned to two adjacent odd harmonic frequencies ■ Possible under- and overcompensation	1 phase $200/freq. 3 phase $30/freq.

SOURCE: Jih-Sheng Lai and Thomas Key. "Effectiveness of Harmonic Mitigation Equipment for Commercial Office Buildings." *IEEE 1996 Industry Applications Society Annual Meeting,* October 8–10, 1996,

TABLE 8.8 Cost of Active Harmonic Elimination Equipment Options

Mechanism	Equipment	Features	Approximate price/kVA
Cancel harmonic currents	Parallel filter at or near load commonly used topology	■ Suitable for current source converters or current harmonic loads ■ Compensates harmonic currents in real time	$500
Reshape voltage fundamental	Series filter at load center requires a current transf.	■ Suitable for voltage source converters or voltage harmonic loads ■ Real-time compensation of voltage	$750
Cancels current and voltage harmonics	Series and parallel active converters at or near load	■ Real-time compensation of both voltages and currents ■ Most expensive, commercial products available	$1000

SOURCE: Jih-Sheng Lai and Thomas Key. "Effectiveness of Harmonic Mitigation Equipment for Commercial Office Buildings." *IEEE 1996 Industry Applications Society Annual Meeting,* October 8–10, 1996,

impact includes the cost to purchase, install, operate, and maintain a static VAR (volt-amperes reactive) compensator at the flicker source to keep the voltage steady under varying load conditions and thus solve the flicker problem. Static VAR compensators to eliminate flicker produced by large arc furnaces can cost $1 to 2 million or more.

Utility-side power quality improvements

Utilities can provide various power quality improvements on their systems. If the disturbance is originating from the utility, the most obvious solution is to eliminate the disturbance. They can minimize faults by trimming trees, installing animal guards, coordinating the switching of their capacitors with their customers, grounding their distribution towers better, using arresters to divert the fault away from the end user's facilities, and improving their maintenance practices. They can also raise the terminal voltage so that when disturbances occur the voltage does not drop below the sensitive level of the end-use equipment (usually 75 to 80 percent).

Interruptions and voltage sag power conditioning technologies. One study of a large industrial end user experiencing power quality problems due to interruptions and voltage sags looked at four basic power conditioning technologies. These four options were current-limiting feeder reactors, primary static switches, dynamic voltage restorers (DVRs), and static voltage regulators (SVRs). Table 8.9 provides a summary of the features and costs of each of these devices.

Flicker and harmonics power quality improvements. The source of flicker and harmonics is usually the end user rather than the utility. Consequently, any power quality improvements to flicker and harmonics needs to be made in the end-user facilities rather than on the utility side of the meter.

Before beginning the economic evaluation, analysts need to determine the benefit of each power quality improvement. This basically involves determining the reduced power quality problem cost resulting from each power quality improvement.

Step 3: Determine Reduced Power Quality Problem Cost

Different power quality improvement technologies have different effects on reducing the cost impact of particular disturbances on specific types of loads. In the case of interruptions and voltage sags, the different power quality improvements have varying degrees of effectiveness.

TABLE 8.9 Cost of Utility Power Quality Improvement Technologies for Interruptions and Voltage Sags

Device	Equipment	Features	Approximate price
Current-limiting feeder reactor	May need to replace main substation transformer	■ Limits voltage sag during parallel distribution faults ■ Not effective for transmission faults	$0.5–1 million
Primary static switch	Requires parallel feeders and split bus	■ Protects for all distribution faults and outages ■ Switches to alternative source in 4 ms ■ Not effective for transmission events	$600,000 for 10-MW load
Dynamic voltage restorer (DVR)	Uses storage capacitor	■ Provides voltage correction for 3-phase sags of 50% ■ Corrects for distribution and transmission faults	$3 million
Static voltage regulator (SVR)	Uses SCRs to switch taps on an autotransformer	■ Provides voltage boost for sags as low as 50% ■ New technology with little history of experience	$150/kVA

SOURCE: Siddharth Bhatt. 1998. "Economic Decision Making Methodology for Power Quality Costs and Solutions Applicable to Both Sides of the Meter." *Proceedings of PQA '98 Southern Hemisphere Conference,* November 9–11, Cape Town, South Africa.

Weighting factors for various types of technologies provide a means to evaluate the effectiveness of these technologies.

Interruption and voltage sag reduction technologies

A study of a plastic extruder plant experiencing interruptions and voltage sags looked at four technology alternatives for reducing voltage sags and interruptions. These technologies included controls protection, service entrance energy storage devices, installation of a primary static switch, and a combination of controls protection and a static switch. Table 8.10 compares the effectiveness and cost of these alternatives to mitigate various types of voltage sags.

Benefits of filters to reduce or eliminate harmonics

Quantifying the benefits of filters to reduce or eliminate harmonics includes calculating the energy savings and slowing down of equipment aging. Most studies evaluate only the energy savings that result from reduced harmonics. One study analyzed the effect of various types of filters and their location in a building's power system serving a 60-kW load to determine their impact on reducing harmonic losses. Table 8.11 summarizes the reduction in cable and transformer losses resulting from passive filters located in the branch circuit and load center.

Benefits of reducing flicker

Quantifying the benefits of reducing flicker is subjective. It primarily involves reducing customer complaints to a reasonable level. It is difficult to put a clear monetary value to customer complaints. Most flicker studies focus on reducing flicker to a level that no longer is visible to the human eye.

Step 4: Determine Economic Analysis Method and Assumptions

An understanding of power quality economics is necessary to weigh the power quality problem cost against the power quality solution benefits. As with all economic analysis, the time value of money over the life cycle of the alternatives needs to be evaluated. Power quality solutions and loss production savings occur over time and must somehow be compared to the initial cost of purchasing and installing the power quality improvements. There are basically three standard methods for

TABLE 8.10 Weighting of Power Quality Improvement Alternatives by Disturbance Intensity

Type of condition affecting customer	Weighting	Base performance, events/year	Reduction with controls protection	Reduction with service entrance energy storage	Reduction with primary static switch	Reduction with static switch and controls protection
Interruption	1.0	6.0	0%	60%	100%	100%
Sag below 50%	1.0	0.0	0%	90%	90%	90%
Sag between 50 and 70%	0.5	6.5	50%	90%	50%	70%
Sag between 70 and 80%	0.2	8.5	85%	95%	30%	90%
Sag between 80 and 90%	0.1	42.5	95%	95%	10%	92%
Total events affecting plant		63.5	15.2	5.6	47.45	6.2
Total events weighted		15.2	8.4	3.0	6.6	1.5

SOURCE: *Courtesy of Electrotek Concepts, Inc.*

TABLE 8.11 Energy Savings Benefits of Harmonic Filters in an Office Building

Location	Branch circuit passive filter	Load center passive filter
Cables, W	1993	1191
Transformer, W	2591	986
Total savings for 60-kW load, W	4584	2177
Percent savings for 60-kW load	7.6	3.6

SOURCE: Jih-Sheng Lai and Thomas Key. "Effectiveness of Harmonic Mitigation Equipment for Commercial Office Buildings." *IEEE 1996 Industry Applications Society Annual Meeting,* October 8–10, 1996.

evaluating alternative power quality improvement choices. These three methods are

1. Equivalent investment cost
2. Present worth
3. Benefit to cost

Each one of these methods will be discussed as it applies to the initial cost of the power quality solution and the cost to operate and maintain the power quality improvement. Each method must be applied to the power quality economic formula in such a way that the various parts of the formula are compared on an equitable basis. How to determine the power quality economic method as it relates to the power quality economics formula is a matter of company policy and personal preference. The economic method should have no effect on the decision as to what power quality improvement to make. In evaluating power quality improvements, the power quality economic method is applied to the components of the total power quality improvement cost:

$$\text{TPQIC} = \text{PQIB} - \text{PQIC} \tag{8.1}$$

where TPQIC = total cost of the power quality improvement
PQIB = power quality improvement benefit in \$/year
PQIC = power quality improvement cost in \$/year

The resulting numbers derived from any one of these methods applied to the TPQIC formula can be used to choose the most cost-effective power quality improvement. Some power quality improvement purchasers, because of the uncertainty of the assumptions and values used in the TPQIC formula, choose to use sensitivity analysis. A sensitivity analysis involves determining how sensitive the TPQIC results are to assumptions and values in the TPQIC formula. How to perform a sensitivity analysis will be discussed later in this chapter.

Power quality improvement—purchaser perspective

The purchaser, whether a utility or an end user of power quality improvement, such as power conditioning equipment, wants to make a decision as to what power quality improvement to make with a minimum amount of difficulty. An understanding of the various methods for performing an economic analysis of a power quality improvement purchase is essential to reducing that difficulty. The purchaser needs to decide on an economic methodology that he or she is most comfortable in using. It is important the purchaser consistently use the same economic method throughout the purchasing decision making process. This allows the supplier of the power quality improvement to provide one that meets the needs and values of the customer and takes into consideration the life cycle of the improvement.

Life cycle

In performing any kind of economic analysis it is necessary to take into account the life cycle cost of the power quality improvement. Life cycle costing is the fundamental concept used in deriving the TPQIC formula. It involves calculating the total cost of ownership over the life span of the power quality improvement. Only then can the reduced cost of power quality problems be compared to the cost of purchasing, operating, and maintaining the power quality improvement. Then what is the life span of a power quality improvement? Is it based on its expected life before failure? Or is it based on its expected life before replacement?

Most utilities and some commercial and industrial users use the expected life before replacement to evaluate the power quality improvement. This is because the improvement will probably have to be replaced because of changes in the facility requirements before it fails.

There are many factors that affect a power quality improvement's life. Anything that affects the effectiveness of the power quality improvement reduces its life. Such things as overloading the facilities, transients, poor wiring and grounding, and extreme temperatures affect the life of power conditioning equipment and other power quality improvements. High voltage is one of the major causes of reduced power conditioning equipment life.

Most power quality improvement studies assume a 5-year time frame. This time frame allows for the uncertainty of future changes in the facility that may reflect on the effectiveness of the power quality improvement.

Time value of money

Each one of the economic analysis methods involves the time value of money. What is the time value of money? Time value of money means that money increases in value over time, depending on the return on the investment. If money is deposited to an interest-bearing savings account or money market, it could appreciate in value to the tune of about 5 to 6 percent a year, while if money is invested in a stock or mutual fund, it could appreciate 10 to 20 percent or more a year. Each company or person has an expectation on the time value of money. But to ignore its value over time is not practical. Both the cost and the benefit of the power quality improvement have time value.

An understanding of simple interest rate is essential to an understanding of the time value of money. Simple interest rate, carrying charge rate, minimum acceptable rate of return, cash flow diagrams, and the various engineering economic factors necessary to perform power quality economic analysis are discussed in various engineering economic books and in *Energy Efficient Transformers,* McGraw-Hill, 1997, by Barry W. Kennedy. This chapter compares the various methods for evaluating a power quality improvements benefits and initial cost and explains how to use these methods starting with the equivalent first cost method.

Equivalent first cost

The equivalent first cost method is probably the most popular method. This is because it is the most straightforward of all the methods. This method involves taking the TPQIC formula and adding the various components without any additional modifications to those components. The price is the bid price of the power quality improvement supplied by the power conditioning manufacturer or the power quality expert. This price requires no modification. The other methods for evaluating power quality improvements are a modification of the equivalent first cost method, starting with the present worth method.

Present worth method

The present worth method requires referring each component of the TPQIC formula back to a common date. This provides a comparison of the cost and benefits of various power quality improvement alternatives. The life of the study and the carrying-charge rate (fixed-charge rate) remain constant in this method.

In the present worth method the annual cost and benefits are multiplied by the uniform series present worth (USPW) factor. This converts the equal annual cost values into a present worth value. Thus the levelized annual price and the levelized annual costs of operation and maintenance of the power quality improvement are each converted into present worth values. This can best be seen by looking at a harmonic filter example. In this case, the following assumptions are made: Cost of energy is $0.10/kWh, filter equipment life is 12 years, the discount rate per year is 8 percent, 240 personal computers operate 12 hours per day, 365 days per year, and maintenance and repair cost applies only at the subpanel. Table 8.12 summarizes the results of this economic analysis.

Again, because all the components of the TPQIC are multiplied by the uniform series present worth factor, the relative relationship between power quality improvement alternatives is not changed. The cost-effectiveness of the various alternatives will be the same in the present worth method as in the equivalent initial cost method and the benefit-to-cost method. The USPW can be determined from a table, assuming a minimum acceptable rate of return and the life of the study.

Benefit-to-cost method

The benefit-to-cost method involves taking the ratio of the annual benefit of the power quality improvement to the annual cost of the power quality problem. This provides a comparison of the cost and benefits of various power quality improvement alternatives. The life of the study and the carrying-charge rate (fixed-charge rate) remain constant in

TABLE 8.12 Present Value of Different Filter Options in an Office Building

Location	Branch circuit passive filter,	Load center $ passive filter, $
Purchase cost	12,000	1800
Floor space cost	0	1000
Installation cost	0	500
Maintenance/repair	0	462
Operating cost at $0.10/kWh	4,038	4,038
Life cycle cost	16,038	7,800
Life cycle energy savings	14,941	7,800
Present value	−1,097	−552
Daily cost	0.16	0.13

SOURCE: Jih-Sheng Lai and Thomas Key. "Effectiveness of Harmonic Mitigation Equipment for Commercial Office Buildings." *IEEE 1996 Industry Applications Society Annual Meeting,* October 8–10, 1996.

this method. This can best be seen by looking at the TPQIC formula for this method:

$$\text{Benefit/cost} = \text{PQIB/PQIC} \tag{8.4}$$

The power quality improvements with benefit-to-cost ratios greater than 1 and with the greatest benefit-to-cost ratio are the preferred alternatives. This method can best be understood by applying it to a large customer served by San Diego Gas and Electric that was experiencing outages and voltage sag power quality problems. In this case, the following assumptions were made: Total load equals 1000 kVA, life of the study is 5 years, and the discount rate per year is 10 percent. Table 8.13 summarizes the results of this economic analysis.

Step 5: Perform Economic Analysis

Before performing the economic analysis, analysts must take into consideration the uncertainty of the assumptions. This can be accomplished by performing an uncertainty analysis.

Uncertainty

Uncertainty as to the validity of the TPQIC values is always a concern of the utility or end user. No matter what method is used to calculate TPQIC, its value is uncertain. This uncertainty is increased by to the lack of stability of rates in the utility industry and the changes in power quality. This increased uncertainty causes a concern about the various assumptions required in calculating the TPQIC. The uncertainty is compounded by the need to evaluate over the life of the study. Reliance on the assumed future value of the cost of energy and capacity, escalation and discount rates, and load can effect the value of the TPQIC and the consequent decision to buy the most cost-effective power quality improvement. How does the power quality improvement purchaser deal with these uncertainties? Rather than rely on the absolute TPQIC values, many decision makers can use the sensitivity analysis approach.

Sensitivity analysis

Sensitivity analysis is a method for determining the effect of changes in the components in the TPQIC formula on the overall TPQIC results. It is accomplished by assuming small changes in those components and calculating the resulting change in the TPQIC. These components include fixed-charge rate, minimum rate of return, system energy and capacity cost, magnitude of operation and maintenance cost, the

TABLE 8.13 Benefit/Cost Comparison of Power Quality Improvement Alternatives

Power conditioning technology	Expected savings, XCi	Expected savings, $	Cost for solution, $/kVA	Size required, kVA	Total solution cost, $	Annual operating cost, % of total cost	Total annual cost, $	Benefit/ cost ratio
Feeder reactor	1.7	170,000		1,000,000	2	283,797	0.6	
Primary static switch	2.2	220,000	60	10,000	600,000	5	188,278	1.17
Electronic voltage regulator	4.10	410,000	200	10,000	2,000,000	10	727,595	0.56
UPS	5.40	540,000	800	2,000	1,600,000	25	822,076	0.6 6
Synchronous motor-generator with flywheel	5.15	515,000	400	2,000	800,000	25	411,038	1.25
Energy storage technologies	5.15	515,000	800	2,000	1,600,000	15	662,0 76	0.78
Secondary static switch	2.20	220,000	100	2,000	200,000	5	62,759	3.51
Protect controls with CVTs	3.26	326,000			50,000	5	15,690	20.78
Protect controls and selected drives	3.90	390,000			150,000	8	$15,570	7.56

SOURCE: McGranaghan, et al. "Economic Evaluation Procedure for Assessing Power Quality Improvement Alternatives." *Proceedings of PQA '97 North America,* March 3–6, 1997, Columbus, Ohio.

weighting factors of various levels of power quality, and projected inflation. The price of the power quality improvement is usually based on bid prices and needs to be evaluated for its sensitivity to possible change.

The first step in performing a sensitivity analysis is develop a base case of TPQIC, based on the mostly likely TPQIC component values. The next step is to vary the value of the TPQIC components from the base case. Then the changes in TPQIC can be plotted on a graph as one TPQIC component changes while the others remain the same. At the point where a particular TPQIC component incremental change results in a change in the decision as to which power quality improvement is cost-effective, the decision maker can decide if this change is likely to occur.

One method of performing a sensitivity analysis is to calculate the parameter sensitivity. Parameter sensitivity is defined as the percent input parameter variation required for a 1 percent change in the total levelized cost of ownership output parameter. Large numbers are an indication that the TPQIC is insensitive to change in parameters. The parameter sensitivity ratio can be determined from the following formula:

$$\text{Parameter sensitivity ratio in percent} = \frac{\text{percent change input} \times 10^2}{\text{percent change output}} \qquad (8.5)$$

There are other factors that need to be considered in evaluating the TPQIC of a power quality improvement. They include the environmental effects of equipment, equipment reliability, and the effect of operating temperature on TPQIC.

Computer Programs

Several custom power manufacturers have developed computer programs for calculating TPQIC. In addition to the programs developed by the custom power manufacturers, EPRI has developed a computer program for evaluating power quality improvements. The EPRI program is called the Economic Assessment Module of EPRI's Power Quality Diagnostic System. It is available to EPRI members of the power quality business unit. Other programs are available from various manufacturers of power conditioning and customer power equipment. They are all modifications to the TPQIC method. They are usually run on Windows. Some custom power manufacturers have developed computer programs.

References

1. Bhatt, Siddharth. 1998. "Economic Decision Making Methodology for Power Quality Costs and Solutions Applicable to Both Sides of the Meter." *Proceedings of PQA '98 Southern Hemisphere Conference,* Cape Town, South Africa, November, 9–11.
2. Dugan, R. C., et al., 1994. "Impact of Fast Tripping of Utility Breakers on Industrial Load Interruptions." *IEEE 1994 Industry Applications Society Annual Meeting,* October 2–6.
3. McGranaghan, et al. 1997. "Economic Evaluation Procedure for Assessing Power Quality Improvement Alternatives." *Proceedings of PQA '97 North America,* Columbus, Ohio, March 3–6.
4. Wagner, John P. 1992. "Cost of Power Quality in the ITE Industry." *Proceedings of Second International Conference on Power Quality End-use Applications and Perspectives,* vol. 1, Atlanta, Georgia, September 28–30.
5. Billmann, Jennifer. 1995. "Good Power-Quality Service Is Achievable." *Electric Light & Power,* vol. 73, no. 7, July, p. 23.
6. Sullivan, M. J., T. Vardell, and M. Johnson, 1996. "Power Interruption Costs to Industrial and Commercial Consumers of Electricity. *1996 Industrial and Commercial Power Systems Technical Conference,* May 6–9, p. 23–35.———, B. Noland Suddeth, Terry Vardell, and Ali Vojdani. 1996. "Interruption Costs, Customer Satisfaction and Expectations for Service Reliability." *IEEE Transactions on Power Systems,* vol. 11, no. 2, May, pp. 989–995.
7. Dugan, R. C., et al. 1999. "Using Voltage Sag and Interruption Indices in Distribution Planning." *IEEE Transactions on Power Delivery,* vol. 2, Singapore, January 31–February 4, pp. 1164–1169.
8. Dougherty, Jeff G., and Wayne L. Stebbins. 2000. "Power Quality: A Utility and Industry Perspective." *Energy User News,* March, vol. 26, no. 3, p. 12–15.
9. Muller, Dave. 1999. "Analyzing the Economics of Customer Power Solutions." *IEEE/PES 1999 Winter Meeting,* New York, NY, January 31–February 4, 1999.
10. Roettger, Bill, et al. 1998. "Evaluating Power Quality Solutions with PQDS Economic Assessment Module." *Proceedings of PQA '98 North America,* Phoenix, Arizona, June 8–11.
11. Key, Thomas, and Jih-Sheng Lai. 1995. "Costs and Benefits of Current Reduction for Switch-Mode Power Supplies in a Commercial Office Building." *IEEE 1995 Industry Applications Society Annual Meeting,* October 8–12, Orlando, Florida, 1995.
12. Lai, Jih-Sheng, and Thomas Key. 1996. "Effectiveness of Harmonic Mitigation Equipment for Commercial Office Buildings." *IEEE 1996 Industry Applications Society Annual Meeting,* October 8–10, 1996.

Chapter

9

Future Trends

Many future trends on both sides of the revenue meter will affect power quality issues. The increased use of computers and the deregulation of the electric utility industry are both future trends that will have a significant impact on power quality. These trends invoke several questions. How much will these trends affect power quality? Will power quality deteriorate and become more costly? Will there be more power quality services available to the end user? Will the end user have to become more knowledgeable about power quality? What new technologies are likely to develop in the future? Will new technologies become more sensitive to the quality of power? Will there be increased or reduced use of sensitive electronic equipment? Will the future use of more sensitive electronic equipment result in more power quality problems between end users?

Both utilities and their customers have questions about how to prepare for changes in power quality service. How can utilities or residential, commercial, and industrial end users of power respond in a way that takes into account the future trends of power quality? How can utilities provide power quality services without losing customers? What will end users do if their local utility does not care about power quality?

Many utility customers have questions about the effect of utility deregulation on the quality and reliability of the power they receive from their utilities. How will deregulation affect the utility industry's ability to provide reliable and quality power? Will the utilities continue to provide power quality services? Will they reduce or increase their power quality research and development efforts? Will they continue to develop ways to reduce the cost of power at the expense of power quality? How will they respond to the changes occurring in the electric utility industry?

The power quality industry has several concerns about the future of research and development. How will research organizations, like the Electric Power Research Institute (EPRI), respond to the changes in the utility industry? Will utilities continue to fund research and development for power quality products? How will their research and development priorities be affected by the changes in the electric utility industry?

This chapter will attempt to answer these and other questions about the future trends in power quality. Future trends are driven by the three factors of competition, technology, and deregulation. The purpose of this chapter is to examine how the future trends of competition, technology changes, and deregulation will affect the availability of power quality in the foreign and domestic markets. One of today's major uncertainties and a future trend that could have the greatest effect on power quality is the deregulation of the electric utility industry in the United States.

United States Electric Utility Deregulation

The deregulation of the utility industry blurs the roles of the utilities and their customers. It redefines who is responsible for delivering power quality. How do utility customers determine not only who is responsible for power quality but what level of power quality they should expect? The characteristics and sensitivity of end-user equipment within customer facilities ultimately define power quality requirements. Improving the energy efficiency and productivity of industrial and commercial facilities can sometimes result in the use of technology that either causes power quality problems or is sensitive to power quality variations. Historically, utilities have concerned themselves only with power quality problems that they cause to their customers.

Deregulation will change utilities from full-service to specific-service companies. Transmission, distribution, and generation functions in a utility could become separate companies. Several issues related to the roles of these new companies will become apparent. What are the power quality requirements at the interface between the transmission company and the distribution company? What is the base level of power quality that must be supplied by the distribution company to its end-use customers? What kinds of enhanced power quality services can the energy service company offer to end-use customers? How will these changes affect the market for power conditioning and measuring instruments? How will deregulation affect the reliability and power quality of the power system? The answers to all of these questions come from an examination of the new structure for the utility industry.

The electric utility industry will probably undergo the same radical change that occurred in the gas, trucking, telecommunications, and airline industries. Several factors contribute to causing the electric utility industry to change. The variation of electric rates from one state to another throughout the United States provides a strong impetus for the utility industry to change. As shown in Figure 9.1, Hawaii had the highest rate of 12.12 cents per kilowatt-hour in 1996, while Wyoming had the lowest rate at 4.31 cents per kilowatt-hour. The electricity consumer sees this inequity of rates as unfair. Why should an end user in one state pay more for electricity than an end user in another state? Why shouldn't all end users of electricity have the same access to lower rates? Yes, the consumer's desire for lower rates provides one of the main driving forces to deregulate the electric utility industry. But how will it affect power quality? Won't deregulation result in cheaper but lower-grade power? How will end users get the power quality they need? The more they understand

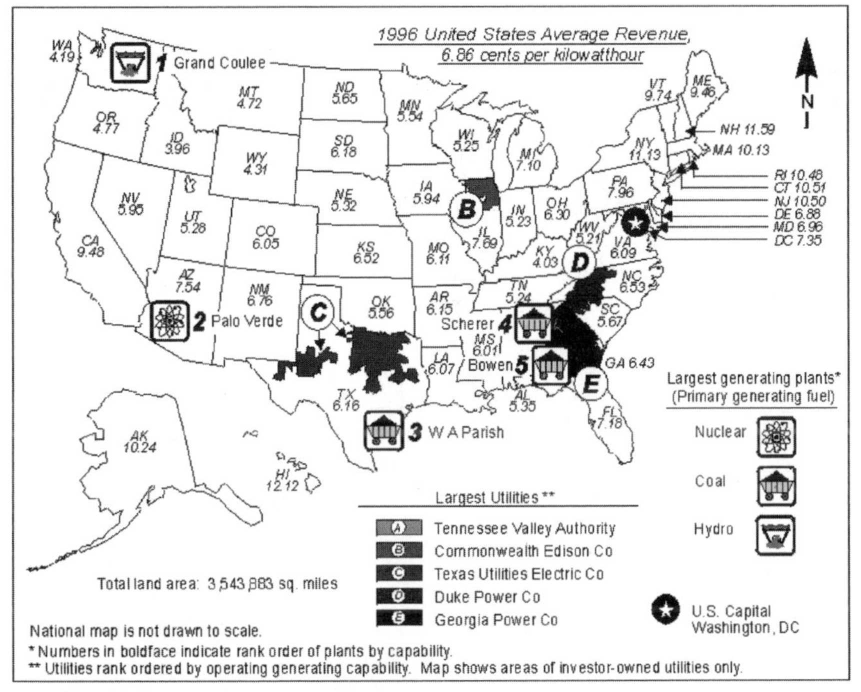

Figure 9.1 Electric utility rates by states. (*Courtesy of Energy Information Administration.*)

the electric utility industry and how it will change, the better they will be able to obtain the power quality they need.

U.S. electric power industry

The U.S. electric power industry has included traditional electric utilities, power marketers, and nonutility power producers. In 1996, approximately 3200 traditional utilities included investor-owned, publicly owned, cooperatives, and federal utilities. Investor-owned utilities represented 8 percent of the total number of utilities and 75 percent of the generation, sales, and revenue. The approximately 2000 publicly owned utilities represented 62 percent of the total utilities, provided 10 percent of the generation, accounted for 15 percent of retail sales and 13 percent of revenues, and included municipalities, public power districts, state agencies, and irrigation districts. The approximately 950 cooperative utilities represented 29 percent of the utilities, approximately 8 percent of sales and revenues, and about 4 percent of generation and generating capability. The 10 federal electric utilities included the U.S. Army Corps of Engineers in the Department of Defense, the Bureau of Indian Affairs and Bureau of Reclamation in the Department of the Interior, the International Boundary and Water Commission in the Department of State, the Power Marketing Administration in the Department of Energy (Bonneville, Southeastern, Southwestern, and Western Area), and the Tennessee Valley Authority (TVA). The three federal agencies that own and operate generation facilities include TVA, the U.S. Army Corps of Engineers, and the U.S. Bureau of Reclamation. Even though power marketers have bought and sold electricity, they have not owned or operated generation, transmission, or distribution facilities. Nonutilities include owners of qualifying facilities and wholesale exempt generators, cogenerators, and independent power producers. Figure 9.2 is a bar graph of the 1996 composition of the electric power industry in the United States.

Local, state, and federal agencies presently regulate these utilities. Federal agencies, like the Federal Energy Regulatory Agency (FERC), Nuclear Regulatory Commission (NRC), and the Environmental Protection Agency (EPA), regulate interstate activities and wholesale rates (sales and purchases between utilities); license hydroelectric facilities, nuclear safety, and waste disposal; and oversee environmental concerns. Each state regulates intrastate activities, plant and transmission line construction, and retail rates.

As a result of the 1965 power blackout in the northeast, the electric utility industry in 1968 formed the North American Electric Reliability Council (NERC). As shown in Figure 9.3, the NERC con-

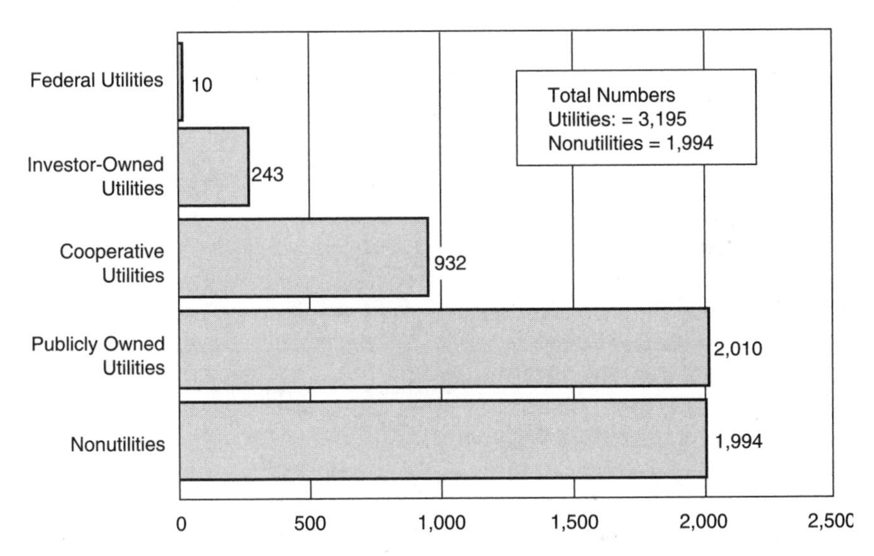

Notes:
• Data are preliminary. • Power marketers, Puerto Rico, and U,S, Territories are not included.
• Nonutilities represent the number of generating facilities, as these facilities are generally incorporated, and each is required to file Form EIA-867.

Sources:
Energy Information Administration, Office of Coal, Nuclear, Electric, and Alternate Fuels.

Figure 9.2 Composition of the U.S. electric power industry in 1996. (*Courtesy of Energy Information Administration.*)

sists of 10 regional councils in the 48 contiguous states, a portion of Baja California, Mexico, and portions of Canada bordering the United States. The councils coordinate the bulk power policies that affect the reliability and adequacy of electrical service of the interconnected power systems in their areas. NERC continues to function even though the electric utility industry began to restructure with the passage of the 1992 Energy Policy Act.

1992 Energy Policy Act

With the passage of the 1992 Energy Policy Act, the U.S. Congress set in motion the process of deregulating the electric power industry. The purpose of this legislation was to bring competition to an industry dominated for many years by monopolistic vertically integrated utilities. Congress initiated the following three steps to encourage competition in the electric power industry. The first step required utilities to provide open access to "wheeling" on their transmission and distribution (T&D) systems. The second step required utilities to separate their power business from their transmission and distribution business. The third step

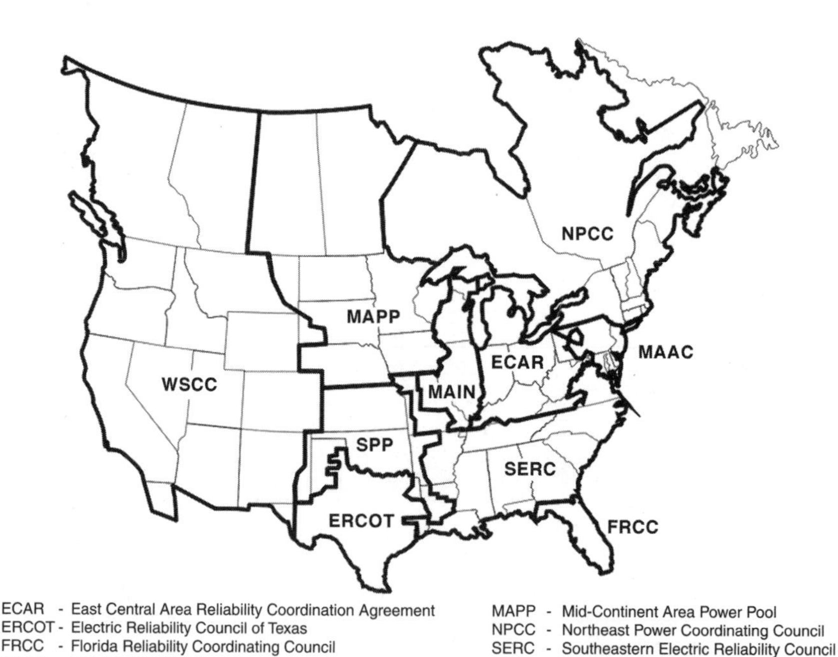

Figure 9.3 NERC regional councils and Alaska affiliate. *(Reprinted with permission © 2000 by the North American Electric Reliability Council. All rights reserved.)*

provided end users an opportunity to choose the electrical supplier regardless of who provided the transmission and distribution service to them. This last step implemented retail wheeling. As shown in Figure 9.4, retail wheeling allows an electrical supplier to wheel (transfer) power on a local distribution company's system to deliver power to the end user. Congress designed the deregulating process to encourage electrical suppliers to compete with one another and supposedly reduce the price of electricity. This deregulation process has caused the electrical power industry to make business decisions similar to those that occurred during the deregulation of the telecommunication and airline industries.

The reaction by some utilities to the 1992 Energy Policy Act has been to make decisions regarding additions or improvements to their transmission and distribution power systems and generation resources based on short-range factors. Because of the uncertainties about the effect of deregulation, these utilities' primary concern is to keep capital expenditures to a minimum. Many utility analysts think that new generation resources are being delayed to the last possible moment. This has resulted in a decrease in generation capacity margin (power supply versus demand) from 20% in 1990 to approximately 13% in 1996, as shown in Figure 9.5. This will result in utilities relying more on switching breakers generation to obtain

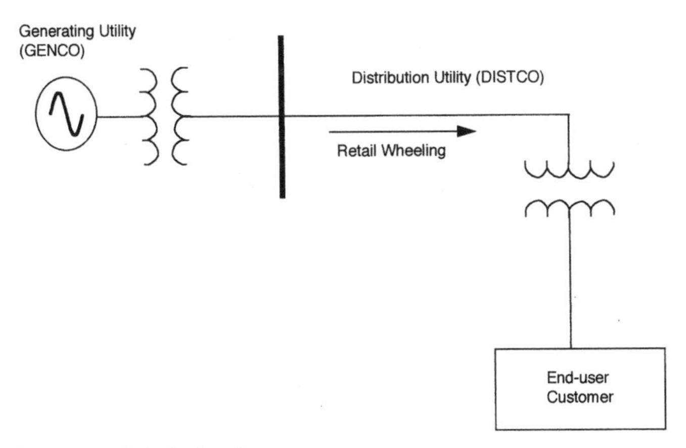

Figure 9.4 Retail wheeling.

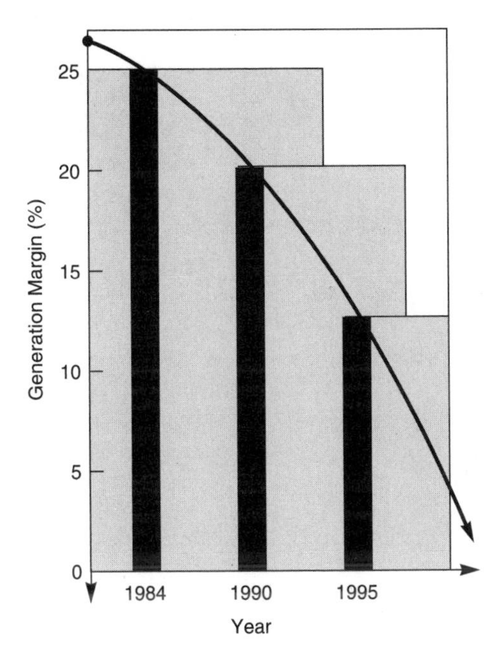

Figure 9.5 Decreasing generation margin. (*Source*: Cambridge Energy Research Associates and Venture Development Corporation.)

power from alternative sources. More utility switching will probably cause more power quality problems from switching surges and voltage sags. On the end-user side of the meter, increased use of sensitive power electronic equipment, rising from 30% of the electric power in 1995 to 50% after the turn of the century, will result in more quality problems. This will result in the increased need for power quality products and services.

The 1992 Energy Policy Act set in motion, at the national level, the changes needed to start deregulation of the U.S. electric power industry.

However, since each state regulates the utilities within its jurisdiction, each state will have to pass legislation to deregulate electric utilities. Rhode Island and California started the process of deregulation in 1996. Many states are considering the consequences of deregulating the electric utilities in their states. As shown on the map in Figure 9.6, as of July 1, 1999, almost half of the states enacted restructuring (deregulation) legislation, and the remainder are in various stages of considering it. Meanwhile, utilities, in anticipation of being deregulated, have begun to unbundle their electric services. What is unbundling?

Unbundling

Unbundling is the buzz word in the utility industry for separating the various services that the regulated, all-service utility provided to its

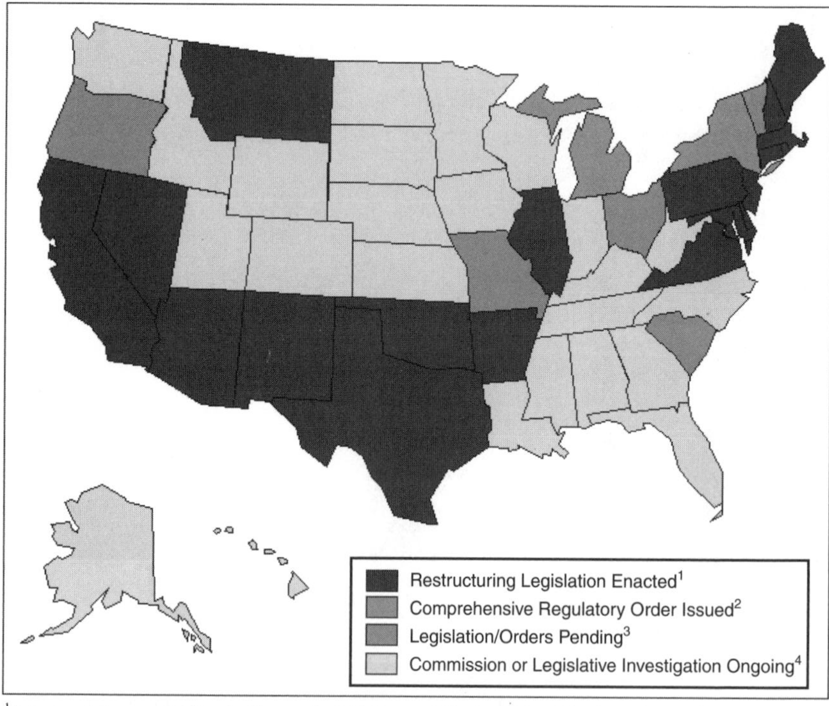

[1]Arizona, Arkansas, California, Connecticut, Delaware, Illinois, Maine, Maryland, Massachusetts, Montana, Nevada, New Hampshire, New Jersey, New Mexico, Ohio, Oklahoma, Pennsylvania, Rhode Island, Texas, and Virginia.
[2]Michigan, New York, and Vermont.
[3]Missouri, Oregon, and South Carolina.
[4]Alabama, Alaska, Colorado, District of Columbia, Florida, Georgia, Hawaii, Idaho, Indiana, Iowa, Kansas, Kentucky, Louisiana, Minnesota, Mississippi, Nebraska, North Carolina, North Dakota, South Dakota, Tennessee, Utah, Washington, West Virginia, Wisconsin, and Wyoming.

Figure 9.6 July 1, 1999, state-by-state status of electric utility restructuring. (*Courtesy of Energy Information Administration.*)

customers. Those services, including power quality, were bundled together under one rate. It was like buying a postage stamp. Everyone paid the same price for the stamp no matter how far the letter was sent. So everyone paid postage stamp rates for their electricity. In the new deregulated utility industry, the utility customer will no longer get all those services under one postage stamp rate but will have to pay for those services separately, as shown in Figure 9.7.

As discussed in Chapter 1, electrical customers will pay for electrical energy provided by unregulated GENCOs or independent power producers (IPPs) and delivery of the energy from regulated transmission companies (TRANSCOs) and distribution companies (DISTCOs). Either regulated DISTCOs or unregulated energy service companies called ESCOs will provide power quality services for an extra charge, as shown in Figure 9.8. The production companies (GENCOs or IPPs) will sell power under contracts to bulk power traders. A separate company called a meter service provider (MSP) will provide revenue meters. Regional transmission network operators (TRANSCOs) will provide transmission access to get the power to the distribution systems supplying the customers. Finally, DISTCOs will provide the final delivery of the electricity to individual end-use customers. These distribution companies, or "wires" companies, will likely supply the only delivery services. Regulators will regulate access charges and terms for these services.

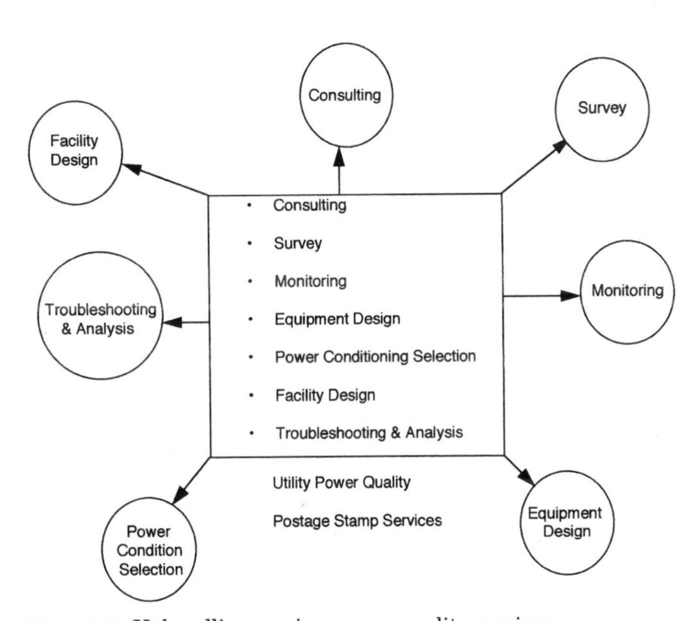

Figure 9.7 Unbundling services power quality services.

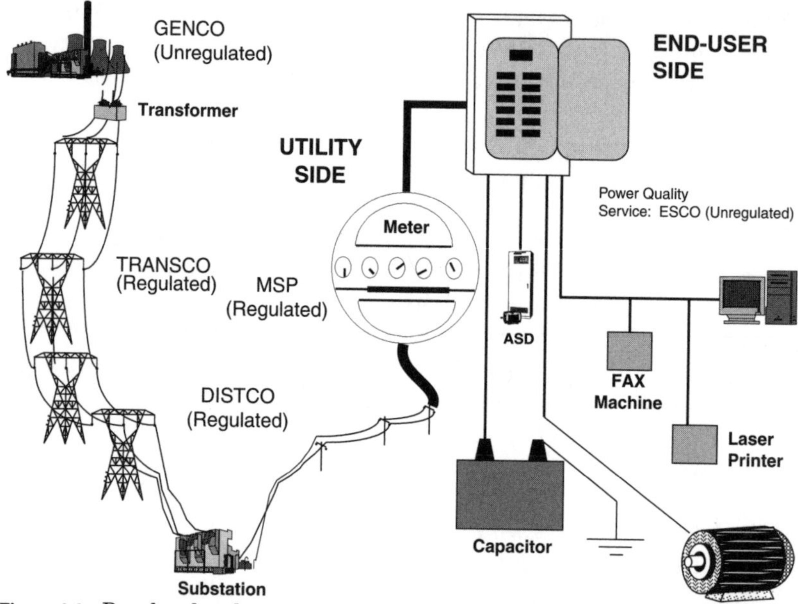

Figure 9.8 Regulated and unregulated power quality services.

Utility power quality business opportunities. The new utility structure provides business opportunities for unbundled utility customer services. Utilities have created unregulated subsidiaries at a tremendous pace in order to tap this potential market for customer services. They have formed retail energy marketing businesses or separate energy service companies (ESCOs). These businesses have the opportunity to provide services that include evaluation of power quality concerns and implementation of power quality improvement technologies. These businesses have no geographic boundaries. Once they acquire expertise and products, they can offer services worldwide, unrestricted by traditional service territory. One of these services includes power quality enhancement.

Power quality enhancement programs can include one or several elements. A power quality program could include power quality training, monitoring, surveys or audits, and the selling or leasing of power conditioning equipment. Many utilities have tried selling or leasing power conditioning equipment and found it to be a low profit producer, while others have found that offering emergency backup service at a premium cost can be a worthwhile business. Many have found that offering power quality consulting and monitoring service improves their relationships with their customers. Some utilities are learning to look at power quality services as not only a way to improve customer relations but also as a new revenue source. As shown in Figure 9.9,

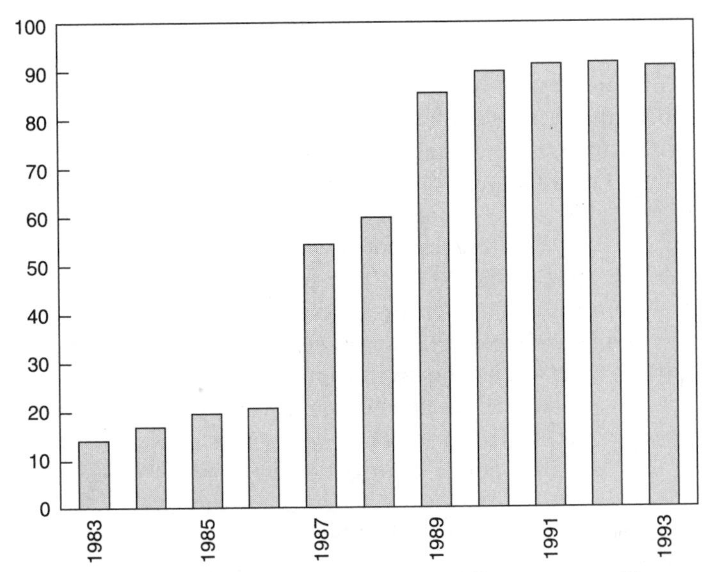

Figure 9.9 Estimated percent of utilities providing power quality program. (*Courtesy of EPRI.*)

since 1993, most utilities provide some type of power quality service to their customers.

Several utilities offer a variety of power quality services. For example, Consolidated Edison Co. (Con Ed), the New York City utility, created a Power Quality Service Center (PQSC). The PQSC provides training and technical information on power quality in a deregulated utility environment to Con Ed's customers and other interested parties in its service region. It has already presented several 1-day seminars on wiring and grounding, site surveys, and power measurement. Con Ed has also installed 100 permanent power quality monitors throughout its service territory, using the Internet to access the power quality status of its customers. Many utilities offer power quality services and products to their industrial, commercial, and residential customers. In addition to utilities, power conditioning equipment and power quality measuring instruments manufacturers see a growth trend in the power quality market.

Power quality market growth. The July 1999 issue of *Power Quality Assurance* magazine contained several articles by experts from Frost and Sullivan, an international marketing and training company, on the increased need for power quality products and services. In these articles, Frost and Sullivan experts segmented the power quality product market into test and measurement instruments, UPSs, TVSSs, and other power conditioning equipment. Frost and Sullivan expects the total North American power quality market in 1999 of $5.13 billion to increase to $8.37 billion by the year 2003, as shown in Figure 9.10. One article projected that the market for power quality test and measurement instrumentation will increase from $214.6 million in 1999 to $280.4 million by 2003, as shown in Figure 9.11. Meanwhile, the market for UPSs increased from $2.4 billion in 1999 to $5.0 billion by 2003, as shown in Figure 9.12. The market for TVSSs and other power conditioning equipment grew from $1.89 billion in 1999 to $5.03 billion by 2003, as shown in Figure 9.13. Another article indicated that the market for market quality services will increase from $372.1 million in 1999 to $497 million by 2003, as shown in Figure 9.14. One power quality service that utilities as well as power quality equipment manufacturers offer is power quality training. They see power quality training as a marketing tool for other power quality services and products.

Power quality training. Utilities; consultants; universities; professional organizations like IEEE and the National Electrical Contractors Association (NECA); electrical research organizations like EPRI; utility associations like the Northwest Power Quality Service Center (NWPQSC), American Public Power Association (APPA), and Western Area Power Institute (WAPI); and manufacturers of power conditioning

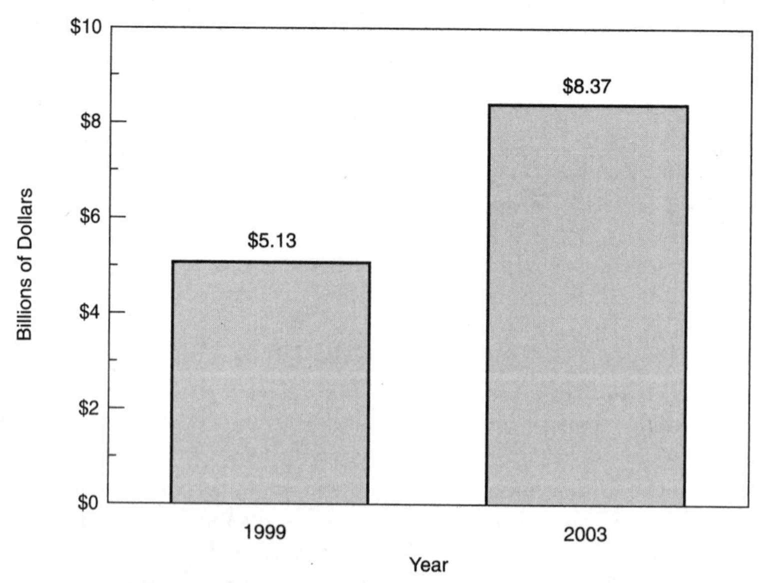

Figure 9.10 North American power quality market segment growth. (*Courtesy of Power Quality Assurance magazine.*)

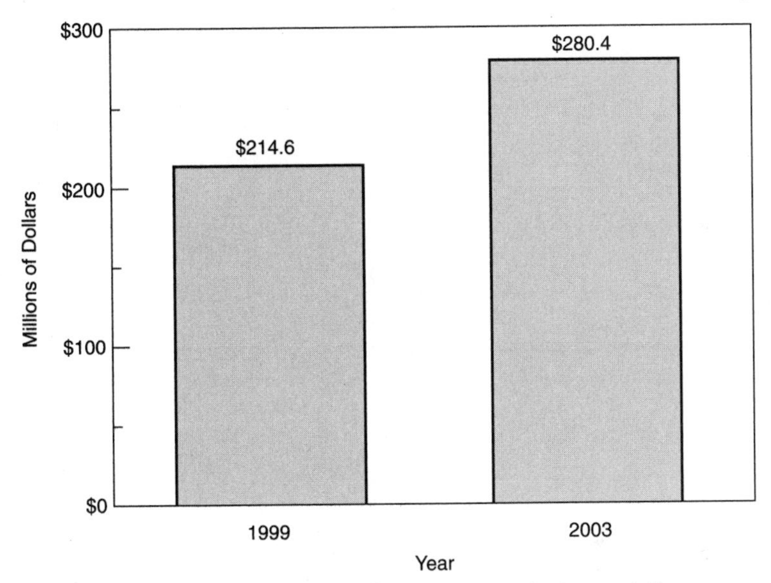

Figure 9.11 North American test and measurement instrumentation power quality market. (*Courtesy of Power Quality Assurance magazine.*)

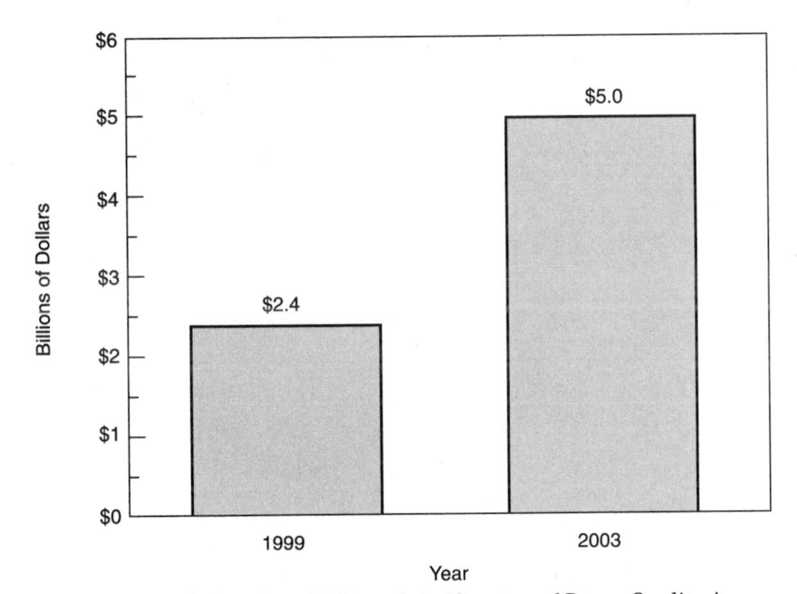

Figure 9.12 North American UPS market. (*Courtesy of Power Quality Assurance magazine.*)

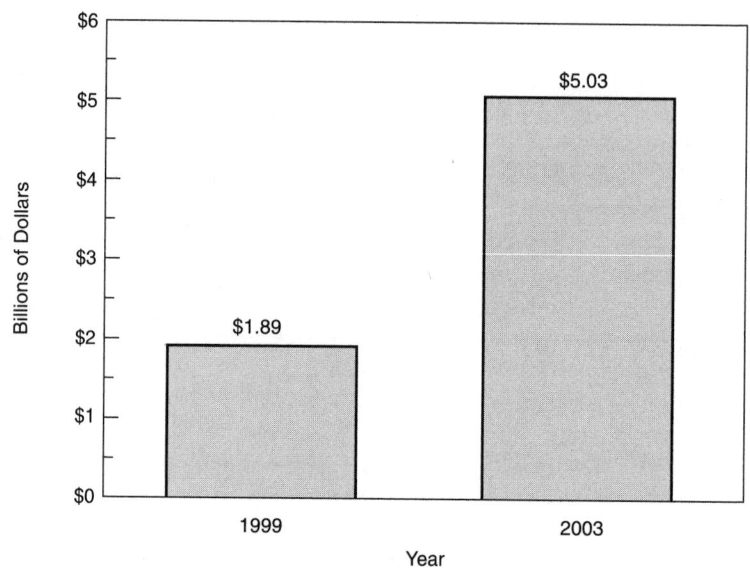

Figure 9.13 North American TVSS and power conditioning equipment market. (*Courtesy of Power Quality Assurance magazine.*)

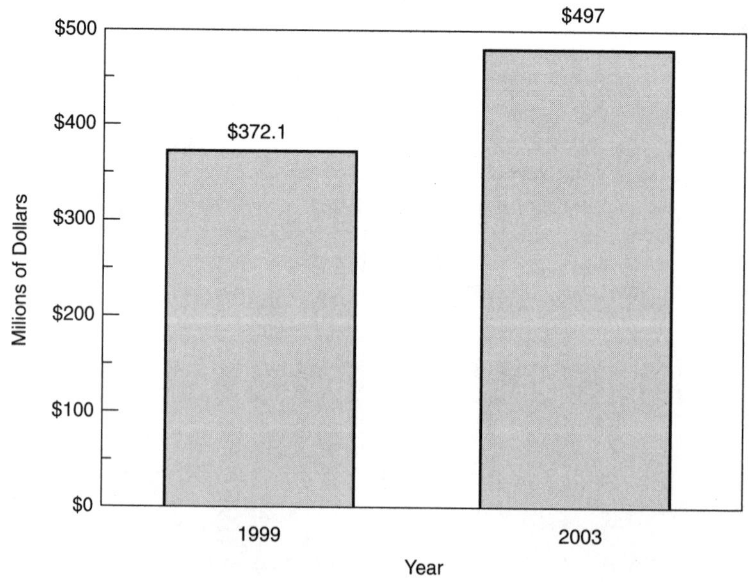

Figure 9.14 North American power quality services market. (*Courtesy of Power Quality Assurance magazine.*)

equipment and power quality instruments provide various types of power quality training. Training in the last few years has progressed from classroom lectures to a combination of classroom lectures and hands-on classes that involve the use of power quality instruments. They offer classes not only on the technical aspects of power quality but also on how to develop a power quality program and business. The hands-on classes, as shown in Figure 9.15 and lecture classes, as shown in Figure 9.16, provide participants essential experience in how to use power quality measurement instruments and troubleshoot problems. However, they have disadvantages for the presenters and the participants. The presenters must provide and set up expensive equipment. The participants usually must travel a long distance to the location of the class. New computer-based courses provide the hands-on experience without these disadvantages.

Computer-based courses may come on CD-ROM or even on the Internet. They offer the same hands-on experience available from actual instruments by the use of virtual power quality instruments and power quality problem situations displayed on a computer screen. These computer-based interactive courses avoid the need for presenters to provide expensive power quality laboratory exercises and participants to pay for expensive trips to the classes. They also have the advantage of allowing the participants to control when they want to access the course. Power quality training will become even more important to electricity end users as they experience more choices in the deregulated utility environment.

End-user choices and concerns. While the new utility structure will provide opportunities for new businesses, it will also cause new problems for unprepared end users. They will need to know how to access companies that can help them solve power quality problems. Like good Boy Scouts, end users will need to be prepared. Yes, electricity consumers will have the opportunity to choose who sells them electrical

Figure 9.15 Hands-on power quality class. (*Courtesy of PowerCET.*)

Figure 9.16 Power quality lecture class. (*Courtesy of PowerCET.*)

energy. They may even have a choice of who provides them their electric revenue meters. With choice comes responsibility. When the local utility becomes primarily a "wires" company, it will most likely not offer power quality services. Consequently, end users will probably not be able to rely on their local utility to help them solve their power quality problems. And if the local utility does provide power quality services, it will probably charge for those services.

End users can prepare for deregulation by taking several precautions, as shown in Figure 9.17. First, they can request that their local utility perform a power quality survey. Many utilities offer this service free or at a minimal cost. Next, they can keep good records of their maintenance practices and any electrical failures or equipment malfunctions. This information will help solve power quality problems. They can keep current on the changes happening in the electric utility industry and how they might affect their local servicing utility. The old adage of "Caveat emptor—Let the buyer beware" applies to the electrical consumer in the deregulated utility environment. As in any market, informed buyers make better choices. They should consider taking power quality training classes. From these classes, they can learn how to access good power quality service, buy appropriate products, and possibly solve their own power quality problems. If they have sensitive equipment in their facilities, they should consider the benefits of installing permanent power quality monitoring instruments along with their revenue meters at the point of common coupling with the utility. Finally, they should have readily available the telephone numbers of electronic equipment vendors, power conditioning equipment vendors, local utility customer service representatives, and power quality experts. Certainly, if they arm themselves with the knowledge and procedures contained in this book, they will have the tools that will prepare them for the changes in the utility industry. They will know how to obtain the power quality they need in the deregulated utility industry environment, as shown in Figure 9.17.

Many end users and utilities have learned to reach across the meter into each other's electrical power systems. Many have learned the value of establishing power quality standards and contracts. They have learned how to deal with power quality issues that affect both sides of the meter before the electric utility industry becomes more complicated. They understand that they need contracts that protect the power quality of one end user from impacting the power quality of another end user.

For example, a municipal utility in the Pacific Northwest recently became concerned about the effects of a new load. The utility was concerned that the new load might inject harmonics into the utility's distribution system and into customers adjacent to the new load. The new load melts titanium in an electron beam furnace. Electron beam furnaces generate harmonics that can get onto the utility system and affect the operation of other utility customers' facilities. In order to protect its own distribution system voltage from harmonic distortion caused by the electron beam furnace, it prepared a contract between the utility and the electron beam furnace customer. In the contract,

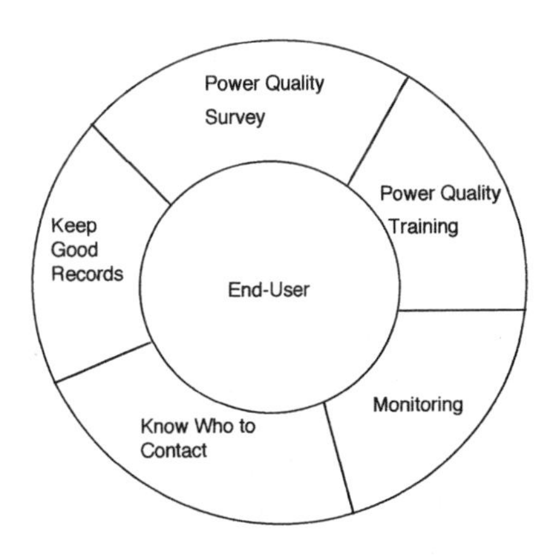

Figure 9.17 End-user power quality precautions.

the electron beam furnace customer agreed to keep its harmonics from exceeding IEEE 519-1992 standards. The utility installed a permanent monitor at the site to record and display any harmonics. The utility connected the monitor to the Internet and utilized a system that allows access of the monitor data via the Internet.

The new deregulated utility structure creates the need for contracts between companies other than just between the local utility and its customer. The new deregulated model includes GENCOs, TRANSCOs, DISTCOs, and ESCOs as well as end-use customers. End-use customers will have to deal with all these companies. The contracts will have to address issues of reliability and power quality, as well as the obvious issues of prices and delivery requirements. Power quality contracts will help establish responsibilities for various types of power quality problems and procedures for addressing those problems when they occur.

Requirements for power quality contracts

Parties involved in the contract and the characteristics of the power system affect the requirements of particular power quality contracts. These contracts address several areas, including the following:

- Reliability/power quality concerns to be evaluated
- Performance indices to be used
- Expected level of performance (baseline)
- Penalty for performance outside the expected level and/or incentives for performance better than the expected level (financial penalties, performance-based rates, shared savings, etc.)
- Measurement/calculation methods to verify performance
- Responsibilities for each party in achieving the desired performance
- Responsibilities of the parties for resolving problems

The following sections present summaries of the most important of these concerns for each type of contract. Each concern contains a description of important factors included in the contracts. Figure 9.18 illustrates some of the important contractual relationships for power quality considerations.

Contracts between TRANSCO and DISTCO or direct-service customer

Contracts between transmission companies and distribution companies (or large direct-service customers) define the power quality

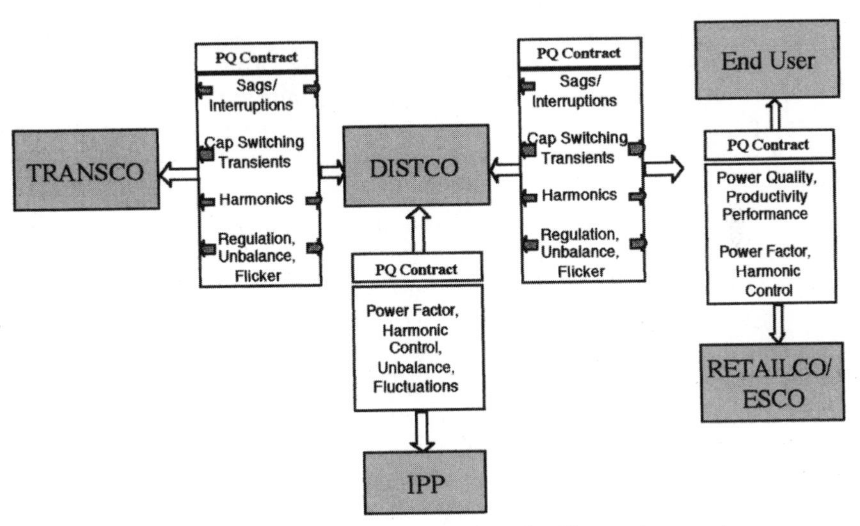

Figure 9.18 Contract relationships in the restructured utility environment.

requirements and responsibilities at the distribution substation interface between the two systems. Power quality disturbances provide an effective way to categorize these requirements and responsibilities.

Voltage regulation, unbalance, flicker. The contract between the two companies should contain a description of the steady-state voltage characteristics that the TRANSCO supplies to the DISTCO or to a direct-service customer. It should also describe the responsibilities for voltage regulation between the two companies. Control of flicker levels requires limits on both parties. Responsibilities of the TRANSCO include the overall flicker levels in the voltage. However, the DISTCO or the direct-service customer has the responsibility to control fluctuating load characteristics. This becomes especially important for contracts between the transmission company and a large arc furnace customer.

Harmonic distortion. The transmission supply company, or TRANSCO, has the responsibility to supply quality voltage to the DISTCO or direct-service customer, while the DISTCO or direct-service customer has the responsibility to minimize the harmonic loading on its system. IEEE 519-1992, or a similar standard describes the harmonic current limits at the point of common coupling.

Transient voltages. Many utilities use capacitor banks at the transmission level for system voltage support and to improve power transfer capabilities. When they switch these capacitor banks, it creates transient voltages. The transient voltage can impact distribution systems and end-use customers' sensitive loads. After deregulation in England, utilities and their customers experienced this problem with no clear definition of responsibility for controlling the transient voltages. Everyone pointed the responsibility finger at the other company. The transmission company declared that the transient voltages were not excessive The distribution companies that served the customers declared that the transients were caused by the transmission system. In the United States, vertically integrated utilities solved these problems by controlling the switching of transmission capacitor banks (synchronous closing or closing resistors). Power quality contracts should define the requirements for control of switching transients at the point of common coupling (the supply substation). These requirements should limit transients from capacitor switching to very low levels at this point because of their potential for causing problems at lower voltages.

Voltage sags and interruptions. Contracts should define the expected voltage sag and interruption performance at the point of common connection. It is important to recognize that faults on the transmission system or faults on the downline distribution system can cause voltage sags. Utilities have already implemented contracts that set voltage sag limits that they will supply to their large customers. For example, Detroit Edison, Consumers Power, and Centerior Electric have voltage sag contracts with their large automotive manufacturing customers. They have agreed to compensate their customers with payments or reduced rates when the voltage sag and interruption performance falls outside of specified levels.

Contracts between DISTCO and end users (or end-user representative)

The power quality requirements at the point of common coupling between the distribution system and end-use customers require definition. In some cases, the end users might act as customers of the distribution company. In other cases, the end users might have retail marketers or energy service companies represent them to distribution companies. Regulations will probably define the basic power quality requirements at this interface. However, opportunities for performance-based rates or enhanced power quality service from the distribution system will create the need for more creative contracts.

Voltage regulation, unbalance, flicker. These contracts will include definitions of the steady-state characteristics of the voltage DISTCOs supply to their end-use customers. They will also require end-use customers to control fluctuating loads, unbalanced loads, and motor starting.

Harmonic distortion. IEEE 519-1992 describes the split of responsibility between the customer and the distribution system supplier in controlling harmonic distortion levels. The distribution companies have the responsibility to limit the voltage distortion they supply their customers. End-use customers have the responsibility to limit harmonic currents created by nonlinear loads within their facilities.

Transient voltages. Contracts between the distribution companies and their customers can limit the impact of transient voltages. The importance of capacitor switching transients results from their impact on sensitive loads. Distribution system suppliers should control their capacitor switching to minimize transient voltage magnitudes. However, customers should avoid magnifying transients by controlling their use of power factor correction capacitors within their facilities. The contracts should define the basic requirements and responsibilities for surge suppression to avoid problems with high-frequency transients associated with lightning.

Voltage sags and interruptions. These contracts should define expected voltage sag and interruption performance. Utilities might offer enhanced performance options in cases where economically they can improve performance through modifications, or they might use power conditioning equipment at the distribution system level.

Contracts between RETAILCO or ESCO and end user

The retail energy marketers (RETAILCOs) or the energy service companies (ESCOs) will have separate contracts with their end-use customers. These complex contracts will probably require more creativity to write than the contracts between the DISTCOs and their end-use customers. ESCOs may offer a wide range of services for improving the power quality, efficiency, and productivity of their customers. These diverse services will dictate the contract requirements. One of these services includes enhanced power quality requirements to improve productivity.

Enhanced power quality requirements to improve productivity

The characteristics of the facility equipment define the power quality requirements of the facility. Typically, voltage and current define power quality requirements. However, performance of the process can instead provide a useful way to establish power quality requirements. ESCOs instead of DISTCOs can guarantee power quality to their customers by providing power conditioning equipment to protect the facility from power quality problems. End-user customers can pay for these services in terms of shared savings from improved productivity (similar to many contracts that specify payments to energy service companies from the shared savings of energy efficiency improvements). They also can make fixed payments based on the power quality improvement requirements.

Power factor and harmonic control. The DISTCOs will require their customers to meet certain harmonic control requirements. They will probably include in their tariffs power factor penalties. The RETAILCOs or the ESCOs will have to deal with these requirements. They could offer to integrate harmonic control and power factor correction along with power conditioning equipment for voltage sag and transient control.

Contracts between DISTCO and small IPP

Deregulation also creates more opportunities for small independent power producers (IPPs) to generate and sell electricity. Many of these smaller producers may locate their generators on distribution systems and create a need to define the power quality requirements for this interface (along with protection and reliability requirements). The power quality contracts will define the expected power quality that the IPP can expect at the interface (similar to the contract with end users) and will define the requirements for the IPP in terms of the quality of the generated power. Important areas to consider for the IPP requirements are the power fluctuations (e.g., start-up for motor-generator systems, power fluctuations for wind or photovoltaic systems), harmonic characteristics of the generated current, power factor characteristics, balance.

Deregulation versus Regulation

Will the utilities responsible for T&D systems be deregulated or remain regulated? The emphasis up to now in the deregulation process is to deregulate the generation portion of the utility business. Electric utility industry analysts seem to think that T&D systems, or "wires" companies, will remain regulated. The need for power quality services will continue whether the wires utilities are regulated or deregulated.

If the wires utilities are deregulated, they will have to compete among themselves to participate in the wheeling of power to the end user of electricity. Utilities that provide reliable and quality service can compete better in the long run than utilities that provide poor power quality and unreliable service. When deregulated, the telecommunications industry had a similar experience. Because of modern society's dependency on electricity for health and safety, electric utilities have even more concern about power quality than telecommunication utilities.

Most likely, federal and state regulators will continue to regulate the wires utilities. They need to regulate monopolies, like the wires utilities. If the federal and state regulators do regulate the wires utilities, they will insist on the utilities designing reliable systems high in power quality. They will either institute incentives or set standards that encourage wires utilities to keep their power systems reliable with good power quality.

One way the regulators may encourage utilities to keep systems high in reliability and power quality is to require that they meet certain reliability and power quality standards. In a regulated environment, utilities often maintained high standards of reliability and power quality. In a deregulated environment, specific standards of power quality will become essential.

Power Quality Standards

In a deregulated utility environment, power quality standards will likely become mandatory. There is a great deal of concern that the restructured utility industry will not provide the same reliable, high-quality power that it has supplied in the past. In order to prevent the quality of power and reliability of individual utility power systems from deteriorating, regulators will most likely require utilities to adhere to power quality and reliability standards. This will cause an increased interest in power quality standards. Specific benchmarks of power quality in various utility systems will be required to determine whether a particular utility has met these standards. Utilities and their customers will need to install permanent monitoring to develop benchmark indices similar to those developed in the 2-year EPRI monitoring project. Many utilities, including TVA, Duke Power, and Consolidated Edison, have already developed bench marking indices based on monitoring critical loads over a long period of time. The U.S. utilities will need to develop standards similar to the Euronorms (EN50160) in Europe for harmonics, flicker, regulation, unbalance, and disturbances and NRS 048 in South Africa for voltage sag.

International Utility Competition

Increased utility competition has not confined itself to the United States. A recent and continuing future trend is increased international utility competition. Deregulation and privatization of utilities throughout the world has caused and will continue to cause increased international utility competition . The deregulation process in other countries is similar to the process in the United States. Privatization is the process of transferring the production and delivery of electricity from the public sector to private ownership and operation. Private utilities are prevalent in the United States but are a relatively new phenomenon in many other countries. In recent years, utilities in South America, Europe, Central America, and the Middle East have been privatized. This has resulted in many mergers of utilities across international boundaries and the so-called globalization of the utility industry. Examples include at least 16 U.S. utilities deciding to expand into international operations and, for the first time, a foreign utility bought a U.S. utility. This has resulted in concerns about the ability of utilities throughout the world to deliver power high in quality and reliability.

To alleviate these concerns, utilities and regulators use three strategies to prevent the deterioration of power quality and reliability that may come from utility deregulation and privatization. These strategies are: (1) investment in new and existing technology, (2) development of performance standards, and (3) entering into new contractual arrangements, as shown in Figure 9.19.

An example of a recent investment in technology to improve power quality is the installation of a high-voltage direct-current (HVDC) submarine link from the Swedish mainland to the island of Gotland. The wind generators on the island of Gotland cause the lights to flicker on the Swedish mainland. The HVDC allows the GEAB (Gotland's local utility) to use the ac-to-dc converters to control the reactive power into its system and thus prevent the light flicker caused by the wind generators on the island.

An example of using standards to maintain power quality is that of the Office of Electricity Regulation (OFFER) in the United Kingdom publishing two types of power quality standards for the 12 regional

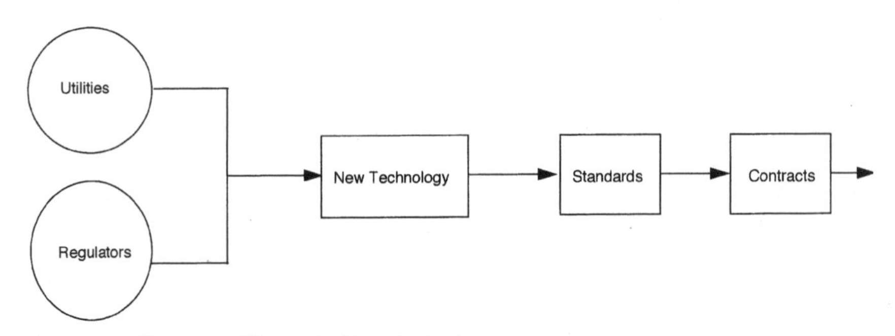

Figure 9.19 Power quality protection strategies.

electric utilities in England, Wales, and Scotland. These standards require the utilities to provide their customers a minimal level of service. If these utilities fail to meet these standards, they have to pay their customers a penalty based on the schedule shown in Table 9.1. As a result of not meeting these standards, these utilities paid their customers $150,000 in penalties in 1996.

An example of the utilization of contracts to maintain an agreed level of power quality occurred in 1995 in France. Electricité de France entered into contracts with its customers with a commitment to reduce the average interruption of service. These contracts require the utility to pay its customers a penalty if the contractual requirements are not met. The French utility was able to reduce its average interruption of service from a high of 113 min in 1992 to 78 min in 1995, as shown in Figure 9.20. Increased utility competition, both international and national, has brought attention to the future of utility research and development.

Research and Development

Research and development (R&D) in the electric utility industry has shifted from technologies that help meet regulators' requirements to technologies that will help utilities compete in the deregulated market. This shift in emphasis should not affect R&D for power quality. R&D for improving power quality needs to continue. The wires utilities that need to provide service high in power quality and reliability will most likely continue to be regulated. There will be an increasing need for industrial and commercial end users to have electricity high in power quality and reliability. The state and federal regulators will continue to set standards and regulations that will encourage power quality. In the past, EPRI collaborated with standards organizations, like IEEE, IEC, and ANSI, to develop and improve power quality standards. This research has resulted in the development of reliability indices and benchmark data for power quality. Research needs to continue in expanding power quality standards. State agencies will need to take a stronger role in power quality research to ensure that deregulated utilities do not neglect power quality research. For example, the state of California collected $62.5 million of research funds from a rate surcharge in the first year of deregulation.

EPRI and its member utilities see electromagnetic compatibility (EMC) as a major emerging power quality challenge. EPRI, therefore, plans to identify and assess EMC interactions, develop EMC applications and solutions, and incorporate EMC into its power quality research. Research into new tools for analyzing power quality problems, like EPRI's Power Quality Tool Box, need to be expanded. New methods of power quality analysis using artificial intelligence and expert systems

TABLE 9.1 United Kingdom Utility-Guaranteed Standards of Performance*

Service	Performance level	Penalty payment†
1 Respond to failure of a supplier's fuse	Within 4 hours of any notification during working hours‡	£20
2. Restoring electricity supplies and faults	Must be restored within 24 hours	£40 (domestic customers), £100 (nondomestic customers) for not restoring supplies within 24 hours plus £20 for each additional 12 hours
3. Providing supply and meter	Arrange an appointment within 3 working days for domestic customers (and five working days for nondomestic customers)‡	£20–£100
4. Estimating charges	Within 10 working days for simple jobs and 20 working days for most others‡	£40
5. Notice of supply interruption customers	Customers must be given at least 2 days' notice‡	£20 domestic customers, £40 nondomestic
6. Investigation of voltage complaints	Visit or substantive reply within 10 working days‡	£20
7. Responding to meter problems	Visit within 10 working days‡ or substantive reply within 5 working days	£20
8. Responding to customers' queries about charges payment queries	A substantive reply within 5 working days	£20
9. Making and keeping appointments	Companies must offer and keep a morning or afternoon appointment, or a timed appointment if requested by the customer	£20
10. Notifying customers of payments owed under standards	Write to customer within 10 working days of failure‡	£20

*Details of the standards are set out in regulations made by the director general. Companies are required to send an explanatory leaflet to customers at least once a year. Companies may not have to make payments if failure is caused by severe weather or other matters outside their control, but this depends on the particular circumstances and companies must make all reasonable efforts to meet the standards. The standards apply to tariff customers and those marked ‡.
†One English pound (£) = 1.6412 U.S. dollar ($)
‡Varies among companies.
SOURCE: Courtesy of *Transmission and Distribution World* magazine.

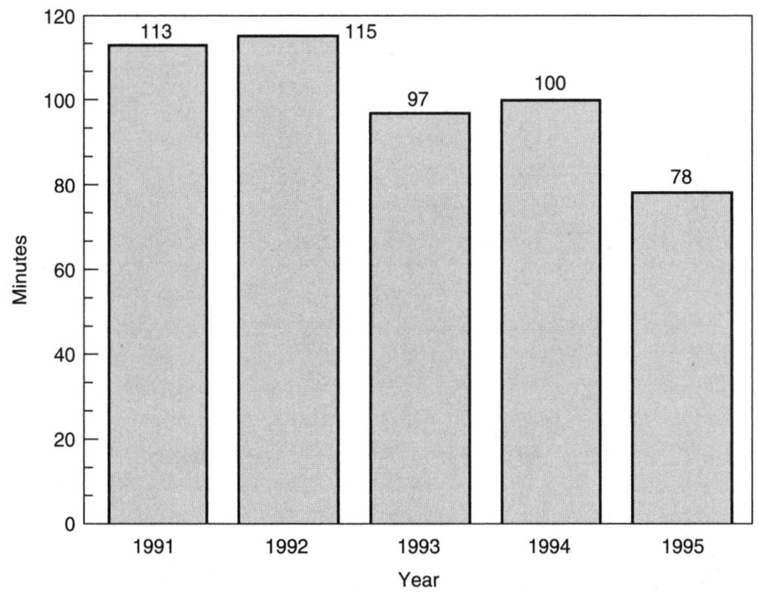

Figure 9.20 Electricité de France improvement in interruption of service. (*Courtesy of Transmission and Distribution World* magazine.)

will need funding. Future research and development will need to focus on developing products and services that prevent power quality problems. An example of such products/services is power quality parks.

Power Quality Parks

Power quality parks are industrial and commercial parks especially designed to give the level of power quality that certain industrial and commercial customers, like hospitals, semiconductor chip fabricators, and stock broker offices, need to run their facilities efficiently and effectively. Energy producers design these parks to attract customers that need power high in quality. Utilities supplying electricity to power quality parks would provide a service high in power quality. This service could include redundant feeders and a power quality park substation with an active filter to filter out harmonics, a standby generator to prevent interruptions to critical loads, and a custom power device, like a dynamic voltage restorer (DVR), to prevent voltage sags. The electrical distribution system design in the park would prevent industrial and commercial power users from affecting each other's power quality. These parks will contain the latest state-of-the-art power quality monitoring and conditioning equipment. They are a logical extension of utilities providing their customers choices in who provides them electrical energy.

Utilities need to provide their customers not only a choice of who provides them electricity but a choice of the level of power quality. As the electrical utilities become more competitive, they should see the need to provide their customers the opportunity to chose a power quality park that guarantees them the level of power quality that some of their customers need. Power quality parks will act as an important marketing tool that would benefit energy providers as well as their customers. Figure 9.21 shows a schematic of the concept of a power

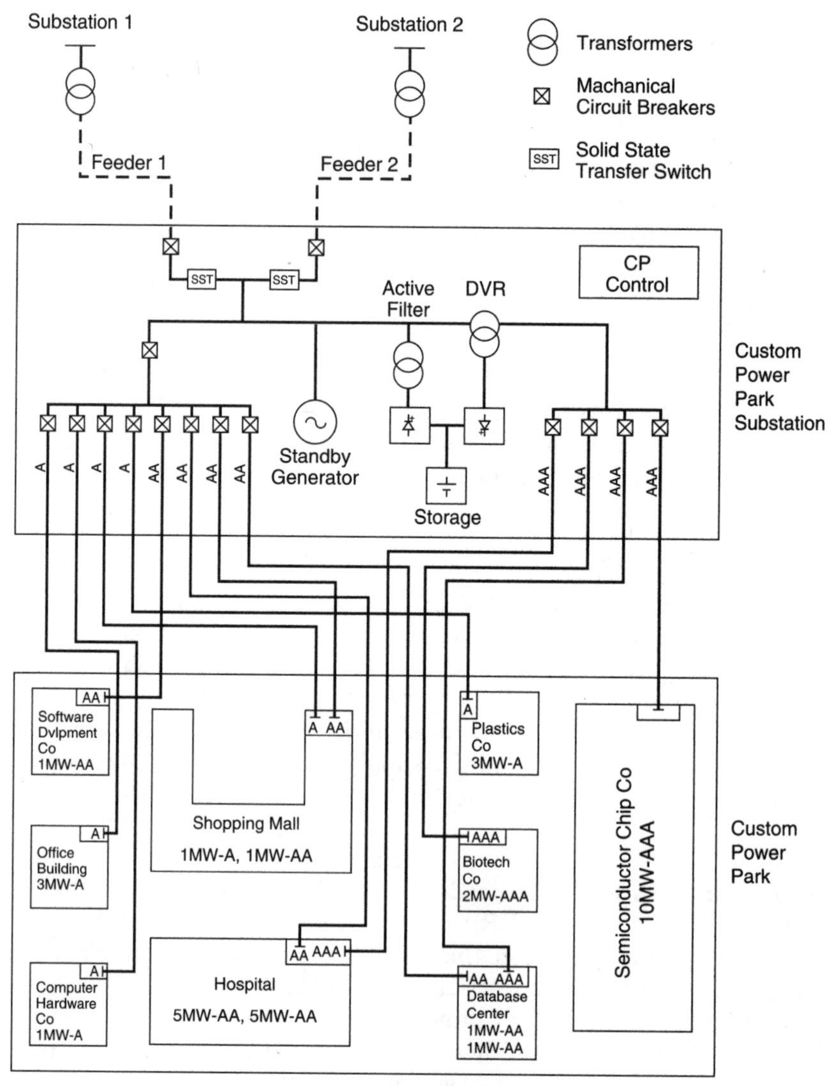

Figure 9.21 Power quality park concept schematic. (*Courtesy of Narian Hingorani.*) The Custom Power Park concept was conceived by Narian Hingorani.

quality park showing various levels of power quality service, i.e., low (A), medium (AA), and high (AAA).

References

1. Porter, Gregory J., and Andy Van Sciver. 1997. "Deregulation Simplified." *Power Quality Assurance,* vol. 8, no. 6 November/December, pp. 12–18.
2. "Power Quality Outlook." 1996. URL address: *http://www.apc.com/english/about/finan/reports/1996/anrep008.htm.* Available from American Power Conversion.
3. Kennedy, Barry W., and M. F. McGranaghan. 1998. "Power Quality Contracts in a Restructured Competitive Electricity Industry." *Proceedings of the 8th International Power Quality Applications Conference—PQA '98 Southern Hemisphere,* November 9–11, Cape Town, South Africa.
4. Simmons, Nelson W. 1998. "Services in a Deregulated Environment." *Proceedings of PQA '98 North America,* June 8–11, Phoenix, Arizona.
5. Mark F. McGranaghan. 1998. "Deregulation and the Need for Power Quality Standards." *Proceedings of PQA '98 North America,* June 8–11, Phoenix, Arizona.
6. Bell, Robert A., and Wayne H. Seden. 1995. "Utility R&D: The Cutting Edge of Competition." *Public Utilities Fortnightly (1994),* vol. 133, no. 15, August, pp. 29–32.
7. Morcos, M. M., and W. R. Anis Ibrahim. 1999. "Electric Power Quality and Artificial Intelligence: Overview and Applicability." *IEEE Power Engineering Review,* vol. 19, no. 6, June, pp. 5–10.
8. Bates, Jennifer L. 1999. "The Lucrative North American UPS Market Continues to Present Vendors with Opportunities." *Power Quality Assurance,* vol. 10, no. 4, July.
9. Clemmensen, Jane, and Susan Tonkin. 1999. "Competition Heats Up in the Power Quality Services Market." *Power Quality Assurance,* vol. 10, no. 4, July.
10. Fong, Dora. 1999. "Prospectives on Market Demand for TVSS and Power Conditioning Equipment." *Power Quality Assurance,* vol. 10, no. 4, July.
11. Hingorani, Narain G. 1998 "Overview of Custom Power." Paper presented at Panel Session on Application of Custom Power Devices for Enhanced Power Quality. July 14, San Diego, CA.
12. Hazan, Earl. 1997. "Reliability: Stacking the Deck in Your Favor." *Transmission & Distribution World,* vol. 49, no. 1, January, pp. 45–48.
13. Kennedy, Barry W., and D. Sabin. 1999 "Use the Internet for Power Quality Reporting." *Electrical World,* vol. 213, no. 5, September/October, p. 64.

Glossary

ac utility power (commercial power) The electric power that an electric power utility company supplies its customers.

ACSR (aluminum conductor, steel reinforced) A conductor that has a steel core surrounded by stranded aluminum wires.

active filter A power electronic device that eliminates harmonics.

aerial Describes power lines or telephone lines or cables installed above ground on poles or overhead structures.

alternating current (ac) An electric current that reverses (or alternates) its direction at regular intervals. Its magnitude begins at zero, increases in value and reaches a maximum value before returning to zero and continuing downward to a negative peak value before increasing in value again. Since the current flows in the positive direction for the same amount of time that it flows in the negative direction, the average value of the current flow equals zero. This cycle repeats itself continuously. The number of such cycles per second equals the frequency in hertz, i.e., 60 cycles per second in the United States and 50 cycles per second in Europe and Asia.

ammeter An electrical test instrument that measures current in a circuit.

ampacity The amount of current expressed in amperes that a device or conductor can carry.

ampere (AMP, amps, *I*, or A) The unit of measurement for the rate of flow of electrons (coulombs per second), or current, through a wire, similar to the flow of water through a pipe. One ampere equals a group of electrons whose total charge equals 1 coulomb that passes a point in a conductor in 1 second. Mathematical formulas represent it by the symbol *I*. Amps is a shortened version of the term.

analog A physical system that represents data by measurement of a continuous physical variable, such as voltage or pressure, or a readout display on a dial rather than by numerical digits.

ANSI American National Standards Institute.

apparent (demand) power The total power a device uses plus the power stored in the device as magnetic and electric fields. It is measured in kilovolt-amperes, or kVA, and is the vector sum of reactive and active power, as shown in the formula: $(kVA)^2 = (kW)^2 + (kVAR)^2$.

arc or arcing Usually an undesirable phenomenon of electricity flowing through the air from a conductor of one potential to another conductor of a different potential. Arcing can produce visible flashes and flames and cause crackling sounds. Overloaded electrical equipment, load conductor connections from one power source to another, insulation leakage, conductor contamination, and electrostatic discharges can cause arcing.

arrester A nonlinear impedance device that protects equipment from high-voltage transients by limiting the amplitude of voltage on a power line. Usually connected between the power conductors to suppress transients larger than a selected voltage, as in lightning protection. The term implies that the device stops overvoltage problems (e.g., lightning). In actuality, it limits or clamps the voltage to a specified level.

attenuation The reduction of a signal or electrical surge from one point to another. Wire resistance, arresters, and power conditioners attenuate surges to varying degrees.

autotransfer switch (ATS) A device that automatically transfers power from one source to another.

autotransformer A transformer that steps voltage up or down and has primary and secondary windings that share common turns but provide no isolation.

auxiliary source A power source that provides emergency power to a critical load during an interruption. Typical auxiliary sources include battery backup or diesel generators that provide uninterruptible power supply to computers and other sensitive loads during a power outage.

AWG American wire gauge, refers to the U.S. standard for wire size.

balanced load An alternating current power system that has more than two current-carrying conductors that carry equal, i.e., balanced, currents.

ballast A device that starts a fluorescent or high-intensity discharge (HID) lamp and maintains the proper operating current and voltage.

battery A combination of wet or dry cells that cause a chemical reaction that produces an electric current.

battery charger A device or a system that converts ac to dc power in order to keep the battery backup fully charged.

battery disconnect switch A switch that disconnects a battery reservoir from a uninterruptible power supply (UPS) and protects personnel when batteries or UPSs require service.

battery reservoir A combination of cells or batteries that provides power to a UPS's inverter (direct current to alternating current converter) when operating in the emergency mode.

bidirectional converter A device that changes (or converts) power in both directions, i.e., from alternating current to direct current and vice versa.

blackout Describes the total loss of electrical power for more than 1 minute.

blink A nontechnical, ambiguous term that usually refers to a short-duration voltage sag or outage.

bonding An interconnection between two or more points (usually grounding systems) that reduces any voltage difference.

branch circuit An electrical circuit individually protected by a fuse or circuit breaker that starts at the service panel and ends at the electrical outlets.

break-before-make A switch or relay operational sequence that requires breaking (opening) the existing connection before making the new connection.

brownout An extended, i.e., longer than a few cycles, voltage reduction of more than 10 percent. Utilities sometimes deliberately cause brownouts to reduce the load on their electrical systems when the demand for electricity exceeds the generating capacity. Generally, utility customers do not notice the reduction, except when it affects their sensitive electronic equipment.

BTU British thermal unit; the standard unit for measuring the quantity of heat energy, such as the heat content of fuel. One Btu equals the amount of heat energy necessary to raise the temperature of 1 pound of water 1° Fahrenheit (3412 BTU = 1 kW-h).

buck-boost transformer (regulator) A small, low-voltage transformer that uses the principle of changing the voltage by adding transformer windings that either reduce (buck) or increase (boost) the voltage.

building transformer A transformer that reduces the utility high voltage to a lower voltage that the customer can use in the building.

bump A nontechnical, ambiguous, and undefined term used to describe a short-duration rise in voltage that may or may not affect equipment operations.

bus bar (bus) A heavy, rigid conductor that serves as a common connection for two or more circuits.

bypass A circuit that provides an alternative path for the electrical power to go around (or bypass) its normal path and allows maintenance personnel to service equipment, like UPSs, without interrupting service.

cable A fully insulated wire or bundle of wires installed underground or overhead.

calcium An element in lead-acid batteries that hardens the plate material.

calibration The procedure that determines the location of scale graduations on an instrument by comparison to a standard series of values.

capacitor An electrical device that contains two conductive plates separated by a dielectric (insulator or nonconductor of electricity) substance. Installed in substations, on poles, and inside facilities to correct unwanted conditions in an electrical system. It stores small amounts of energy, "detours" high-frequency transients to ground, and resists changes in voltage. It helps improve the efficiency of the flow of electricity through distribution lines by reducing energy losses and increasing the power factor.

capacity The maximum load of electricity that equipment can carry.

carbon block Slice of carbon typically used as a resistor.

CBEMA The former Computer Business Equipment Manufacturers Association; replaced by the Information Technology Industry Council (ITIC).

CBEMA curve A set of curves developed by the Computer Business Equipment Manufacturers Association (CBEMA) that represents the withstand capabilities of computers in terms of the magnitude and duration of a voltage disturbance. It was a standard for measuring the performance of all types of equipment and power systems until replaced by the Information Technology Industry Council (ITIC) curve.

charge voltage The voltage required for storage batteries to maintain their maximum charge.

charger An ac-to-dc converter that provides power to a UPS inverter and maintains the battery reservoir charge.

check meter Installed as a temporary instrument to measure the amount of current used by a particular appliance or piece of equipment.

chip (microchip) A tiny slice of semiconductor material that provides the "brains" for a computer to operate.

choke An inductor that filters (chokes) out undesirable flow of current at specified frequencies; usually used to block high-frequency transients or harmonics that might damage sensitive equipment.

circuit A path through which electric current travels.

circuit breaker (CB) A device that opens (breaks) or closes a circuit to interrupt or apply electric power to an electrical apparatus. An operator can open it manually. It opens a circuit automatically when it senses an overload. Located in utility substations and on transmission and distribution lines as well as in the home, factory, and office.

clamp-on CT A current transformer that clamps around a current-carrying conductor and picks up current from the conductor. Used to monitor current at several points for short periods of time.

clamp-on voltage The maximum voltage that protective devices (surge protection) allow into an electric circuit.

class A, B, C Categories of location for transient suppression within a facility. Class A refers to outlets and long branch circuits. Class B refers to major feeders and short branch circuits near the distribution panel. Class C refers to the commercial power service entrance and outside the facility.

clean ground An undefined term that describes an earth connection that does not cause electrical equipment to malfunction. See *computer ground*; or *isolated ground.*

clean power An undefined, imprecise term sometimes used to describe electrical power that does not cause power quality problems.

clearance Distance required between conductors of various voltages and the ground.

coaxial cable A cable that contains two concentric conductors separated by a tube of dielectric substance (the outer conductor acts as a shield/ground).

commercial power See *ac utility power.*

common mode (CM) Refers to electrical interference measured as a ground reference signal common to both of two current-carrying conductors.

common-mode noise (voltage) Abnormal signals (undesirable voltage) that appear between a current-carrying line and associated ground, i.e., between neutral or phase and ground.

compensating winding A winding in the primary of a ferroresonant transformer that compensates for changes in voltage.

conductor An object or substance, usually a wire, cable, bus bar, rod, or tube, that conducts, i.e., provides a path for, electric current.

conduit A tube, duct, or pipe that provides a metallic or nonmetallic tubular raceway for wires and cables that carry data or power.

converter Refers to either a rectifier that changes alternating current (ac) to direct current (dc) or an inverter that changes direct current (dc) to alternating current (ac).

core The ferrous center of a transformer or choke used to increase the strength of the magnetic field.

core saturation Condition where a transformer or inductor core reaches its maximum magnetic strength and can no longer support any additional increase in magnetic flux density (field).

corona A visible or audible discharge of electricity from an object caused by the ionization of the air around the surface of the object.

coulomb The combined negative electrical charge of 6.24×10^{18} electrons or the quantity of electric charge per second that equals 1 ampere.

counterpoise A buried wire grounding system connected to transmission and distribution tower or pole footings to provide a low resistance path to earth.

coupling The means by which power or signals transfer from one circuit element or network to another. Also, the effect of a power or signal source interfering with a signal transmission system.

CPU Central processing unit; master control ("intelligence") circuitry of a computer system that performs all of the logic functions, arithmetic functions, program execution, and certain input/output control functions.

crest factor Ratio of a quantity's maximum value to its effective value. The crest factor for commercial ac voltage in the United States equals 1.414 (the crest factor for a sine wave), while a switched-mode power supply nonlinear current waveform usually has a crest factor equal to 3 or 4.

critical Circuit or feeder that has an important or costly operation associated with it.

critical load Electrical devices and equipment that provide essential electrical service whose failure to operate satisfactorily jeopardizes the health or safety of personnel, and/or results in loss of function, financial loss, or damage to property critical to the user. Electrical equipment that requires an uninterrupted power input to prevent damage or injury to personnel, facilities, or itself.

cross arm The crossing member of a transmission or distribution line wood pole or steel tower that supports the line's insulators.

cross talk The unwanted transfer of energy from one communication circuit to another by means of mutual inductive, capacitve, or conductive coupling.

crowbar A circuit that protects sensitive equipment by becoming a "crowbar" (or short circuit) between power conductors when conditions require it and temporarily shunting all current to ground and clamping the voltage to zero.

CSA Canadian Standards Association.

current The amount of electricity (movement of electrons) that flows through a conductor, such as a wire, measured in amperes and identified by the symbol *I*.

current balance The equal flow of current in each phase of a three-phase power system.

current carrying A circuit component carrying current.

current distortion Distortion in the ac waveform. See *distortion*.

current transformer (CT) A device that provides a means for measuring current beyond the range of a meter and uses the strength of the magnetic field around the conductor to induce a current in its secondary.

customer service charge That portion of the utility customer's bill that remains the same from month to month independent of the amount of energy the customer, uses. It recovers the costs associated with connecting a customer to the company's distribution system, including the service connection and metering equipment, operating expenses, such as meter reading, and the overhead cost of customer accounting and collections.

cutout A piece of easily melted metal inserted in an electric circuit that melts when the current becomes too large for the circuit to carry because of too much demand.

cycle A complete sequence of a wave pattern that repeats itself at regular intervals.

cycles per second The frequency of alternating current measured in hertz.

decibel (dB) The standard unit that expresses relative power of signal levels in terms of the ratio of the logarithm of the power output to power input; i.e., decibel (dB) = $10 \log_{10} (P_1/P_2)$ for power.

delta A standard three-phase circuit connection configured such that the ends of each phase winding connected in series form a closed loop with each phase 120 electrical degrees from the other. It appears as a triangle and looks like the Greek letter delta (Δ).

delta connection A method of connecting a three-phase source or load in series to provide a closed circuit (three-wire, plus ground).

delta-delta Three-phase transformer circuit created by a delta source and a delta load with both the primary and secondary windings connected in a delta configuration.

delta-wye Three-phase transformer circuit created by a delta source and a wye load with the transformer primary connected in a delta configuration and the transformer secondary connected in a wye configuration.

demand (kW) The total amount of electrical power at any given time that a utility customer requires from the utility.

demand billing Billing based on the electrical demand, as specified in the rate schedule or contract, that a customer requires from the utility.

demand charge The specified charge or rate that the utility charges its customers to pay for the amount of maximum power they use in a specified time period, usually a month.

derating factor A factor in percent for reducing the capacity of electrical equipment, like transformers.

differential amplifier One that has two input signal connections and a zero signal reference lead and an output equal to the algebraic sum of the instantaneous voltages appearing between the two input signal connections.

differential-mode voltage The voltage (noise) that appears across two specified sets of active conductors. See *transverse mode noise.*

digital-to-analog converter A circuit or device that changes digital input data to an analog form.

diode A two-terminal device that conducts current better in one direction that the other. Its uses include rectification (ac to dc conversion) and detection (retrieving an information signal from a modulated carrier wave).

diode/capacitor input A power supply that uses a full-wave bridge rectifier connected to a capacitor to produce a pulse current.

dip See *sag.*

direct coupling Two circuits connected to each other through an inductor, resistor, or wire.

direct current (dc) A type of current that never reverses its direction but flows in only one direction and whose average value does not equal zero but a constant number.

dirty power An undefined, imprecise term sometimes used to describe the electric power that causes power quality problems, especially to electronic equipment operation.

disk A nonvolatile mass memory storage device for computers that uses a magnetic medium to store and retrieve computer data.

displacement power factor Refers to the power factor of an undistorted fundamental sine wave and equals the ratio of the active power, in watts, of the fundamental wave to the apparent power, volt-amperes, of the fundamental wave.

DISTCO A distribution company, or "wires" company, that only provides electrical distribution service.

distortion Any deviation from the normal sine wave for an ac quantity; usually describes the abnormal and undesirable waveshape of voltage or current.

distortion factor The ratio of the root mean square (rms) of the harmonic content of a waveform to the rms of the fundamental quantity, expressed in percent of the fundamental. See *total harmonic distortion* (THD).

distribution In utility power system usage, the transport of electricity to the ultimate usage points, such as homes and industries, directly from nearby generators or from interchanges with higher-voltage transmission networks. In consumer power system usage, the transport of electricity inside a home, building, office, or factory.

distribution line A line or system that distributes power from a transmission system to an end-use customer with a nominal operating voltage of less than 69,000 V.

disturbance Any event that adversely affects the normal power flow in a system, such as lightning or a short circuit.

DMM (digital multimeter) An instrument that measures voltage, current, and resistance and displays measurements on a digital readout.

dropout Loss of equipment operation due to noise, sag, or interruption. A discrete voltage loss. Or a voltage sag (complete or partial) for a very short period of time (milliseconds).

dropout voltage The voltage at which a device fails to operate.

duct Enclosure through which cables and wires pass.

dynamic braking resistor A device that helps maintain power system stability during a disturbance.

earth ground A low-impedance path to earth that discharges lightning, static, and radiated energy, and keeps the main service entrance at earth potential.

earthing electrode A ground electrode, water pipe, building steel, or some combination of these, that establishes a building's earth ground.

ECL Emitter-coupled logic; extremely high-speed electronic circuitry where very fast switching between specific voltage levels rather than semiconductor saturation and cutoff determine changes in binary logic.

eddy currents Induced currents in transformer windings and core caused by the magnetic field from the normal alternating current and harmonics.

effective value For any time-variant voltage or current waveform, the constant value that gives the same average power. For sinusoidal waveforms, the effective value equals 0.707 times the peak value.

efficiency The percentage of input power that a device uses expressed mathematically as the ratio of output power to input power: efficiency = P_o/P_i, where P_o is power output in watts, P_i is power input in watts.

EHV Extra-high voltage; covers voltages from 345 to 800 kV.

electromagnetic compatibility The ability of a device or system to operate satisfactorily without causing electromagnetic disturbances to another device or system.

electric field Describes forces associated with electrical charges.

electrical degree Allows mathematical relationships between various electrical quantities. One cycle of ac power passes through 360°.

electrode Generally, a conducting medium through which an electric current enters or leaves a different medium, such as an electrolyte, gas, or vacuum.

electrolyte Any substance that in solution separates into ions and conducts electric current, such as the plates of a battery in an acid or alkaline solution.

electromagnetic A magnetic field caused by an electric current. For example, the electric current in energized power lines causes electromagnetic fields that can interfere with nearby data cables.

electromagnetic compatibility The ability of a device, equipment, or system to function satisfactorily in its electromagnetic environment without introducing intolerable electromagnetic disturbances to anything in that environment.

electromagnetic disturbance Any electromagnetic occurrence that may degrade the performance of a device, equipment, or system.

electromagnetic environment Electromagnetic phenomena that occur within a given location.

electromagnetic susceptibility The inability of a device, equipment, or system to function without degradation in the presence of an electromagnetic disturbance.

electromechanical device A mechanical device that an electric device controls, such as a solenoid or shunt trip circuit breaker.

electrostatic potential A potential difference (electric charge), usually in kilovolts, that occurs between two points and pertains to stationary electrically charged bodies.

electrostatic shield A metallic barrier or shield located between a transformer's primary and secondary windings that reduces capacitive coupling and high-frequency noise.

EMF Electromotive force, or voltage.

EMI Electromagnetic interference.

EMP Electromagnetic pulse.

end user Residential, commercial, or industrial consumer of electricity.

energy The capability of doing work over time, expressed in kilowatt-hours for electrical energy.

engine-generator Combination of an internal combustion engine and an electrical generator driven by the engine.

equipment event log A record of equipment problems and activities that provides diagnostic information along with power monitor data to correlate equipment problems with power events.

equipment grounding conductor The conductor that connects the non-current-carrying parts of conduits, raceways, and equipment enclosures to the grounded conductor (neutral) and the grounding electrode at the service equipment entrance (main panel) or secondary of a separately derived system (e.g., isolation transformer). See Section 100 in ANSI/NFPA 70-1990.

error burst A large number of errors within a given period of time compared to the preceding and following time periods.

ESCO Energy service company.

ESD Electrostatic discharge. The effects of static discharge can range from simple skin irritation for an individual to degraded or destroyed semiconductor junctions for an electronic device.

event summary A plot of recorded power monitor events over time.

failure mode The observed effect of failure.

farad Unit of measurement for capacitance.

faraday shield A grounded metallic barrier that improves the isolation between the windings of a transformer by reducing the leakage capacitance between the primary and secondary transformer windings.

fast tripping (fuse saving) The common utility protective relaying practice in which the circuit breaker or line recloser operates faster than a fuse can blow. It clears transient faults without a sustained interruption but subject's industrial loads to a momentary or temporary interruption.

fault An unintentional short circuit that causes a failure or interruption in an electrical circuit or a power system.

fault locator A device or system that indicates where a fault occurred on a power system.

fault, transient A short circuit on the power system, usually caused by lightning, tree branches, or animals and cleared by momentarily interrupting the current.

feedback The return of a portion of a device's output to its input.

feeder An electrical supply line, either overhead or underground, that connects from a generating plant or an interchange point to a load or distribution system.

feeder lockout Describes the situation when automatic protective devices disconnect a feeder from the power system until manually reconnected, and indicates a serious problem on the power system, usually equipment failure or a broken conductor.

FERC Federal Energy Regulatory Commission.

ferroresonance Resonance resulting when the iron core of an inductive component of an inductive/capacitive circuit becomes saturated and increases the inductive reactance with respect to the capacitance reactance.

ferroresonant transformer A voltage-regulating transformer that uses core saturation and output capacitance to maintain a stable output voltage even when the input voltage fluctuates.

FFT (fast fourier transform) A set of algorithms that speeds up the calculation of a Fourier series; used in equipment that measure harmonics.

field effects The effects of electric and magnetic fields on objects near current-carrying conductors.

filter A device that contains selective combinations of resistors, inductors, or capacitors to block certain frequencies or direct current, while allowing other frequencies to pass.

FIPS Pub. 94 Federal Information Processing Standards publication 94 (September 21, 1983), produced as an official publication of the National Bureau of Standards (since renamed National Institute for Standards and Technology). It gives guidelines for federal agencies in the use of automatic data processing (ADP) facilities in an electrical environment.

firm energy Electric energy that a utility has assured a customer is available to meet all, or any agreed-on portion, of the customer's load requirements over a defined time period.

firm power Power that the supplier has guaranteed to provide its customers at all times except when certain uncontrollable forces, agreed on by the supplier and its customers, occur.

fixture A complete lighting unit that includes one or more lamps connected to a power source.

flashover The flow of a high current between two points of different potential, usually due to insulation breakdown or a lightning surge on a transmission line.

flicker Slow variations in voltage that cause the light intensity in a fluorescent light to vary and give the impression of unsteadiness of visual perception.

fluctuation A change, i.e., surge or sag, in voltage amplitude, often caused by load switching or fault clearing.

fluorescent lamp A lamp that uses less energy than an incandescent lamp by passing a current through a mercury vapor to generate ultraviolet energy that becomes visible light when it hits the phosphorus coating inside the bulb.

flux A magnetic field's lines of force.

flywheel A heavy disk or wheel that rotates on the shaft of a motor-generator to control its rotational speed.

forced outage A power system interruption caused by the improper operation of equipment or human error.

forward transfer impedance The amount of impedance between the source and load that affects the transfer of power, including harmonics and inrush current, to the load.

fossil fuel Nonrenewable fuel, such as coal, oil, and natural gas, found in the ground, made up of decayed plant and animal life and used to generate electricity.

Fourier series An infinite series developed by Francois Fourier to approximate a given function on a specified domain using a linear combination of sines and cosines.

FPN Fine print note; National Electrical Code (NEC) explanatory material.

frequency The number of times in a specific period (how frequently, usually in cycles per second) alternating current reverses its direction, measured in hertz (Hz). Each reversal from one direction to another and back again constitutes a cycle. In North America, utilities provide power with a frequency of 60 cycles per second, or 60 hertz. In ac circuits, designates number of times per second that the current completes a full cycle in positive and negative directions. See also *alternating current.*

frequency deviation A variation, i.e., increase or decrease, in the nominal power frequency that can occur from several cycles to several hours.

frequency modulation (FM) A method to change a carrier frequency so that it varies above and below a center or resting frequency in step with the signal it transmits.

frequency response A measurement of how well a device or circuit transmits various frequencies.

fuel See *fossil fuel.*

fuel cell An electrochemical cell that produces electrical energy directly from the chemical reaction of a fuel and an oxidant on a continuous basis.

full load The greatest amount of load that a piece of equipment, circuit, or conductor can carry without overloading.

full-wave bridge rectifier A configuration of power diodes that changes alternating current into pulsating direct current.

fundamental (component) The component of order 1 (50 to 60 Hz) of the Fourier series of a periodic quantity.

fundamental frequency The lowest-frequency component of a periodically recurring, multifrequency wave, i.e., 60 Hz in the case of U.S. power frequency.

fuse A piece of metal that, when placed in an electric circuit, melts when the current becomes too great and opens the circuit, thereby protecting the circuit and loads from damage caused by an overload.

gain An increase in an electrical quantity that equals the ratio of output to input voltage, current, or power of a signal as it passes through an electronic device; usually expressed in decibels (dB).

galvanometer A device that measures small electrical currents by means of a mechanical motion produced from the electric current.

gauss The unit that measures the strength of a magnetic field.

GENCO Generating company.

generate To produce electrical energy.

generating station (power plant) A place that contains prime movers, electric generators, and auxiliary equipment to convert mechanical (water, wind), chemical (fossil fuels), solar (sun), and/or nuclear energy into electric energy.

generation The act or process of converting other forms, e.g., mechanical, chemical, solar, or nuclear of energy, into electric energy.

generator (electric) A machine that transforms mechanical energy into electrical energy.

GFI (ground fault interrupter) A device that interrupts the flow of electric current in an electric circuit when the fault current to ground exceeds some predetermined value.

gigawatt One billion (10^{12}) watts; describes the capacity of large electrical systems.

glitch An undefined, imprecise power quality term that describes a voltage variation (usually of very short duration) that causes electronic equipment to malfunction.

grid An interconnected system of electric transmission lines and associated equipment that moves or transfers bulk electric energy from the generators to the loads.

ground The connection between earth and an electric circuit that causes electric current to flow out of the circuit and into the earth. Sometimes confused with bonding. Should always conform to the National Electrical Code.

ground electrode A conductor or group of conductors in contact with the earth that provides a low-impedance connection to the ground.

ground fault Any undesirable current path from a current-carrying conductor to ground.

ground grid A system of interconnected bare conductors arranged in a pattern over a specified area, on or buried below the surface of the earth, that provides safety for workers by limiting potential differences within its perimeter to safe levels. It does not act as a signal reference grid.

ground impedance tester An instrument that measures the impedance of a circuit from the point of test to the bond between the neutral and ground bond. Some can handle 120 V ac single-phase voltage while others can handle 600 V ac three-phase voltage. It also measures voltage and determines the presence of neutral-to-ground connections, isolated ground shorts, reversed polarity, and an open equipment grounding conductor.

ground loop A undesirable loop formed when two or more points that have nominally the same ground potential in an electrical system connect to the same conducting path such that either or both points no longer have the same ground potential. It causes an unexpected current to flow in a non-current-carrying conductor when the two or more points of an electrical system connect to the earth at different points.

ground mat See *ground grid.*

ground noise An undefined, imprecise term that describes unwanted electrical signals appearing between the earth conductor and any other conductor.

ground, radial Single conductor to ground.

ground window The opening through which all grounding conductors, including metallic raceways, enter a specific area. Provides the building grounding system a connection to an area that would otherwise have no grounding connection.

grounded Connected to earth or to some conducting body that substitutes for earth.

guy wire A cable that supports a pole.

hardware In computer usage, the physical components of the computer system, such as circuits, cabinets, racks, and tape readers, as contrasted with software.

hard-wired Applied to equipment that connects its power source with wiring (generally customer- or contractor-supplied) attached directly to terminal blocks or distribution panels rather than via an input line cord and output receptacle.

harmonic A sinusoidal wave, such as ac voltage, current, or power, that has a frequency equal to an integral multiple of the fundamental frequency.

harmonic content The waveform that remains after subtracting the fundamental component from a harmonic waveform.

harmonic distortion Waveform changes caused by the power supplies of certain electrical or electronic appliances. Periodic distortion of the sine wave. See *distortion* and *total harmonic distortion* (THD). Regularly appearing distortion of the sine wave whose frequency is a multiple of the fundamental frequency. Converts the normal sine wave into a complex waveform.

harmonic filter On power systems, a device for filtering one or more harmonics from the power system. Most are passive combinations of inductance, capacitance, and resistance. Newer technologies include active filters that can also address reactive power needs.

harmonic neutralization A process to cancel harmonics at the output of a circuit by inverting them and feeding them back in their opposite phase.

harmonic number The whole number that equals the ratio of the frequency of a harmonic to the fundamental frequency.

harmonic resonance The power quality term that describes the condition that sometimes occurs in electrical systems in which high currents flow through and damage capacitors or clear fuses in connecting circuits. A condition in which the power system resonates at one of the major harmonics produced by nonlinear elements in the system and increases the harmonic distortion.

heat pump A device that takes heat from one source and transfers it to another place. In a building, it heats the air in the winter and cools the air in the summer. Several different types of heat pumps exist, but the most common are air to air heat exchangers.

henry Unit of measurement for inductance.

hertz (Hz) Unit of frequency; 1 hertz (Hz) equals 1 cycle per second.

high-intensity discharge (HID) lamp A type of lamp that uses mercury, metal halide, or high-pressure sodium to illuminate.

high-pass filter A filter that passes all frequencies above a certain level, and stops all lower frequencies.

horsepower A unit that measures the power of motors or engines. One horsepower equals 746 watts. However, for all practical purposes, one horsepower almost equals 1000 watts or 1 kilowatt, taking into consideration the starting load and motor inefficiency. For example, a 5-horsepower motor has a rating of approximately 5000 watts or 5 kW, or a $^1/_3$-horsepower furnace motor has a rating of 333 watts.

HV High voltage.

hybrid A device that combines different technologies to improve its function.

hydroelectric Refers to a type of generating station where mechanical energy produced by falling water converts into electrical energy by turning a turbine-generator.

hysteresis loss The transformer core loss due to the friction caused by the molecules' resistance to being magnetized and demagnetized.

I^2R The expression for power that comes from the flow of current through a resistance: $P = I^2R$.

IEEE Institute of Electrical and Electronics Engineers.

ITIC Information Technology Industry Council, which replaced the former Computer Business Equipment Manufacturers Association.

ITIC curve Replaces the former CBMA curve and provides a set of curves developed by the Information Technology Industry Council (ITIC) that represents the withstand capabilities of computers in terms of the magnitude and duration of the voltage disturbance.

impedance Forces that resist current flow in ac circuits; i.e., resistance, inductive reactance, and capacitive reactance.

impulse (notch) A disturbance of the voltage waveform that lasts less than $\frac{1}{2}$ cycle (1 millisecond) and occurs in opposite polarity to the waveform (see *transient*).

impulsive transient A sudden nonpower frequency change in the steady-state condition of voltage or current that occurs in unidirectional polarity (primarily either positive or negative).

incandescent lamp A type of light that uses electric current flowing through a tungsten filament to heat the filament to the point where it glows (incandescence) and produces illumination.

inductance A coil's ability to store energy in the form of a magnetic field and oppose changes in current flowing through it. The value of inductance is a function of the cross-sectional area, number of turns of coil, length of coil and core material.

inductive reactance The impedance to alternating current produced by inductors.

inductor (choke) A coiled conductor usually wrapped around a ferrous core; tends to oppose any change in the flow of current.

input line cord The power cord connected to the input terminals of the UPS that plugs into an ac utility outlet to supply power to the UPS.

input plug A plug that connects to the end of the input line cord at one end and into a receptacle connected to an ac utility outlet at the other end.

inrush The high initial large current that a device, like a motor, requires when it first starts up.

inrush current The initial large current demand needed to start up certain types of electrical equipment, like motors, before their resistance or impedance increases to their normal operating value.

instantaneous Describes the short time range from $\frac{1}{2}$ cycle to 30 cycles of the power frequency.

instantaneous reclosing A term that describes the time (18 to 30 cycles) to reclose (reconnect) a utility breaker as quickly as possible after the interrupting fault current.

insulation (electrical) A non-electrical-conducting material, like rubber or polyethylene, that resists electrical flow and covers an electric wire.

insulation (thermal) A non-heat-conducting material, like fiberglass, that resists heat flow and provides a barrier to heat loss.

insulator A device made of porcelain, glass, rubber, or wood that prevents electricity from flowing, like a porcelain support that insulates conductors from a pole or tower.

interconnection system A connection between two electrical systems that allows the transfer of electric energy in either direction.

interharmonic (component) A frequency component of a periodic quantity that is not an integer multiple of the power supply frequency (e.g., 50 or 60 Hz).

interharmonics A power quality term that describes the waveform distortion at frequencies other than an even multiple of the fundamental frequency. For example, a harmonic frequency may equal 2, 3, or 4 times the fundamental frequency while an interharmonic may equal 1.5, 2.5, or 6.3 times the fundamental.

internal impedance The inherent impedance inside a device or circuit.

interruptible power Power that the supplier and the customer have agreed can be stopped by the supplier.

interruptible rate A lower rate to large industrial and commercial customers who agree to reduce their electricity use in times of peak demand.

interruption A complete stop in the flow of electricity, lasting from a fraction of a second to hours (see also *power outage, momentary*).

interruption, momentary (electrical power systems) The time, not to exceed 5 minutes, required to restore interrupted electrical service by automatic or supervisory-controlled switching operations or by manual switching at locations where an operator is immediately available.

interruption, momentary (power quality monitoring) A complete loss of voltage (<0.1 pu) on one or more phase conductors for a short time period between 30 cycles and 3 seconds.

interruption, sustained (electrical power systems) Any interruption not classified as a momentary interruption of greater than 5 minutes.

interruption, sustained (power quality) The complete loss of voltage (<0.1 pu) on one of more phase conductors for a long time period greater than 1 minute.

interruption, temporary The complete loss of voltage (<0.1 pu) on one or more phase conductors for a short time period between 3 seconds and 1 minute.

inverter A machine, device, or system that changes direct current (dc) power into alternating current (ac) power.

investor-owned utility A company that pays taxes, makes a profit, and private investors (stockholders) own.

isolated ground An insulated equipment grounding conductor that runs in the same conduit or raceway as the supply conductors and is insulated from the metallic raceway and all ground points throughout its length. It originates at an isolated ground-type receptacle or equipment input terminal block and terminates at the point where neutral and ground bond at the power source. See NFPA 70-1990, Section 250-74, Exception 4, and Section 250-75, Exception.

isolation Separation between electrical input and output, such as an isolation transformer or optical coupler or separation of one section of a system from the undesired influences of other sections.

isolation transformer A multiple-winding transformer with physically separate primary and secondary windings that allows the magnetic field in the windings of the primary to create (induce) electrical power in the secondary winding but minimizes electrostatic transfer to the secondary windings. This way the electrical power available at the input transfers to the output, but some of the unwanted electrical effects in the input power do not reach the transformer's output. See also *electrostatic shield.*

joule A unit of energy or work that equals 1 watt-second or 0.0002778 watt-hours (1 kilowatt-hour equals 3,600,000 joules).

jumper A short length of conductor that connects two points in a circuit.

junction box (J box) A box with a blank cover that provides space to connect and branch enclosed conductors.

***K*-factor** A number determined from the amount of harmonics in the load current that provides a means to derate equipment to carry the extra harmonic load.

***K*-rated transformer** A transformer specially designed to handle harmonics.

kilo (k) A metric prefix meaning 1000, or 10^3.

kilohertz One thousand hertz.

kilovolt (kV) One thousand volts.

kilowatt (kW) One thousand watts, or the number of watts divided by 1000. Kilowatts are real power and are important in sizing UPSs, motor generators, and other power conditioners. See also *power factor.*

kilowatt-hour (kWh) The basic unit of electric energy that equals 1000 watts of power used for 1 hour. The amount of power the customer uses is measured in kilowatt-hours (kWh). A measurement of power and time used by utilities for billing purposes. For example, a 75-watt light bulb that burns for 10 hours consumes 0.75 kWh (75 watts $\times$ 10 hours) or 750 watt-hours.

Kirchhoff's laws of electric networks The sum of the electrical currents flowing to a point in a network equals the sum of the currents flowing away from that point.

kVA (1000 VA) Kilovolt amperes (volts times amperes) divided by 1000; provides the actual measured power (apparent power). Used for circuit and equipment sizing.

kW One thousand watts.

kWh One thousand kilowatt-hours.

lag The time delay between two events (usually electrical quantities), such as the delay of current behind voltage.

lagging load An inductive load that resists changes in current in which current lags voltage. The time lag between current and voltage is measured in electrical degrees and is known as the *phase angle.* The cosine of this angle equals the power factor (linear loads only).

lattice tower A transmission tower or substation structure that contains skew as well as horizontal and vertical members.

***LC* circuit** An electrical network that contains both inductive and capacitive elements.

leading load A capacitive load that resists changes in voltage with current leading the voltage.

LED Light-emitting diode; a semiconductor that emits light when current passes through it.

lighting efficiency A comparison of the amount of light a lamp emits to the amount of energy it uses, measured in lumens per watt. For example, a lamp that yields 100 lumens per watt is twice as efficient as a lamp that yields 50 lumens per watt.

lightning arrester A lightning arrester protects lines, transformers, and equipment from lightning surges by transferring the charge to the ground. Serves the same purpose on a line as a safety value on a steam boiler.

line A system of poles, conduits, wires, cables, transformers, fixtures, and accessory equipment used for the distribution of electricity to the public.

line crew A team of highly trained workers who service and repair lines and equipment.

line filter A filter in series with a transmission line that removes unwanted electrical signals.

line imbalance Unequal loads on the phase lines of a multiphase feeder.

line loss The heat loss in a line caused by the flow of the current through the resistance in the line, usually called I^2R loss.

line person A person who repairs and maintains power and telecommunication lines.

line to line Describes an electrical quantity, such as voltage, between two conductors in a multiphase line.

line to neutral Describes an electrical quantity, such as voltage, between one phase of the line and the neutral.

line trap A device that confines and carries a communication signal transmitted over a telephone line, radio circuit, or power line.

linear load An electrical load device, which, in steady-state operation, provides an essentially constant load impedance to the power source throughout the applied voltage cycle and in which the current relationship to voltage remains constant for a relatively constant load impedance.

load The amount of electric power or energy delivered or required at any specified point or points on an electrical system. It originates primarily at the energy-consuming equipment of the electrical customer.

load balancing Switching the various loads on a multiphase feeder to equalize the current in each line.

load factor The ratio of average load in kilowatts to the peak load during a specified period of time.

load fault A malfunction, like a tree touching a power line that causes the load to demand abnormally high amounts of current from the source.

load following The practice of automatic load shaping on a second-by-second basis to maintain a continuous balance between loads and generation.

load interrupter switch A power system switch that can interrupt a circuit under load or a limited amount of fault current.

load management Controlling the level and shape of the demand for electrical energy so that the demand matches the existing and future availability of electrical energy.

load regulation A term used to describe the effects of low forward transfer impedance.

load shaping Either the arrangement and operation of generating resources to meet a given load or the arrangement of (interchange) load to meet a given generation over specified time periods, such as hourly, weekly, monthly, or yearly.

load shedding The dropping of loads in isolated areas by the use of automatic relays to protect the bulk power system from collapse. This could occur when the generated amount cannot meet the load requirements or transmission lines are in danger of overload.

load switching Transferring the load from one source to another.

load unbalance Unequal loads on the phase lines of a multiphase system.

logger A device that stores and prints out important information on the operation of power systems.

long-duration variation A change of the rms value of a voltage from nominal voltage for a time greater than 1 minute. Usually further described by a modifier indicating the magnitude of the voltage change (e.g., undervoltage, overvoltage, or voltage interruption).

loop A set of branches that form a closed circuit where the omission of any branch would eliminate the closed-circuit path.

loss The power dissipated in a power system circuit expressed in watts. In communications, the ratio of the signal power delivered by a device under ideal conditions to the signal power actually delivered, expressed in decibels (dB).

low-pass filter A filter that passes all frequencies below a certain designated cutoff point and blocks all frequencies above that point.

low-side surges A term started by distribution transformer designers to describe the current surge that appears across the transformer secondary terminals when lightning strikes grounded conductors nearby.

lumen A lumen is a unit of light output from a lamp, measured in foot-candles.

luminaire A light fixture.

magnetic field An area where magnetic forces emanate from a magnet.

magnetic synthesizer A device made of resonant circuits including nonlinear inductors and capacitors to store energy, pulsating saturation transformers to modify the voltage waveform, and filters to filter out harmonic distortion to protect sensitive loads from voltage sags, transients, overvoltage, undervoltage, and voltage surges.

main service entrance The enclosure that contains connection panels and breakers, located at the point where the utility power lines enter a building.

maintenance bypass A circuit that allows maintenance personnel to repair or service equipment without affecting supply of electricity to other equipment.

make-before-break Operational sequence of a switch or relay where the new connection occurs prior to disconnection of the existing connection.

manual bypass switch (MBS) A manually operated transfer switch that allows maintenance personal to bypass the major electronics in the UPS so the UPS can be safely serviced.

mega (M) A metric prefix meaning 1,000,000, or 10^6.

megahertz (MHz) One million hertz or cycles per second.

megawatt (MW) One million watts or 1000 kilowatts.

metal oxide varistor (MOV) A component of a surge suppressor that limits overvoltage conditions (electrical surges) on power and data lines. Its resistance varies according to the rate of applied voltage and current and decreases from a very high level (thousands of ohms) to a very low level (a few ohms) when the applied voltage exceeds the breakdown point.

meter board The board on which the main switch and associated equipment mount.

meter constant The ratio of the output of an instrument transformer (current transformer, power transformer) to the input of the meter to determine the difference between meter readings to determine the kilowatt-hours used.

meter inspection An examination of the meter to determine its accuracy.

meter loop (meter socket) The necessary equipment and wiring that the customer's electrician installs prior to the meter installation.

meter test An instrumental test of meter accuracy under all load conditions.

micro (μ) A metric prefix that means one-millionth of a unit, or 10^{-6}.

micron One-millionth of a meter.

microwaves Radio frequencies of 1000 megahertz or greater that have very short wavelengths and exhibit some of the properties of light and allow communication signals sent point-to-point in a concentrated beam.

mil A unit of length equal to one-thousandth of an inch, or 10^{-3} inch.

milli (m) A metric prefix that equals one thousandth of a unit, or 10^{-3}.

minimum ground clearance The least distance between a conductor and ground level under selected design-loading conditions.

mitigate (power quality) The reduction of power quality disturbances by the use of power conditioning equipment.

mobile substation A movable substation that substitutes for a fixed substation when a substation is not working or additional power is needed.

modem An abbreviation for units or equipment panels containing both a modulator and a demodulator that allows data equipment to connect to a communication line, for example, to connect computer equipment to telephone lines.

module A subunit of an electronic system that can be plugged in or otherwise easily replaced.

momentary (power quality monitoring) An undefined, imprecise term used to describe a short-duration (usually 30 cycles or 3 seconds) power quality event, such as a voltage sag or surge.

momentary interruption See *interruption, momentary.*

Moore's law The principle by which, in the computer industry, each new chip contains roughly twice as much capacity as its predecessor released 18 to 24 months previously. Named after an Intel founder, Gordon Moore, who made this observation in a 1965 speech.

motor alternator A machine that contains an ac generator mechanically connected to utility power or a battery-driven electric motor.

motor generator See *motor alternator.*

MOV See *metal oxide varistor.*

MTBBF Mean time between failure, the probable length of time that a component will survive when taken from a particular batch if operated under the same conditions as a sample from the same batch.

MTTR Mean time to repair.

nameplate rating The full-load continuous rating of a specific electrical apparatus, like a transformer or generator, under specified conditions set by the manufacturer.

nano (n) A metric prefix meaning one-billionth of a unit, or 10^{-9}.

National Electrical Code (NEC) A national code of standards and practices for the electrical and electronics industry.

NEC National Electrical Code.

negative resistance The characteristic of a circuit in which current varies inversely with applied voltage.

negatively sequenced Occurs when a three-phase electrical quantity, such as voltage, current, or power, crosses the zero line in the order of ACB rather than ABC.

NEMA National Electrical Manufacturers Association.

network A system of transmission and distribution lines cross-connected and operated to permit multiple power supply to any principal point on it. It provides a high level of reliability by restoring power quickly to customers during an outage by switching them to another circuit.

neutral The grounded junction point of the legs of a wye circuit or the grounded center point of one coil of a delta transformer secondary.

neutral conductor The conductor in a three-phase wye system that acts as the return conductor by conducting the resultant current in an unbalanced three-phase system. It bonds to ground on the output of a three-phase delta-wye transformer.

neutralizing winding An extra winding that cancels harmonics developed in a saturated secondary winding and produces a sinusoidal output waveform from a ferroresonant transformer.

node A terminal of any branch of a network or a terminal common to two or more branches of a network.

noise Any unwanted extraneous electrical quantity superposed on a useful signal that tends to obscure the signal's information content, or any undesired audible signal; undesired sound; more specifically, sound that constitutes a hazard or annoyance.

nominal value A specified value that has become the standard. For example, nominal voltage for a home circuit equals 120 V.

nominal voltage The normal or specified voltage level. For three-phase wye systems, nominal voltages are 480/277 V (600/346 V in Canada) and 208/120 V, where the first number expresses phase-to-phase (or line-to-line) voltages and the second number equals the phase-to-neutral voltage. The nominal voltage for most single-phase systems is 240/120 V.

nondamaging distortion of electricity Disturbances from other appliances and electronic lighting that interfere primarily with communications equipment.

nonfirm energy (or power) Energy (or power) supplied or available under an arrangement that does not guarantee continuous availability of firm power (see *firm power*).

nonlinear load A load in which the current varies with the voltage in a nonlinear fashion. For example, in a switched-mode power supply or almost any other electronic power supply, the current does not vary in direct proportion to the voltage because it uses power in pulses or other waveforms that do not track the sine wave.

nonlinear load current Current drawn by a load that does not have a direct relationship to the voltage waveforms.

nonrenewable resource A natural fuel, such as a mineral that exists in finite supply, like oil, gas, and coal that cannot renew itself once used.

nonsinusoidal A waveform that does not conform to the shape of a sine wave (see *sine wave*).

nonspinning reserve That reserve generating capacity that provides generation to a load within a specified time period.

normal mode (NM) Refers to electrical interference measured between line and neutral (current-carrying conductors). The operation of lights, switches, and motors generates normal-mode interference.

normal-mode noise (voltage) A voltage that occurs between energized conductors in a circuit, but not between the grounding conductor and an energized conductor.

notching A negative or positive change in the waveshape that repeats cycle to cycle; caused by high peak currents of variable-speed drives or other phase loads.

NRC Nuclear Regulatory Commission, the federal agency that licenses nuclear facilities and oversees these facilities to make sure they follow regulations and standards.

off-line Describes the operation of a standby utility power system that supplies power directly to the load and then transfers to battery power supplied through an inverter when the voltage drops below a specified level. The time required between the loss of utility power and inverter start-up can disrupt sensitive loads.

off-peak energy Period of relatively low system demand for electrical energy as specified by the supplier.

ohm (Ω) The unit of measurement for electrical resistance or opposition to current flow that equals the amount of resistance of a conductor such that a constant current of 1 ampere produces a voltage of 1 volt across it.

Ohm's law The relationship between voltage (pressure), current (electron flow), and resistance. The current in an electrical circuit is directly proportional to the voltage and inversely proportional to the resistance: $E = IR$, $I = E/R$, and $R = E/I$, where E = voltage, I = current, and R = resistance.

on-line Describes the operation of an uninterruptible power system which supplies conditioned power through an inverter or converter to the load and supplies the backup power to the load without delay when a utility power outage occurs.

on-peak energy Electric energy supplied during periods of relatively high system demand as specified by the supplier.

open circuit Describes the condition of a disconnect in a circuit caused intentionally or by a fault.

orderly shutdown The shutdown of units within a computer system in a step-by-step fashion to prevent damage to the system and avoid data loss or corruption.

oscillation The change with time of a quantity's value from the maximum and minimum values of one cycle.

oscillator An electrical device that sets up and maintains oscillations at a frequency determined by the electrical constraints of the system.

oscillatory transient A power quality term that describes a voltage or current transient that rises suddenly and sharply to some level and then degrades over time to a waveform that decreases in frequency and amplitude.

oscillogram A record of an oscilloscope's display.

oscilloscope An electronic instrument that produces a visible graphical display on a cathode-ray tube of the instantaneous value of one or more rapidly varying electrical quantities as a function of time or other electrical or mechanical quantity.

OSHA Occupational Safety and Health Administration.

out of phase A condition that occurs when two waves have the same frequency, but their maximum values occur at different times.

outage In a power system, the state of a component, such as a generating unit or transmission line, that no longer performs its function because of some event directly associated with the component. See *blackout*.

outlet A point on an electric circuit where loads can access electric power.

output The current, voltage, power, or driving force that a circuit or a device delivers.

output impedance A measure of a source's ability to supply current to a load.

overhead ground wire A protective wire strung above the conductors to shield the conductors from lightning.

overload A condition when the flow of electricity exceeds the rated capacity of a device or system or when the load wants more from the power source, utility, or UPS than the power source can supply.

overload capacity The maximum load that a machine, apparatus, or device can carry beyond its normal nameplate rating without damaging itself.

overshoot The tendency of a control circuit's output to continue to momentarily increase after the input signal has stopped increasing.

overvoltage An increase in the normal voltage level lasting for seconds or minutes greater than the rating of a device or component. The term can also apply to transients and surges. When applied to a long-duration variation it refers to a voltage with a value at least 10 percent above the nominal voltage for a period of time greater than 1 minute.

pad-mount transformer A transformer inside a large metal cabinet on a pad outside a house or building that converts the utility voltage supplied via an underground line to a level that a customer can use (120/240 V for a home), similar to a pole-mounted transformer. The cabinet has a lock and a "DANGER" sign on it.

panelboard A single panel or group of panel units assembled into a single panel, containing buses and, overcurrent protection devices (with or without switches) to control power circuits.

parallel operation The connection of two or more components side by side to the same pair of terminals to increase the capacity of the power system or provide power redundancy. For example, paralleling for capacity two transformers means that two 50-kVA units in parallel have the sum of their individual ratings, i.e., the power of a single 100-kVA transformer. Paralleling two UPS units for redundancy means that if one fails, the other unit will provide back-up power.

parallel resonance See *resonance, parallel.*

parity error An unintentional change in the bit structure of a data word due to the presence of a pulse or transient.

passive filter A combination of inductors, capacitors, and resistors that eliminates one or more harmonics. The most common type simply uses an inductor in series with a shunt capacitor, which diverts the major distorting harmonic component from the circuit.

peak current Maximum instantaneous current during a cycle.

peak demand The maximum amount of power required to supply customers.

peak load The maximum electrical demand (instantaneous or average) in a stated period of time.

peaking capacity Generating capacity available to meet that portion of the load above the base load.

periodic waveform A waveform that repeats itself after a period of time.

peripheral Any device that processes data put into or taken out of a computer.

phase The stage or progress of a cyclic movement, such as a current or voltage wave. Also a conductor that carries one of three separate phases (designated A, B, and C) of power in an alternating current system. Almost all residential customers use single-phase service. Large commercial and industrial customers use either two-phase or three-phase service.

phase angle In a power system, the displacement, in time, of the phase of one quantity from the phase of another, at power system frequency.

phase balancing Connecting loads in a three-phase power system so that all three phases carry the same current.

phase changer A utility-owned device that changes the phase of the service supplied to meet the equipment needs of the customer.

phase compensation Switching capacitors into or out of a power distribution system to reduce the phase difference between the current and voltage to keep the power factor close to unity.

phase conductor The wire cable in each phase of a transmission or distribution line.

phase rotation The sequence of electrical quantities, like voltage or current, in all three phases of a three-phase system.

phase shift The displacement in time of one voltage-waveform relative to other voltage waveforms.

photovoltaic generation A method to convert solar energy directly into electrical energy that uses specially designed semiconductors called photovoltaics.

pico (p) A metric prefix meaning one millionth of a millionth, or 10^{-12}.

planned electric outage An interruption of service to electric lines to permit work on the electric lines or equipment served by the electric lines.

points of common coupling (PCC) Points where the electric utility connects to its end-user customer.

polarity An electrical condition that determines the direction in which current tends to flow.

pole-mounted transformer A distribution transformer, mounted on a pole, that steps primary distribution voltage down to a level that customers can use.

pole structure In a transmission or communications system, a column (or columns) of tapered wood or steel, supporting overhead conductors.

polyphase An alternating current supply with two or more hot conductors. Voltages between the conductors and the voltage waveforms for each conductor are usually displaced 120°. The voltage from each hot conductor to neutral is equal.

positively sequenced A three-phase electrical quantity that has all three phases cross zero in the order ABC.

potential transformers (PTs) Transformers that reduce high voltage to within the range of a meter.

pothead A flared, pot-shaped, insulated fitting that connects underground cables to overhead lines.

power The time rate of transferring or transforming energy, expressed in watts or kilowatts, that equals (1) the product of applied voltage and resulting in-phase current for ac circuits or (2) volts times amperes for dc circuits. For single-phase ac circuits, watts equal volts times amperes times power factor.

power amplifier An electronic device that contains a local source of power that can increase the input power signal significantly.

power conditioner A device that modifies the voltage magnitude or frequency or provides alternative sources of energy.

power distribution unit (PDU) A portable device that provides power to computer equipment.

power disturbance A disturbance, like a surge or sag, that originates from the utility's power system.

power factor (PF) The ratio of total watts (the real power) to the total root mean square (rms) volt-amperes (apparent power); W/VA = power factor. Inductive loads cause leading power factor and capacitive loads cause lagging power factor; nonlinear loads cause harmonic power factor.

power factor (true) The ratio of active power (watts) to apparent power (volt-amperes).

power-factor-corrected power supplies (PFCs) Power supplies with capacitors that increase the power factor to unity.

power factor, displacement The power factor of the fundamental frequency components of the voltage and current waveforms.

power flows Studies of line and equipment power loadings on transmission or distribution networks for specific conditions of system generation, load, and line configuration to provide information for planning future system additions to ensure power high in quality and reliability.

power grid A network of power lines and associated equipment used to transmit and distribute electricity over a geographic area.

power interruption alert The notification of the public that load shedding or rolling blackouts are imminent.

power line carrier A voiceband or narrow-band signal communication system that depends on imposing a carrier with information on high-voltage power transmission lines.

power marketing agencies United States federal agencies in the Power Marketing Administration in the Department of Energy (Bonneville, southeastern, southwestern, and western areas), and the Tennessee Valley Authority (TVA).

power plant A generating station or a place where electricity is produced.

power pool Two or more electric systems interconnected and coordinated to supply power in the most economical manner for their combined load requirements and maintenance program.

power quality It depends on your perspective. If you are an electrical engineer, power quality expert, or electrician, you may tend to look at power quality as a problem that must be solved. If you are an economist, power marketer, or purchaser of electrical power, you may look at power as a product and power quality as an important part of that product. End users see power quality as a measure of how the power affects the operation of their equipment.

power quality problem The difference between the quality of electricity at an electrical outlet and the quality of the electricity required to reliably operate an appliance, resulting in a malfunction or damage.

power supply A device that converts available electric service energy (alternating current) into direct current energy at a voltage suitable for electronic components.

power surge An undefined, imprecise term sometimes used to describe a transient that damages equipment.

power warning An appeal to the public that an immediate reduction in power usage is necessary to avert overload of the electrical system. Public appeals are made when other efforts such as emergency purchases, voluntary curtailment, contracted curtailment, and voltage reductions are unsuccessful in supplying the demand.

power warning fault (PWF) An option in a UPS that supplies a warning signal to some computer systems that the UPS may shut down. Some computers can take advantage of this signal to automatically back up and shut down before the UPS shuts down.

power watch An announcement made when conditions are such that further steps to manage capacity may affect the public.

power-line monitor An instrument that monitors the condition of the power supplied to a given load.

preference customers In accordance with Congressional directives, cooperatives and public bodies (states, public utility districts, counties, municipalities, and federal agencies) have preferential rights to federally generated hydropower.

primary circuit The distribution circuit (less than 69,000 V) on the high-voltage side of the transformer.

primary distribution feeder (primaries) Distribution lines that carry the highest distribution voltage. They are usually located at the topmost position of the utility pole.

private utility See *investor-owned utility.*

program As applied to digital computers, the set of instructions to perform the sequence of operations to solve a particular problem or a related group of problems, or to perform a particular computer task.

propagation The travel of an electrical waveform along a medium, as in a surge passing along a power cord to a system.

protective scheme A group of interrelated devices that prevent damage to equipment caused by very high voltages and currents.

protector Another name for an arrester or diverter.

pulsating dc A voltage that remains at the same polarity while it goes up and down.

pulse An abrupt variation of short duration of a physical quantity followed by a rapid return to the initial value.

pulse modulation Describes methods of transmitting information on a pulse train.

pulse-width modulation A method for varying a pulse train width in relationship to another signal's characteristics.

Q A quality factor that indicates the electrical quality of a coil, capacitor, or circuit and equals the ratio of reactance to resistance. The higher the Q, the greater the selectivity of the circuit.

radial array A group of earthing electrodes or conductors of equal length and ampacity, connected at a central point and extending outward at equal angles, spoke fashion, to provide a low-earth-impedance reference.

radial ground See *ground, radial.*

radiation Transmission of energy by means of electromagnetic waves. Radiant energy of any wavelength, when absorbed, may become thermal energy and result in an increase in the temperature of the absorbing body.

random error An error that does not repeat itself, like noise.

rate base The value, specified by a regulatory authority, that a utility can earn to gain a certain rate of return.

rates The prices a utility charges for different types of electrical service.

reactance A physical property of a circuit component that tends to hinder the flow of alternating current.

reactive power The out-of-phase component of the total volt-amperes in an electric circuit that represents the power required to provide a magnetic field to drive reactive loads, like motors, in a circuit. Expressed in VARs (volt-amperes reactive).

reactor, current-limiting An inductor connected in series in a circuit to limit short-circuit current to a predetermined value to protect equipment from damage due to excessive current.

real power The in-phase component of volt-amperes in an electric circuit that does useful work, expressed in watts or kilowatts.

receptacle The contact in an electrical outlet, often referred to as a wall socket.

receptacle tester Three-lamp circuit testers that plug into a receptacle and measure wiring connections to the receptacle. They indicate the wiring errors in the receptacles by a combination of lights. "IEEE Recommended Practice for Power and Grounding Sensitive Electronic Equipment" (*The Emerald Book*), page 32, says "These devices have some limitations. They may indicate incorrect wiring, but cannot be relied upon to indicate correct wiring."

recloser A circuit-interrupting device that recloses the circuit after an automatic trip.

reclosing The common utility practice on overhead lines of closing the breaker within a short time after clearing a fault, taking advantage of the fact that most faults are transient, or temporary.

reclosure The automatic closing of a circuit-interrupting device after automatic tripping.

recorder, strip-chart A device that simultaneously makes one or more permanent records of varying quantities (voltage, megawatts, etc.) as a function of time.

recovery time Time required for the output voltage or current of a power system to return to normal operating condition after a load or line change or interruption.

recovery voltage The voltage that occurs across the terminals of a circuit-interrupting device when it interrupts the flow of electricity.

rectifier An electrical device, like a battery charger, that converts ac power into dc power.

redundancy The addition of extra components in an electrical power system that provide backup in the event of loss of any of those components.

reflection The return wave created when a traveling wave encounters a load, a source, or a junction that has a change in line impedance.

regulated power supply A power supply that keeps the output voltage constant when the load changes.

relay An electromagnetic device that interprets input conditions (which reflect the operation of another piece of equipment) in a prescribed manner, and, after specified conditions occur, responds to cause contact operation or similar abrupt change in a circuit controlling the equipment.

relay differential A relay that responds by design to the difference between incoming and outgoing electrical quantities associated with the protected electrical apparatus.

reliability Generally, the ability of an item to perform a required function under stated conditions for a stated period of time.

relief valve A device that reduces pressure quickly.

repeater A station between terminals of a microwave system that receives a signal from a distant station, amplifies the signal, and retransmits it to another distant station.

reserve capacity The capacity in excess of that required to carry peak load.

resistance A conductor's characteristic that retards the flow of electrons (current).

resistance value The electrical quantity that impedes current flowing in a conductor, expressed in ohms (Ω).

resistor A device that introduces resistance into a circuit.

resonance The condition that occurs when the capacitive reactance equals the inductive reactance of a circuit.

resonance, parallel For capacitive reactance and inductive reactance in parallel, a condition that occurs when the capacitive reactance equals the inductive reactance of the circuit. It maximizes the impedance and minimizes the current flow in the circuit.

resonance, series For capacitive reactance and inductive reactance in series, a condition that occurs when the capacitive reactance equals the inductive reactance of the circuit. It minimizes the impedance and maximizes the current in the circuit.

response time The time required, after the initiation of a specified disturbance to a device or system, for an output to reach a specified value.

restoration time The time required in relaying and switching to deenergize a transmission line after a fault occurs plus the time to reenergize the line after the fault clears.

RETAILCO Retail marketing company that provides enhanced power quality services as part of its service offerings.

revenue The amount of money a utility collects from its customers.

revenue meter The meter a utility uses to record and display the amount of electrical energy a customer uses in a given period of time.

RF Radio frequency.

RFI Radio-frequency interference.

ride-through The ability of a power conditioner, especially a UPS, to continue to supply power to critical loads when the utility has discontinued power.

right-of-way An easement for a certain purpose over the land of another, such as the strip of land used for a road, electrical transmission and distribution line, ditch, pipeline, etc.

ripple An alternating current (ac) component on a direct current (dc) voltage resulting from incomplete filtering, usually associated with the ac component that appears on the output of the dc power supply.

risers The insulators in a substation that support the bus bar.

rms Root mean square; used for ac voltage and current values that equal the square root of the average of the squares of all the instantaneous amplitudes occurring during one cycle. Referred to as the effective value of ac because it equals the value of ac voltage or current that causes the same amount of heat produced in a circuit containing only resistance from a dc voltage or current of the same value. In a pure sine wave the rms value equals 0.707 times the peak value and the peak value equals 1.414 times the rms value. The normal home wall outlet, which supplies 120 V rms, has a peak voltage of 169.7 V.

rolling blackout A controlled and temporary interruption of electrical service that is necessary when a utility does not have enough power to meet heavy peak demands.

rotating field The electrical field that develops in a multiphase generator caused by the varying currents flowing through parts of a stator winding.

rotor The rotating part of a generator (or motor); usually contains the field winding.

safety ground See *equipment grounding conductor*. An alternate path of return current, during a fault condition, that trips a circuit breaker and establishes a load at earth level.

sag (also called dip or voltage sag) A decrease of the normal voltage level between 0.1 and 0.9 pu in rms voltage or current at the power frequency for durations of 0.5 cycle to 1 minute.

SCADA Supervisory control and data acquisition in a centralized remote control system that includes the transmission of numerical quantities and alarms from substation to a control center.

scheduled outage An outage that results when the utility deliberately removes from service at a selected time a component of the power system, usually to allow construction, maintenance, or test.

schematic diagram A diagram of an electric circuit, with graphical symbols that represent components of the circuit.

SCR Silicon controlled rectifier, a device that acts as an electronic dc switch when triggered to conduct by a pulse or a gate signal, and cuts off the flow of electricity by reducing the main current below a predetermined level (usually zero).

screen freeze (lockup) A term that describes the situation when a computer or computerized equipment stops its operation and its control function.

screen panel (enclosure) An electrical cabinet that houses circuit breakers or fuses for a building or a portion of a building.

secondary The output winding of a two-winding transformer.

secondary circuit The distribution circuit on the low-voltage side of a transformer (usually 120/240 V).

sectionalizer Similar to a reclosure, but opens only when a line becomes "dead" from the operation of a reclosure or breaker upstream. It serves to isolate the section that has the fault and allow the remainder of the circuit to remain energized.

sectionalizing The connecting or disconnecting of sections of a transmission line, distribution line, or substation bus to isolate equipment or line sections for locating problems or doing work.

selectivity In communication, the degree to which a device can accept signals of one frequency or band of frequencies while rejecting all other frequencies. In power system control, the degree to which the interrelated performance of relays and circuit breakers and other protective devices keeps to a minimum the amount of equipment taken out of service for isolation of a fault.

semiconductor An electronic conductor (e.g., silicon, selenium, or germanium) with a resistivity between metals and insulators that allows current to flow through it normally via holes or electrons.

series capacitor In a power system, a capacitor that compensates for voltage drop along a transmission line and reduces the impedance of the transmission line to allow more power to flow in it.

series resonance See *resonance, series.*

service area The territory in which a utility has the responsibility or has the right to supply electric service to ultimate customers.

service channel A band of frequencies, usually including a voice channel, utilized for maintenance and fault indication on a communication system that supports a power system.

service drop The lines running to the customer's house that usually include two 120-V lines and a neutral line, from which the customer can obtain either 120- or 240-V power. When these lines are insulated and twisted together, they become triplex cable.

service entrance equipment The main control and means of disconnection for the supply of electricity to a building that usually contains circuit breakers, switches, and fuses. Newer residential houses usually have 200-A service while older homes generally have 100-A service.

service factor (motor) A measurement of the motor's ability to operate under abnormal conditions. A motor can operate at 1.15 times its rated load continuously when operated at its rated voltage, frequency, temperature, etc. Therefore, a 100-horsepower motor could operate as a 115-hp motor under normal conditions.

sheath A conductive material that surrounds the conductor of a coaxial cable and shields the conductor from outside noise.

shield As normally applied to instrumentation cables, refers to a conductive sheath (usually metallic) applied, over the insulation of a conductor or conductors, to reduce coupling between conductors to prevent them from receiving or generating unwanted electrostatic or electromagnetic fields (noise).

shield wires Grounded wires placed in close proximity to an energized transmission line. Overhead shield (ground) wires protect the transmission line from lightning. Shield wires located beneath the conductors reduce the electric field at ground level.

shielding A barrier to reduce the coupling of undesirable signals. It provides a conducting and/or ferromagnetic barrier between a potentially disturbing noise source and sensitive circuitry. It protects cables (data and power) and electronic circuits. It may include metal barriers, enclosures, or wrappings around source circuits and receiving circuits.

shielding (of utility lines) The construction of a grounded conductor or tower over the lines to intercept lightning strokes in an attempt to keep the lightning currents out of the power system.

short circuit An accidentally established connection between two points in an electric circuit, as when a tree limb or an animal bridges the gap between two conductors. This will cause an overload of current on the line, causing damaged lines, blown fuses, and the faulty operation of protective devices such as reclosers and circuit breakers.

short-duration variation A variation of the rms value of the voltage from nominal voltage for a time greater than $^1/_2$ cycle of the power frequency but less than or equal to 1 minute. Usually further described by a modifier indicating the magnitude of the voltage variation (e.g., sag, swell, or interruption) and possibly a modifier indicating the duration of the variation (e.g., instantaneous, momentary, or temporary).

shunt A device that has resistance or impedance connected in parallel across other devices or apparatus to divert some of the current from it.

shunt filter A filter connected in parallel across a device or circuit to filter out undesirable signals.

shunt trip A device connected in parallel with another device or circuit to disconnect power.

signal A visual, audible, electrical, or other representation that conveys information, or an electronic wave that embodies information.

signal reference grid (or plane) A system of conductive paths among interconnected equipment that reduces noise-induced voltages to levels that minimize improper operation. Common configurations include grids and planes.

signal-to-noise ratio The ratio, at any point of a circuit, of signal power to total circuit noise power, usually expressed in decibels.

silicon avalanche device A semiconductor device that normally acts as an open circuit but changes to a short circuit when the trigger voltage exceeds a certain amount.

simulation The representation of an actual system by analogous characteristics of some device or mathematical equations easier to construct, modify, or understand.

sine wave The sinusoidal form exhibited by alternating current. A graph, with the x axis for time and the y axis for amplitude, depicting ac voltage or current. The center line of the x axis is zero and divides polarity (direction).

single phase A line that carries electrical loads capable of serving the needs of residential customers, small commercial customers, and street lights. It carries a relatively light load compared to heavy-duty three-phase constructs.

With a three-phase source: one or two phase conductors. With a single-phase source: a single output that may be center-tapped for dual voltage levels.

single-phase condition An unusual condition where one phase of a three phase system is lost, causing unusual effects on lighting and other loads.

single-phase line A distribution line energized by a single alternating current; usually serves a residential area.

single-phase power Power provided by a single source with one output. If there is more than one output, the voltages and currents of the outputs are all in phase.

single-point ground A method to avoid differential ground voltage between points in a power system by connecting the power neutral and safety ground at the same single point.

sinusoidal A waveform that can be represented by a sine function.

skin effect The tendency of a high-frequency radio signal current flowing in a conductor to flow near the surface of the conductor.

slew rate The rate of change of ac voltage frequency.

soft-start circuit Circuitry that limits the initial power demand when a UPS is operated in emergency mode and commercial power comes back on. It also controls the rate at which the UPS output increases to normal.

software Programs for directing the operation of computers and computer-controlled equipment, as opposed to *hardware.*

solar heating Heat the sun energy creates. Passive solar heating takes advantage of the heat created through natural means (heat created when sunlight passes through a window and becomes trapped inside a building). Active solar heating systems contain three components: a solar collector, energy storage, and distribution pipes or ducts. Absorber panels collect sunlight, condition it if required, and distribute it through the building by a heat transfer fluid or by air.

solar hot water Similar system to an active solar heating system except it preheats water for normal domestic hot water use. Collectors trap sunlight and create heat. Transfer fluid moves heat from collectors to water holding tanks. This system requires a supplemental heating source for days when the sunlight is not available to heat water to required temperatures.

solenoid An electrical conductor wound as a helix with a small pitch, or as two or more coaxial helixes; or a coil wound in such a manner as to have a movable iron core.

solid-state Describes an electronic device whose electrical functions are performed by semiconductors (as opposed to components that conduct in a vacuum or gas, such as tubes) and otherwise completely static components, such as resistors and capacitors.

spark gap Any short air space between two conductors electrically insulated from each other; or a device that depends on a spark gap for its operation.

spectrum A range of frequencies within which waves have some specified common characteristic, for example, the audio-frequency spectrum.

spikes (voltage) An imprecise, undefined term used to describe the very short duration voltage transients that cause damage to electronic equipment.

spinning reserve A reserve generating capacity connected to an output bus and ready to supply power immediately to the load.

stability The property of a system or element at which its output will ultimately attain a steady state. A power system consists of several generators connected together and to the load by transmission lines. The amount of power that is transmitted from one machine to another after a disturbance, like a line fault, is limited. When this limit is exceeded, the machines become unstable and may lose synchronism with each other. When this happens, relays operate to separate the generators not synchronized. Otherwise, the disturbance would move out over the system, somewhat like a storm moving outward from its center, and result in cascading outages. Also, that attribute of a system that enables it to develop restoring forces equal to or greater than the disturbing forces and thereby remain stable.

standing wave A stationary pattern of waves on conductors or in space created by two waves of the same frequency traveling in opposite directions.

static Audible noise on a radio receiver caused by disturbances, like lightning, motors starting, power line corona, or fluorescent lighting.

static electricity An electric charge that accumulates on an object, usually caused by friction.

static switch A solid-state device that opens and closes circuits without the use of moving mechanical parts.

station service Facilities that provide energy for local use in a generating, switching, converting, or transforming station.

steady state A condition in which circuit values remain essentially constant after all initial fluctuating conditions have stabilized.

stress An external force applied to a component or assembly that tends to damage or destroy it.

substation A small building or fenced-in yard that contains switches, transformers, and other equipment and structures for the purpose of changing voltage, monitoring circuits, switching lines, and performing other service functions. As electricity nears its destination, it goes through a substation where transformers lower the voltage to a level that homes, schools, and factories can use. Substations are also located where high-voltage transmission lines connect to switchgear and stepdown transformers to reduce voltages to lower levels for local distribution networks.

surge A sudden increase of electric current or voltage. A short-duration high-voltage condition that lasts for several cycles, whereas a transient lasts less than $^1/_2$ cycle. Often confused with transient.

surge arresters Electrical devices that limit sudden variations in voltage or current. When connected in series, they limit current and when connected in parallel they limit voltage. They thus protect other electrical equipment and electrical systems.

surge generator An electrical apparatus that produces surges through the use of many capacitor units that store energy and release that energy in the form of surges. It tests in the laboratory various types of electrical apparatus ability to withstand surges.

surge protectors See *surge arresters.*

surge suppressors See *TVSS.*

surplus energy Energy generated beyond what the load requires.

survey A visual and instrument inspection of a facility to determine the causes and solutions to power quality problems.

sustained When applied to quantify the duration of a voltage interruption, refers to the time frame associated with a long duration variation (i.e., greater than 1 minute).

swell A temporary increase in the rms value of the voltage or current of more than 10 percent of the nominal voltage, at the power frequency, for durations from 0.5 cycle to 1 minute.

switched-mode power supply A power supply that uses electronic components to convert ac power into high-frequency dc power.

switchgear A group of switches, relays, circuit breakers, etc., that controls the distribution of power to other distribution equipment and large loads. Also, substation equipment designed and operated to switch electrical circuits and interrupt power flow.

switching station A type of substation that contains various types of switching devices, like breakers and switches, to open and close transmission and distribution lines but contains no transformers to change the voltage.

switchyard The outdoor portion of a substation.

synchronization Maintaining a constant phase relationship between various ac signals.

synchronous Events that have the same period or which occur at the same time. For example, a synchronous transfer mechanism for a standby power generator transfers power to or from the utility in phase. The voltage waveform of the generator and the utility's power system must operate in phase, and the waveforms must occur at the same time and interval during the transfer.

synchronous closing Generally refers to closing all three poles of a capacitor switch in synchronism with the power system to minimize transients.

synchronous condenser A rotating machine that provides variable continuous control of voltage and power factor on transmission lines by operating to either increase (boost) the voltage like a capacitor or decrease (buck) it like a reactor.

synchronous motor An ac motor whose speed varies in proportion to the power input frequency.

system control center A central location that controls and operates the power system.

system frequency Frequency in hertz (cycles per second) of a power system's alternating voltage; equal to 60 hertz in the United States and 50 hertz in Europe and Asia.

systematic error A repeatable portion of an error.

tap A connection point brought out of a transformer winding to permit changing the turns ratio. Also, a terminal where an electric circuit connects to another electric circuit.

tap changer A device that changes the voltage ratio of a transformer or a voltage regulator.

tap switcher A voltage regulator that uses power semiconductors, rated at line voltage and current, to switch taps of a transformer and change the turns ratio and output voltage.

telecommunication equipment Equipment that transmits information, such as words, sounds, data, or images, in the form of electromagnetic signals, as telegraph, telephone, radio, or television signals.

telemetry The transmission of measurements to remote sites or a central location by the use of radio or wire.

temporary When applied to quantify the time of a short-duration variation as a modifier, refers to the time range from 3 s to 1 min.

temporary service Electrical service used for a short period of time, usually at a construction site.

thermal efficiency The ratio of the electric power produced by a power plant to the amount of heat produced by the fuel; a measure of the efficiency of the plant's conversion of thermal to electric energy.

thermocouple A pair of dissimilar conductors so joined at two points that an electromotive force develops when the two junctions at opposite ends experience different temperatures.

three-phase line A line with three conductors that carries heavy loads of electricity, usually to larger commercial and industrial customers.

three-phase power Power from three separate outputs from a single source with a phase differential of 120 electrical degrees between any two adjacent voltages or currents. It has the same phase-to-phase voltage as single-phase power but requires multiplying by the square root of 3, or 1.732, to change from single-phase voltage to line-to-line voltage in a wye-connected three-phase system. The line-to-line voltage equals the phase-to-phase voltage in a delta-connected three-phase system. The power equals 3 times the phase-to-phase power.

thyristor A semiconductor bistable switch (with on and off states) that operates unidirectionally or bidirectionally. A three-terminal device (a controlled rectifier) or a two-terminal device (diode) may trigger it.

tie line A transmission line that connects two or more power systems.

tolerance The allowed change from a specified quantity.

total demand distortion (TDD) The ratio of the root mean square (rms) of the harmonic current to the root mean square value of the rated or maximum demand fundamental current, expressed as a percent.

total disturbance level The level of a given electromagnetic disturbance caused by the superposition of the emission of all pieces of equipment in a given system. Also the total amount of electromagnetic disturbance determined by summing the electromagnetic emissions from each source in a given system.

total harmonic distortion (THD) The ratio of the root mean square (rms) of the harmonic content to the root mean square value of the fundamental quantity, expressed as a percent of the fundamental, that describes a waveshape change caused by the presence of multiples of the fundamental frequency of the ac power. The square root of the sum of the squares of the rms harmonic voltages or currents divided by the rms fundamental voltage or current.

tower A steel structure along transmission lines that supports conductors.

traceability The ability to track a calibration device to a more accurate standard.

transceiver Transmitter and receiver combined together in one cabinet; uses common circuit components to operate the transmitter and receiver.

TRANSCO Transmission company.

transducer A device that senses one form of energy and converts it to another, e.g., temperature to voltage (for monitoring).

transfer switch A switch that transfers load from one source to another.

transfer time The time it takes for a transfer switch to transfer power from one source to another.

transfer trip A relay scheme in which a signal to trigger an operation function transmits from a relay location to a remote location.

transformer A static electrical device that by electromagnetic induction regenerates ac power from one circuit into another. Also used to change voltage from one level to another by the ratio of turns on the primary to turns on the secondary (turns ratio). If the primary windings have twice the number of windings as the secondary, the secondary voltage will have half the primary voltage. A device that changes voltage levels to facilitate the transfer of power from the generating plant to the customer. A step-up transformer increases the voltage (power) of electricity, while a stepdown transformer decreases it.

transformer coupling The linking of separate electric circuits by means of electromagnetic fields, as in a transformer

transformer, current An instrument transformer that gives an accurate low-current (amperes) indication in its secondary winding of the high-amperage current of the power system on its primary winding.

transformer, grounding In a power system, a transformer intended primarily to provide a natural point for grounding purposes.

transformer, isolation See *isolation transformer.*

transformer, potential An instrument transformer that reproduces in its secondary winding a specified portion of the voltage of its primary circuit for control, relaying, or metering.

transient Describes a phenomenon or a quantity that varies between two consecutive steady states during a short time interval. A unidirectional impulse of either polarity or a damped oscillatory wave with the first peak occurring in either polarity. A short duration, fast-rise-time voltage caused by lightning, large motors starting, utility switching operations, and other appliances switching.

transient response The ability of a power conditioner to respond to a change in voltage or power.

transient step load response The ability of a power conditioner to maintain a constant output voltage when sudden load (current) changes occur.

transistor A semiconductor device with three or more terminals that performs functions in an electronic circuit, like amplification and rectification.

transistor-transistor logic (TTL) Electronic circuitry that defines a binary logic state when components saturate or cutoff.

transmission In power system usage, the bulk transport of electricity from large generation centers over significant distances to interchanges with large industries and distribution networks of utilities.

transmission line The conductors that carry electrical energy from one location to another. It has heavy wires that carry large amounts of electricity over long distances from a generating station to the consumers of electricity. They support the conductors high above the ground on tall towers called transmission towers.

transverse mode noise (normal mode) An undesirable voltage that appears from line to line of a power line.

tree crews Teams of utility employees or contractors who clear trees, limbs, and brush from transmission and distribution lines.

tree wire An insulated wire located in heavily treed areas to protect lines from momentary tree limb contact.

triac An electronic device, usually composed of two SCRs connected back to back, that provides switching action for either polarity of an applied voltage and is controlled from a single gate.

trip The opening of a power circuit breaker by protective relays.

trip-out A disconnection of an electric circuit that occurs when the circuit breaker has opened, putting the line out of service; usually refers to an automatic rather than a manual action.

triplen harmonics Odd multiples of the third harmonic, which deserve special attention because of their natural tendency to add to each other.

triplens Harmonics that are a multiple of 3 times the fundamental frequency, for example, third, ninth, fifteenth.

tuned Adjusted for resonance at a specified frequency.

turbine An enclosed rotary wheel turned by water or steam.

turbine-generator A rotary-type unit consisting of a turbine and an electric generator.

TVSS Transient voltage surge suppressor, a device that "clamps" the voltage from a voltage transient and keeps it from damaging sensitive equipment.

UL Underwriters Laboratories.

UHV Ultra-high voltage.

unbalanced load regulation The maximum voltage difference that occurs in the three output phases of an unbalanced power system.

underground (UG) An electrical facility installed below the surface of the earth.

undervoltage A decrease in the normal voltage level lasting for seconds or minutes.

UPS Uninterruptible power supply; contains batteries that store energy, which provides a power source during power interruptions.

utility A company that performs an essential utilitarian service, like providing natural gas and electricity for people.

V ac Volts of alternating current.

VAR Volt-amperes reactive.

varistor A semiconductor device whose resistance varies with the applied voltage.

V dc Volts of direct current.

volt (V) The unit of voltage or potential difference.

volt-ampere Apparent power's unit of measurement.

voltage Electrical pressure, or electromotive force (emf). The force that causes current to flow through a conductor, expressed as a difference of potential between two points, since it is a relational term. Connecting both voltmeter leads to the same point will show no voltage present, although the voltage between that point and ground may equal hundreds or thousands of volts. Thus, most nominal voltages are expressed as phase to phase or phase to neutral. The unit of measurement is volts and the electrical symbol is V.

voltage change The variation of rms or peak voltage for a definite period of time.

voltage dip See *sag.*

voltage distortion Any change from the nominal voltage sine waveform.

voltage drop In an electric supply system, the difference between the voltages at the transmitting and receiving ends of a feeder, main, or service line.

voltage fluctuation A series of voltage changes or a cyclical variation of the voltage envelope.

voltage imbalance (unbalance) A power quality term that describes the difference in voltage between phases in a three-phase system. Determined by measuring voltage in each phase, taking the average of the three phases, and calculating the percentage difference in the phase with the greatest difference. A condition in which the three phase voltages differ in amplitude or are displaced from their normal 120° phase relationship, or both. Frequently expressed as the ratio of the negative sequence or zero sequence voltage to the positive sequence voltage, in percent.

voltage interruption Disappearance of the supply voltage on one or more phases. Usually qualified by an additional term indicating the duration of the interruption (e.g., momentary, temporary, or sustained.)

voltage magnification The amplification of the transient voltage during switching of capacitors on a transformer's primary and secondary side.

voltage regulation Describes the voltage variation from nominal, usually in percent. Also, the degree of control or stability of the rms voltage at the load. Often specified in relation to other parameters, such as input-voltage changes, load changes, or temperature changes. The ability of a power conditioner to maintain a stable output voltage when input voltage changes.

voltage regulator A transformer with windings of the primary and regulated circuits suitably adapted and arranged for the control of the voltage of the regulated circuit.

voltage variation, long-duration A change in the rms nominal voltage for more than 1 minute.

voltage variation, short duration A change in the rms nominal voltage for more than 0.5 cycles and less than 1 minute.

voltage variations Changes in voltage value.

VOM Voltohmmeter.

watt (power) The unit of power that equals 1 joule per second and measures how much electricity an appliance needs to operate satisfactorily. An electrical unit of power often used to rate appliances using relatively small amounts of electricity. Wattage is stamped on light bulbs and all appliances. The mathematical relationship between watts, volts, and amperes is wattage = ampere × voltage. For example, a 120-V, 20-A circuit will carry 2400 watts.

watt-hour The amount of electricity used by one watt in one hour.

watt-hour meter An electric meter that measures and registers the energy (kilowatt-hours) delivered to a circuit.

waveform A graph of a wave that shows its shape and changes in amplitude with time.

waveform distortion A steady-state deviation from an ideal sine wave of power frequency principally characterized by the spectral content of the deviation.

wavetrap See *line trap*.

wheeling The use of the transmission facilities of one system to transmit power of and for another system.

work The transfer of energy from one body to another.

wye A wye connection refers to a polyphase electrical supply where the source transformer has the conductors connected to the terminals in a physical arrangement resembling a Y. Each point of the Y represents the connection of a hot conductor. The angular displacement between each point of the Y equals 120°. The center point provides the common return point for the neutral conductor.

wye-delta Transformer connection with a wye primary and delta secondary.

wye-wye Transformer connection with a wye primary and wye secondary.

zero sequenced All three phases of a power system intersect the zero axis at the same time.

zero signal reference A connection point, bus, or conductor used as one side of a signal circuit that may or may not be designated as ground; sometimes referred to as circuit common.

Zigzag transformer A special type of transformer used to change the phase angle of the transformer primary.

Bibliography

ANSI/IEEE Standard C62.45-1987. "IEEE Guide on Surge Testing for Equipment Connected to Low-Voltage AC Power Circuits." Piscataway, NJ: IEEE.
ANSI/IEEE Standard 142-1991. "IEEE Recommended Practice for Grounding of Industrial and Commercial Power Systems" (*The Green Book*). Piscataway, NJ: IEEE.
ANSI/IEEE Standard 1100-1992. "IEEE Recommended Practice for Power and Grounding Sensitive Electronic Equipment" (*The Emerald Book*). Piscataway, NJ: IEEE.
ANSI/IEEE Standard 519-1992. "Recommended Practices and Requirements for Harmonic Control in Electrical Power Systems." Piscataway, NJ: IEEE.
ANSI/IEEE Standard 1446-1995. "IEEE Recommended Practice for Emergency and Standby Power Systems for Industrial and Commercial Applications." (*The Gold Book*). Piscataway, NJ: IEEE.
ANSI/NEMA Standard C84.1-1995. "Electric Power Systems and Equipment Voltage Ratings." Piscataway, NJ: IEEE.
ANSI/INFPA No. 70-1993. *National Electrical Code.*
B.C. Hydro. 1989. *Power Quality Reference Guide.* Vancouver, B.C., Canada.
B.C. Hydro. 1989. *Solutions—A Power Quality Resource Guide.* Vancouver, B.C., Canada.
B.C. Hydro. "Power Factor." *Guides to Energy Management.* Vancouver, B.C. Canada.
Bender, David. "Clean and Constant: The Basics of Power Quality." *Energy User News,* vol. 23, no. 11, November, pp. 49–55.
Bonneville Power Administration. 1985. *BPA Definitions.* DOE/BP-62. U.S. Department of Energy.
Bonneville Power Administration. 1991. *Reducing Power Factor Cost.* DOE/CE-0380. U.S. Department of Energy.
Bonneville Power Administration and Electrotek Concepts, Inc. 1995. *Industrial Power Factor Analysis Guidebook.* DOE/BP-42892-1. U.S. Department of Energy.
Bush, William. 1991. "Telecom System Fundamentals." *Power Quality,* vol. 2, no. 6, November/December, pp. 30–37.
Carr, Joseph J. 1996. *Elements of Electronic Instrumentation and Measurement.* 3d ed. Englewood Cliffs, NJ: Prentice-Hall.
"Causes of power quality problems." 1997. URL address: *http://biz.fpc.com/pqcauses.htm.* Available from Florida Power Corp.
Christensen, Peter C. 1996. *Retail Wheeling: A Guide for End-Users,* Tulsa, OK: PennWell Publishing.
Clark, H. K. "A New Ball Game." *Power Technology Newsletter.* URL address: *http://www.pti-us.com/pti/.* Available from Power Technologies, Inc.
Clark, O. Melville. 1990. "Data I/O Ports: Your Computer's Achilles Heel." *Power Quality,* vol. 1, no. 2, pp. 94–101.
DeDad, John A. 1993. "The Basics of Voltage Drop Calculations." *EC&M Electrical Construction & Maintenance,* vol. 92, no. 3, May, pp. 90–91.
"Deregulation in the Electric Industry." 1999. URL address: *http://www.emec.com/deregulation/index.html.* Available from Eastern Maine Electric Co-op.
DeNardo, Chuck. 1992. "Don't Let Customers Make a Mystery of Stray Voltages." *Electrical World,* vol. 206, no. 7, July, pp. 53–55.
Devereux, Tony. 1997. "Deregulation—The United Kingdom Outlook." *Power Value Online.* URl address: *http://www.powervalue.com/art0005/art1.html.* Available from *Power Value* magazine.
DeWinkel, Carel, and Jeffrey D. Lamoree. 1993. "Storing Power for Critical Loads." *IEEE Spectrum.* vol. 30, June, pp. 38–42.

Dugan, Roger C., Mark F. McGranaghan, and Wayne H. Beaty. 1996. *Electrical Power Systems Quality,* New York: McGraw-Hill.

——— and B. W. Kennedy. 1995. "Predicting Harmonic Problems Resulting from Customer Capacitor Additions for Demand-Side Management." *IEEE /PES Winter Meeting,* January 29–February 2, 1995, New York.

Dunbar, Mark. 1990." Power Quality Gains in Importance as Use of Electronic Equipment Grows." *Energy User News,* vol. 15, no. 1, January, pp. 1, 3.

Dunklin, Philip I. 1996. *Competition in the Utility Industry.* 2d ed. Atlanta, GA: Chartwell.

Edminister, Joseph A. 1983. *Electric Circuits.* 2d ed. New York: McGraw-Hill.

Electric Transmission and Distribution Reference Book. 4th ed. 1964, Westinghouse Electric Corp., East Pittsburgh, PA.

EPRI and Arthur D. Little, Inc. 1994. "Power Quality Market Assessment." TR-104372 Research Projects 3273-09, -10. Palo Alto, CA: EPRI.

EPRI, Consolidated Edison Co., National Electrical Contractors Association, Northwest Power Quality Service Center and Portland General Energy Systems. 1994. *Power Quality for Electrical Contractors,* Volumes 1 and 2. TR-101536-V1 and TR-101536-V2: Palo Alto, CA: EPRI.

EPRI, Bonneville Power Administration, and Electrotek Concepts. 1995. *Power Quality Workbook for Utility and Industrial Applications.* TR-105500: Palo Alto, CA: EPRI.

EPRI. 1995. *Power Quality Considerations for Power Factor Correction Applications.* Palo Alto, CA: EPRI.

EPRI Power Electronics Application Center. 1993. *Harmonic Elimination Methods for Electronic Power Supplies.* Tech Commentary, vol. 3, no. 4. Knoxville, TN: EPRI.

EPRI Power Electronics Application Center. 1992. *Harmonic Filter for Personal Computers: Passive, Parallel-Connected Resonant.* PQTN Brief 1. Knoxville, TN: EPRI.

EPRI Power Electronics Application Center. 1992. *Residential Service Entrance Meter-Base Surge Protectors.* PQTN Brief 2. Knoxville, TN: EPRI.

EPRI Power Electronics Application Center. 1992. *Plug-In Transient Voltage Surge Suppressors.* PQTN Brief 3. Knoxville, TN: EPRI.

EPRI Power Electronics Application Center. 1992. *Plug-In Transient Voltage Surge Reference Equalizers.* PQTN Brief 4. Knoxville, TN: EPRI.

EPRI Power Electronics Application Center. 1993. *Equalizing Potential Differences.* PQTN Solution No. 1. Knoxville, TN: EPRI.

EPRI Power Electronics Application Center. 1995. *Solving the Clash of Electronics Technologies.* PQTN Solution 4, August. Knoxville, TN: EPRI.

EPRI Power Electronics Application Center. 1995. *Eliminating the Jitters in Computer Monitors.* PQTN Application 4, September. Knoxville, TN: EPRI.

EPRI Power Electronics Application Center. 1995. *Sizing Single-Phase Uninterruptible Power Supplies.* PQTN Application 5. Knoxville, TN: EPRI.

EPRI Power Electronics Application Center. 1994. *Solving the Fast Clock Problem.* PQTN Brief 3. Knoxville, TN: EPRI.

EPRI. 1995. *Power Quality in Commercial Buildings.* BR-105018. Palo Alto, CA: EPRI.

EPRI Power Electronics Application Center. 1996. *Power-Conditioning Performance of Uninterruptible Power Supplies.* PQTN Brief 35. Knoxville, TN: EPRI.

EPRI. 1997. *Power Quality for Healthcare.* Palo Alto, CA: EPRI.

EPRI. 1997. *Power Quality in the Semiconductor Fabrication Industry.* Palo Alto, CA: EPRI.

EPRI. 1997. *Energy Storage in a Restructured Electric Utility Industry: Report on EPRI Think Tanks I and II.* TR-108894. Palo Alto, CA: EPRI.

EPRI. 1998. *Active Harmonic Filter Technology and Market Assessment.* TR-111088. Palo Alto, CA: EPRI.

Fluke. "Understanding Harmonics in Power Distribution Systems." Application Note.

Fulton, Stanley R. 1981. *Basic AC Circuits.* Indianapolis: Howard W. Sams & Co.

Grebe, Thomas. 1993. "How Utilities, Customers Can Tackle Harmonics Problems." *Electrical World,* vol. 207, no. 8, August, pp. 49–51.

Green, Robert. 1985. "For Sensitive Measurements, Use a Good Electrometer." *Research & Development,* vol. 27, January, pp. 88–93.

"Ground Loop Problems and How to Get Rid of Them." 1999. URL address: *htttp://www.hut.fi/Misc/Electronics/docs/groundloop/index.html.*

Hamm, Steven W. "How to Select the Right Surge Suppressor (TVSS)." South Carolina Department of Consumer Affairs Fact Sheet. p. 1.

Herrington, Donald E. 1965. *Dictionary of Electronic Terms.* New Augusta, IN: Editors and Engineers, Ltd.

Heydt, Gerald T. 1992. "Effects of Electronic Equipment on Plant Power Quality." *Plant Engineering,* vol. 46, no. 15, September 17, pp. 79–80.

———. 1992. *Electric Power Quality.* West LaFayette, IN: Stars in a Circle Publications.

Hibig, Rhonda Wright. 1996. *Power Quality Mitigation Reference Guide.* 2d ed. Toronto, Ontario, Canada: Ontario Hydro.

Holt, Mike. 1999. "Neutrals and Grounds—Part 1: The Basics." *Power Quality Assurance,* vol. 9, no. 5, September/October, pp. 66–70.

Hower, Wendy. 1991. "Powerphobia Generates Profits for Chomerics (Concern over Radiation from Power Lines; Manufacturer of High Energy Electromagnetic Interference Insulation for Industrial Machinery)." *Boston Business Journal,* vol. 10, no. 46, January 7, pp. 1–2.

Hunter, Terry. 1996. "B.C. Systems Corporation Case Study: Harmonic Analysis in a Large Commercial Facility." *Electricity Today,* February.

Hurley, Maureen Z., and Kenneth G. Hurwitz. 1997. "Legal Issues and Challenges for Power Quality: An End-User Perspective on System Reliability and Customer Choice." *Future of Power Quality in a Deregulated Marketplace,* October 14.

Ibrahim, A. Rashid, and K. Seshadri. 1995. "Power Quality for Beginners." *IAEEL Newsletter* 3–4/95. URL address: *http://www.stem.se/iaeel/IAEEL/NEWSL/1995/trefyra1995/LiTech_b-3-4-95.html.* Available from Public Utilities Board, Singapore.

IEC 1000-4-7. 1991. "Electromagnetic Compatibility (EMC)—Part 4: Testing and Measurement Techniques Section 7: General Guide on Harmonics and Interharmonics Measurements and Instrumentation, for Power Supply Systems and Equipment Connected Thereto."

IEC 1000-3-5. 1994. "Electromagnetic Compatibility (EMC)—Part 3: Limits—Section 5: Limitation of Voltage Fluctuations and Flicker in Low-Voltage Power Supply Systems for Equipment with Rated Current Greater Than 16 A."

IEEE Standard 100-1988. "IEEE Standard Dictionary of Electrical and Electronic Terms." Piscataway, NJ: IEEE.

IEEE Standard 446-1995. "IEEE Recommended Practice for Emergency and Standby Power Systems for Industrial and Commercial Applications" (*The Orange Book*). Piscataway, NJ: IEEE.

IEEE Standard 1159-1995. "IEEE Recommended Practice for Monitoring Electric Power Quality." Piscataway, NJ: IEEE.

IEEE Standard 1250-1995. "IEEE Guide for Service to Equipment Sensitive to Momentary Voltage Disturbances." Piscataway, NJ: IEEE.

IEEE Standard C62.48-1995. "IEEE Guide on Interactions Between Power System Disturbances and Surge-Protective Devices." Piscataway, NJ: IEEE.

IEEE Standard 1346-1998. "IEEE Recommended Practice for Evaluating Electric Power System Compatibility with Electronic Process Equipment." Piscataway, NJ: IEEE.

Information Technology Industry Council (ITI, formerly CBMA). 1997. "Guidelines for Grounding Information Technology Equipment." pp. 1–11.

Jowett, Jeff. 1998. "Ground Testing Essentials." *EC&M Electrical Construction & Maintenance,* August, p. 82.

Jungreis, Aaron M., and Arthur W. Kelledy. 1995. "Adjustable Speed Drive for Residential Applications." *IEEE Transactions on Industry Applications,* vol. 31, no. 6, November/December, 1995.

Kardon, Redwood. 1998. "Grounding Requirements in the National Electric Code NEC." URL address: *http://www.codecheck.com/grounding.htm.*

Kardon, Redwood. 1998. "What Does Ground Mean?" URL address: *http://www.codecheck.com/grounding2.htm.*

Keller, Peter. 1997. "Use True RMS When Measuring AC Waveforms." *Test & Measurement World,* October, pp. 29–30.

Kennedy, B. W. 1995. "Optimize Placement of In-Plant Power-Factor Correction Capacitors." *Electrical World,* vol. 209, no. 10, October, p. 57–58.

———. 1999. "Application of IEEE 519 Standards in the Restructured Competitive Electricity Industry." *Power Quality '99.* Chicago, November 9–11, 1999.

——— and M. F. McGranaghan. 1997. "Design of a Workbook for Analyzing Utility and End-User Power Quality Concerns." *CIRED 1997.* Birmingham, England.

Lamarre, Leslie. 1998. "The Digital Revolution." *EPRI Journal,* January/February, vol. 23, no. 1, pp. 26–35.

Lamendola, Mark, and Jerry Borland. "Coming to Terms with Power Quality: Understanding What Power Quality Experts Say Can Be a Boon to Your Success in the Electrical Field (Power Quality Advisor)." *EC&M Electrical Construction & Maintenance,* vol. 98, no. 2, February, p. PQ-3.

Lemerande, Cory J. 1998. "Harmonic Distortion: Definitions and Countermeasures." *EC&M Electrical Construction & Maintenance,* vol. 97, no. 3, March, pp. 48–54.

Lenk, John D. 1993. *McGraw-Hill Electronic Testing Handbook.* New York: McGraw-Hill.

Lentz, Roby C., Grank J. Mercede, and Joseph N. Mercede, Jr. 1994. "A Student Design Project to Improve Power Quality for a Commercial Facility." *IEEE / PES 1994 Winter Meeting* January 30–February 3, 1994. New York.

Lewis, Warren. 1986. "Application of the National Electrical Code to the Installation of Sensitive Electronic Equipment." *IEEE Transactions on Industry Applications,* vol. IA-22, no. 3, May/June, pp. 400–415.

Lisker, Lowell, and Gerry Tucker. 1996. "Focus on the Fundamentals of Shielded MV Power Cable (Medium-Voltage Power Cable)." *EC&M Electrical Construction & Maintenance,* vol. 95, no. 10, October, pp. 76–78.

Loper, Orla E., and Edgar Tedsen. *Direct Current Fundamentals.* 3d ed. Auburn Heights, MI: Delmar Publishers.

Marshall, Mike W. 1996. "What Makes a Power Line Vulnerable to Lightning." *Electrical World,* vol. 210, no. 6, June, pp. 22–24.

Martzloff, Francois D., and Thomas M. Gruzs. 1988. "Power Quality Site Surveys: Facts, Fiction, and Fallacies." *IEEE Transactions on Industry Applications,* vol. 24, no. 6, November/December, pp. 1005–1017.

McEachern, Alexander. 1989. *Handbook of Power Signatures.* Basic Measuring Instruments: Foster City, CA.

Moravek, James M. 1993. "Harmonics Terminology: Fact and Fiction." *EC&M Electrical Construction & Maintenance,* vol. 92, no. 2, February, p. 61.

Morinec, Allen G. 1995. "Investigating Power Quality Problems." *PQA '95 Fourth International Conference on Power Quality: Applications and Perspectives,* 9–11 March, New York.

Mosbacher, C. J. 1988. "Use of Electrical and Electronic Instruments." *Research & Development,* vol. 30, no. 4, April, pp. 92–96.

Muller, David. 1996. "Electric Utilities Implement Power-Quality Programs." *Electric Light & Power,* vol. 74, no. 5, May, pp. 9–10.

National Bureau of Standards. 1983. *Guideline on Electrical Power for ADP Installations—FIPS 94,* 24 September, U.S. Commerce Department.

Negley, Michele, and Steven Whisenant. 1998. "The Industry Initiatives." *Proceedings of PQA '98 North America,* June 8–11, Phoenix, Arizona.

Nelson, Robert D. 1992. "Synchronous Condensers Improve Power Factor." *Power Quality.* pp. 18–21.

Newberry, John. 1996. "Deregulation of the Electricity Supply Industry in the United Kingdom and the Effects on Communications Service." *IEEE / PES 1996 Summer Meeting,* 28 July–1 August, 1996, Denver, Co.

Newcombe, Charles. 1994. "Evaluating Harmonics Problems in Commercial and Industrial Facilities." *EC&M Electrical Construction & Maintenance,* vol. 93, no. 9, September, pp. 31–34.

Northwest Power Quality Service Center. 1995. *Uninterruptible Power Supply Specifications and Installation Guide.*

Pansini, Anthony J. 1988. *Electrical Transformers and Power Equipment.* Englewood Cliffs, NJ: Prentice Hall.

Patrick, Dal R., and Stephen W. Fardo. *Understanding AC Circuits—Concepts, Experiments, and Troubleshooting.* 1989. Englewood Cliffs, NJ: Prentice-Hall.

Peterson, Craig. 1994. "Harmonics for Nonelectrical Engineers." *Plant Engineering,* vol. 48, no. 16, October, pp. 88–90.

Peterson, Ivars. 1987. "In Search of Electrical Surges: Sudden Electrical Disturbances Can Severely Damage Electronic Equipment in the Home and in the Workplace." *Science News,* pp. 378–380.

Phipps, James K., John P. Nelson, and Pankaj K. Sen. 1994. "Power Quality and Harmonic Distortion on Distribution Systems." *IEEE Transactions on Industry Applications,* vol. 30, no. 2, March/April, 1994.

Pineda, Juan. 1997. "Power Quality Instruments: Integrating Hardware & Software." *Electricity Today,* September.

Porter, Art. 1989. "Transient Voltage Surge Suppressors (TVSS)." *Public Service Indiana Tech Examination,* pp. 1–8.

"Power Conditioning Issues." 1999. URL address:*http://www.powergy.com/issues.html.* Available from Powergy.

"Power Enhancement and Delivery System (PED)." 1998. URL address: *http://www.pti-us.-com/pti/consult/dist/peds/pedslit.html.* Available from Power Technologies, Inc.

"Power Primer." URL address: *http://www.tsipower.com/mainprimer.htm.* Available from TSI Power Corp.

"Power Quality." 1998. URL address: *http:www.entergy.com/custom/qual.htm.* Available from Entergy Corp.

"Power Quality." 1998. URL address: *http:www.hei.com/*~hecopq/pq00009.htm. Available from Hawaiian Electric Company, Inc.

"Power Quality at a Glance." 1998. URL address: *http://www.bedison.com/products/powersys.htm.* Available from Boston Edison.

"Power Quality Primer." 1999. URL address: *http://www.aimenergyinc.com/p_harmonic_harmref/primer.html.* Available from AIM Energy Inc.

"Power Quality Primer." 1999. URL address: *http://www.copper.org.* Available from Copper Development Association.

"Power Quality: Yes, You Really Do Need It!" 1998. URL address: *http://www.olpsvc.com/circuit%20Breaker/CB598.htm.* Available from *Circuit Breaker.* May/June.

PQAudit™. 1999. URL address: *htttp://www.electrotek.com//PS-STUDY/indust/pqaudit.htm.* Available from Electrotek Concepts.

"PQ 101—A Power Quality Tutorial for the Executive." 1999. URL address: *http://www.powerquality.com/pqpark/pqpk101.htm.* Available from *Power Quality Assurance* magazine.

"PQ 103—Equipment Sensitivity Basics" 1999. URL address: *http://www.powerquality.com/pqpark/pqpk103.htm.* Available from *Power Quality Assurance* magazine.

"Power Quality Problems." 1999. URL address: *http://www.ecsintl.com/primer.htm.*

"Power Quality Outlook." 1996. URL address: *http://www.apc.com/english/about/finan/reports/1996/anrep008.htm.* Available from American Power Conversion.

"Primer for Electric Power Industry." 1999. URL address: *http://www.eia.doe.gov/cneaf/electricity/page/rim2/chapter1.html.* Available from the Electric Power Division of the Energy Information Administration.

"Recognizing Symptoms and Determining Solution." 1999. URL address: *http://www.fpc.com/flpower/biz/pqsol2.htm.* Available from Florida Power Corp.

Riezenman, Michael J. 1995. "Making an Issue of Power Quality." *IEEE Spectrum,* June, p. 81.

Shaughnessy, Tom. 1995. "Harmonics." *PQ Today,* vol. 1, no. 2, Summer, pp. 6–7.

———. 1996. "Grounding: A Historical Perspective." *PQ Today,* vol. 3, no. 1, Summer, pp. 5, 9.

———. 1998. "Types of Grounding Systems." *Power Quality Assurance,* vol. 9, no. 4, July/August, pp. 74–75.

———. 1999. "Grounding Conductor Measurements." *Power Quality Assurance,* vol. 10, no. 3, May/June, pp. 52–61.

Sherr, Sava I. 1985. "Communications Standards and the IEC." *Communications Magazine,* vol. 23, no. 1, pp. 25–27.

"Shielding." 1999. URL address: *http://www.amercable.com/catalog/k5.htm.* Available from AmerCable.

Short, Thomas A. 1992. "Harmonics and IEEE 519." Paper presented for *Electric Council of New England,* September 17, URL address: *http://www.pti-us.com/pti/consult/dist/papers/harmonics/harmonics.htm.* Available from Power Technologies Inc.

Smith, Shawn. 1998. "Uninterruptible Power Supplies: Distributed vs. Centralized?" *EC&M,* vol. 97, no. 3, March, pp. 62–67.

Staff. 1997. "Power Quality in a Deregulated Environment—Part 1: It's More Than Just Price." *Power Quality Assurance Online.* URL address: *htttp://powerquality.com/art0004/staff1.html.* Available from *Power Quality Assurance* magazine.

Staff. 1997. "Power Quality in a Deregulated Environment—Part 2: It's More Than Just Price." *Power Quality Assurance Online.* URL address: *htttp://powerquality.com/art0035/art1.html.* Available from *Power Quality Assurance* magazine.

Staff. 1997. "Power Quality in a Deregulated Environment—Part 3: Differentiate Your Product Through Value-Selling." *Power Quality Assurance Online.* URL address: *htttp://powerquality.com/art0036/art1.htm.* Available from *Power Quality Assurance* magazine.

Stebbins, Wayne. 1994. "Facility Managers Can't Ignore PQ Concerns." *Energy User News,* vol. 19, no. 7, July, pp. 22–23.

———. 1997. "Power Distribution Systems and Power Factor Correction." *Energy User News,* vol. 22, no. 9, September, pp. 36–38.

Steciuk, P. B., and J. Redmon. 1999. "Voltage Sags and Voltage Sag Studies." URL address: *http://www.pti-us.com/pti/consult/dist/papers/voltsag/voltsag.htm.* Available from Power Technologies, Inc.

Stein, Hank. 1996. "FERC (Federal Energy Regulatory Commission)." *Modern Power Systems,* vol. 16, no. 6, June, pp. 19–20.

"Step by Step 9—Power Conditioning." 1999. URL address: *http://www.dansdata.com/sbs9.htm.*

Stevens, Mark. 1993. "How to Beat Lightning Bolts." *Machine Design,* vol. 65, no. 12, June, 25, pp. 51–56.

St. Pierre, Conrad R. 1993. "How Utilities, Customers Can Tackle Harmonics Problems." *Electrical World,* vol. 207, no. 8, August, pp. 49–51.

Studebaker, John M. 1997. *Electricity Purchasing Handbook,* Tulsa, OK: PennWell Publishing.

Sywenky, Andy. 1999. "Power Quality: Basic Investigation." URL address: *http://www.elecleague.ab.ca/basicin.htm.* Available from the Electric League.

"The Law and Power Quality." 1990. URL address: *http://www.powerquality.com/art0017/knox5.htm.* Available from *Power Quality Assurance* magazine.

Thompson, Lawrence M. 1994. *Electrical Measurements and Calibration: Fundamentals and Applications.* 2d ed. Research Triangle Park, NC: Instrument Society of America.

"Understanding Power Quality." 1999. URL address: *http//:www.nimo.com.* Available from Niagara Mohawk.

Underwriters Laboratories Standard for Safety 1449. 1996. *Transient Voltage Surge Suppressors.* 2d ed. Northbrook, IL.

U.S. Navy. 1969. *Basic Electricity.* Washington, DC: Bureau of Naval Personnel.

"Utility Deregulation for Beginners, Part 1." 1999. URL address: *http://www.emec.com/deregulation/dereg1.html.*

Van Valkenburgh, Nooger, and Neville. 1978. *Basic Electricity Revised Edition.* Hasbrouck Heights, NJ: Hayden Book Company.

Waggoner, Ray. 1995. "Power Quality and Good Housekeeping—Part 1." *EC&M Electrical Construction & Maintenance,* vol. 94, no. 2, February, pp. 14–16.

——— 1995. "Power Quality and Good Housekeeping—Part 2." *EC&M Electrical Construction & Maintenance,* vol. 94, no. 3, March, pp. 18–19.

———. 1997. *Practical Guide to Quality Power for Sensitive Electronic Equipment.* Overland, KS: Intertec Publishing.

Wagner, John. 1998. "Power Quality and Info. Technology Equipment." *Energy User News.* vol. 23, no. 5, May, pp. 24.

Waller, Mark. 1988. "Power Protection." *BYTE.* vol. 13, no. 10, October, pp. 270–280.

———. 1992. *Managing the Computer Power Environment.* Indianapolis: PROMPT Publications.

——— 1993. "Power Line vs. Data Line." *Cabling Business.* December, pp. 24–28.

———. 1994. *Harmonics.* Indianapolis: PROMPT Publications.

Warnock, William J. 1993. "Power Pollution Protection." *Security Management,* vol. 37. no. 5, May, pp. 54–58.

Watkins-Miller, Elaine. 1997. "Don't Get Zapped (Office Technology and Power Quality Problems)." *Building,* vol. 91, no. 10, October, pp. 68–69.

Weaver, Thomas. 1992. "Harmonics and Resulting Liability." *EC&M Electrical Construction & Maintenance,* vol. 91, no. 1, January, pp. 29–30.

Weidner, Gary. 1998. "Update: Handheld Multimeters." *Plant Engineering,* vol. 52, no. 11, October, pp. 36–41.

"What Is Power Conditioning?" 1999. URL address: *http://www.onlinepower.com/sidebar.html.*

"What Is Power Quality?" 1998. URL address: *http://www.hlp.com/business/majoraccts/powerqual/PQwhatis.htm.* Available from Houston Light & Power signals). *EC&M Electrical Construction & Maintenance,* vol. 91, no. 4, April, pp. 31–32.

Williamson, Ron. 1992. "True-RMS Meters and Harmonics (Root-Mean Square's Discussion of Digital Multimeter in Measuring Nonsinusoidal Signals)."

Worden, Michael. 1998. "Power Quality—Regulatory Expectations." *Proceedings of PQA '98 North America,* 8–11 June, Phoenix, Arizona.

"Zap Insurance: Surge Suppressors." 2000. *Consumer Reports,* January, pp. 49–50.

Index

About the Author

Barry Kennedy, PE, is CEO of Kennedy Consulting Solutions, contributing editor for *Electronic World T&D Magazine*, and adjunct professor at the University of Portland. He was formerly senior electrical engineer and project manager responsible for transmission and distribution system efficiency, power quality, and power factor research and development for Bonneville Power Administration, a successful supplier of electricity to retail utilities and large industrial companies. Well-known within the industry for his seminars and consulting work preparing power companies and their customers for deregulation, he has developed notable methods, guidebooks, software, training programs, and solutions. Mr. Kennedy is also author of *Energy Efficient Transformers*. He is the project manager and developer of the *Workbook for Utility and Industrial Applications* and the *Industrial Power Factor Analysis Guidebook*. He is a registered professional engineer in Oregon, with a Master's degree in Electrical Engineering from Purdue University. He resides in Sherwood, Oregon.